Plant Tissue and Cell Culture

Plant Biology

Series Editor

Titles in the Series

Vol 1: Phloem Transport / James Cronshaw, William J. Lucas, and Robert T. Giaquinta, *Editors*

Vol 2: Regulation of Chloroplast Differentiation / George Akoyunoglou and Horst Senger, *Editors*

Vol 3: Plant Tissue and Cell Culture / C.E. Green, D.A. Somers, W.P. Hackett, and D.D. Biesboer, *Editors*

Plant Tissue and Cell Culture

Proceedings of the VIth International Congress on Plant Tissue and Cell Culture, held at the University of Minnesota August 3–8, 1986

Editors

C.E. Green
Geo. J. Ball, Inc.
West Chicago, Illinois

D.A. Somers
Department of Agronomy and Plant Genetics
University of Minnesota
St. Paul, Minnesota

W.P. Hackett
Department of Horticulture
University of Minnesota
St. Paul, Minnesota

D.D. Biesboer
Department of Botany
University of Minnesota
St. Paul, Minnesota

Alan R. Liss, Inc., New York

Address all Inquiries to the Publisher
Alan R. Liss, Inc., 41 East 11th Street, New York, NY 10003

Printed in the United States of America

Library of Congress Cataloging-in-Publication Data

International Congress on Plant Tissue and Cell
Culture (6th : 1986 : University of Minnesota)
Plant tissue and cell culture.

(Plant biology ; v. 3)
Includes index.
1. Plant tissue culture—Congresses. 2. Plant cell culture—Congresses. 3. Plant propagation—In vitro—Congresses. I. Green, C.E. (Charles E.), 1943–
II. Title. III. Series.
QK725.I4944 1986 581′.07′24 86–27527
ISBN 0–8451–1802–1

Contents

Section VI. Applications to Crops

Section VII. Technology Development

Sixth IAPTC Congress Organizing Committees

Chairman	C.E. Green
Finance	R.L. Phillips, M.L. Brenner, C.E. Green
Program	B.G. Gengenbach, D.A. Somers, P.E. Read, H.W. Rines, W.P. Hackett, R.L. Phillips, E.J. Staba, I. Rubinstein
Publications	D.A. Somers, W.P. Hackett, D.D. Biesboer, C.E. Green
Local Arrangements	W.R. Bushnell, H.W. Rines, W.P. Hackett
Exhibits	P.E. Read

Contributors

Philip V. Ammirato, Department of Biological Sciences, Barnard College, Columbia University, New York, NY 10027; and DNA Plant Technology Corporation, Cinnaminson, NJ 08077 **[57]**

Nikolaus Amrhein, Lehrstuhl für Pflanzenphysiologie, Ruhr-Universität Bochum, D-4630 Bochum, Federal Republic of Germany **[119]**

Paul G. Arnison, Paladin Hybrids, Inc., Brampton, Ontario L6V 2L2, Canada **[223]**

Hans Becker, Institut für Pharmazeutische Biologie, Universität Heidelberg, D-6900 Heidelberg, Federal Republic of Germany **[199]**

Ludwig Bender, Institut für Pflanzenernaehrung Abt. Gewebekultur, Justus-Liebig-Universität Giessen, 6300 Giessen, Federal Republic of Germany **[151]**

Robert Bernatzky, Department of Plant Breeding and Biometry, Cornell University, Ithaca, NY 14853 **[331]**

Wallace D. Beversdorf, Allelix Inc., Mississauga, Ontario L4V 1P1, Canada **[359]**

M. Bolik, Institute for Resistance Genetics, Federal Biological Research Centre for Agriculture and Forestry, D-8059 Grünbach, Federal Republic of Germany **[343]**

Michel Boulay, AFOCEL domaine de l'Etançon 77370, Nangis, France, and Department of Pomology, University of California, Davis, CA 95616 **[367]**

R.I.S. Brettell, Division of Plant Industry, CSIRO, Canberra, A.C.T., 2601, Australia **[275]**

Danielle Cammaerts, Laboratory of Plant Genetics, Vrije Universiteit Brussel, B-1640 Sint-Genesius-Rode, Belgium **[243]**

Brian J. Cardy, BioTechnica International of Canada, Inc., Calgary, Alberta T2E 7H7, Canada **[223]**

Michael L. Christianson, Department of Molecular Biology, Zoecon Research Institute, Sandoz Crop Protection Corporation, Palo Alto, CA 94304 **[45]**

Edward C. Cocking, Department of Botany, University of Nottingham, Nottingham NG7 2RD, U.K. **[3]**

Friedrich Constabel, Plant Biotechnology Institute, National Research Council, Saskatoon, Saskatchewan S7N 0W9, Canada **[213]**

The numbers in brackets are the opening page numbers of the contributors' articles.

P.A. Davies, Division of Plant Industry, CSIRO, Canberra, A.C.T. 2601, Australia **[275]**

Alicia de la Pena, Max-Planck-Institut für Züchtungsforschung, 5000 Köln 30, Federal Republic of Germany **[303]**

Pierre C.A. Debergh, Department of Horticulture, State University Gent, B-9000 Gent, Belgium **[383]**

S.C. Debnath, Department of Genetics and Plant Breeding, Bangladesh Agricultural University, Mymensingh, Bangladesh **[343]**

S. Deimling, Institute for Resistance Genetics, Federal Biological Research Centre for Agriculture and Forestry, D-8059 Grünbach, Federal Republic of Germany **[343]**

Beate Diettrich, Section of Pharmacy, Martin-Luther-University Halle-Wittenberg, DDR-4050 Halle, Weinbergweg 15, German Democratic Republic **[187]**

Rob Dirks, Laboratory of Plant Genetics, Vrije Universiteit Brussel, B-1640 Sint-Genesius-Rode, Belgium **[243]**

Claire Doré, Station de Génétique et D'Amélioration des Plantes, Institut National de la Recherche Agronomique, 78000 Versailles, France **[419]**

Donald K. Dougall, Department of Botany, University of Tennessee, Knoxville, TN 37996-1100 **[97]**

Udo Eilert, Plant Biotechnology Institute, National Research Council, Saskatoon, Saskatchewan S7N 0W9, Canada **[213]**

Dean Engler, Advanced Genetic Sciences, Inc., Oakland, CA 94608 **[265]**

Erzsebet Fejes, Advanced Genetic Sciences, Inc., Oakland, CA 94608 **[265]**

B. Foroughi-Wehr, Institute for Resistance Genetics, Federal Biological Research Centre for Agriculture and Forestry, D-8059 Grünbach, Federal Republic of Germany **[343]**

Larry C. Fowke, Department of Biology, University of Saskatchewan, Saskatoon, Saskatchewan S7N 0W0, Canada **[17]**

Michael W. Fowler, Wolfson Insitute of Biotechnology, University of Sheffield, Sheffield S10 2TN, U.K. **[459]**

Robert Fraley, Plant Molecular Biology, Corporate Research Laboratories, Monsanto Company, St. Louis, MO 63198 **[317]**

Joyce Fry, Plant Molecular Biology, Corporate Research Laboratories, Monsanto Company, St. Louis, MO 63198 **[317]**

Jo Ann Fujii, Research Department, Plant Genetics, Inc., Davis, CA 95616 **[473]**

Yasuhiro Fujita, Bioscience Research Center, Mitsui Petrochemical Industries, Ltd., Yamaguchi 740, Japan **[169]**

Nancy Hoffmann, Plant Molecular Biology, Corporate Research Laboratories, Monsanto Company, St. Louis, MO 63198 **[317]**

Heike Holländer-Czytko, Lehrstuhl für Pflanzenphysiologie, Ruhr-Universität Bochum, D-4630 Bochum, Federal Republic of Germany **[119]**

Roger Horgan, Department of Botany and Microbiology, University College of Wales, Aberystwyth SY23 3DA, U.K. **[135]**

Robert Horsch, Plant Molecular Biology, Corporate Research Laboratories, Monsanto Company, St. Louis, MO 63198 **[317]**

V.N. Iyer, Biology Department, Carleton University, Ottawa, Ontario K1S 5B6, Canada **[435]**

Michel Jacobs, Laboratory of Plant Genetics, Vrije Universiteit Brussel, B-1640 Sint-Genesius-Rode, Belgium **[243]**

Detlef Johänning, Lehrstuhl für Pflanzenphysiologie, Ruhr-Universität Bochum, D-4630 Bochum, Federal Republic of Germany **[119]**

Barbara Junker, Max-Planck-Institut für Züchtungsforschung, 5000 Köln 30, Federal Republic of Germany **[303]**

K.K. Kartha, Plant Biotechnology Institute, National Research Council, Saskatoon, Saskatchewan S7N 0W9, Canada **[447]**

Wilfred A. Keller, Plant Research Centre, Agriculture Canada, Ottawa, Ontario K1A 0C6, Canada **[223]**

Harry Klee, Plant Molecular Biology, Corporate Research Laboratories, Monsanto Company, St. Louis, MO 63198 **[317]**

Wolfgang Kurz, Plant Biotechnology Institute, National Research Council, Saskatoon, Saskatchewan S7N 0W9, Canada **[213]**

Richard E. Litz, Tropical Research and Education Center, Institute of Food and Agricultural Sciences, University of Florida, Homestead, FL 33031 **[407]**

Horst Lörz, Max-Planck-Institut für Züchtungsforschung, 5000 Köln 30, Federal Republic of Germany **[303]**

Martin Luckner, Section of Pharmacy, Martin-Luther-University Halle-Wittenberg, DDR-4050 Halle, Weinbergweg 15, German Democratic Republic **[187]**

Pal Maliga, Advanced Genetic Sciences, Inc., Oakland, CA 94608 **[265]**

Sheila McCormick, Plant Molecular Biology, Corporate Research Laboratories, Monsanto Company, St. Louis, MO 63198 **[317]**

Brian L. Miki, Genetic Engineering Section, Plant Research Centre, Agriculture Canada, Ottawa, Ontario K1A 0C6, Canada **[435]**

Michael G. Mullins, Department of Agronomy and Horticultural Science, University of Sydney, Sydney, N.S.W. 2006, Australia **[395]**

Ioan Negrutiu, Laboratory of Plant Genetics, Vrije Universiteit Brussel, B-1640 Sint-Genesius-Rode, Belgium; and Friedrich Miescher-Institut, CH-4002 Basel, Switzerland **[243, 289]**

Karl-Hermann Neumann, Institut für Pflanzenernaehrung Abt. Gewebekultur, Justus-Liebig-Universität Giessen, 6300 Giessen, Federal Republic of Germany **[151]**

Jeanne Niedermeyer, Plant Molecular Biology, Corporate Research Laboratories, Monsanto Company, St. Louis, MO 63198 **[317]**

M.A. Pallotta, Division of Plant Industry, CSIRO, Canberra, A.C.T. 2601, Australia **[275]**

Jerzy Paszkowski, Friedrich Miescher-Institut, CH-4002 Basel, Switzerland **[289]**

Ingo Potrykus, Friedrich Miescher-Institut, CH-4002 Basel, Switzerland **[289]**

Keith Redenbaugh, Research Department, Plant Genetics, Inc., Davis, CA 95616 **[473]**

Terry J. Reich, Biology Department, Carleton University, Ottawa, Ontario K1S 5B6, Canada **[435]**

Stephen Rogers, Plant Molecular Biology, Corporate Research Laboratories, Monsanto Company, St. Louis, MO 63198 **[317]**

S.A. Ryan, Division of Plant Industry, CSIRO, Canberra, A.C.T. 2601, Australia **[275]**

Michael W. Saul, Friedrich Miescher-Institut, CH-4002 Basel, Switzerland **[289]**

Jeff Schell, Max-Planck-Institut für Züchtungsforschung, 5000 Köln 30, Federal Republic of Germany **[303]**

R. Schuchmann, Institute for Resistance Genetics, Federal Biological Research Centre for Agriculture and Forestry, D-8059 Grünbach, Federal Republic of Germany **[343]**

Arno Schulz, Lehrstuhl für Pflanzenphysiologie, Ruhr-Universität Bochum, D-4630 Bochum, Federal Republic of Germany **[119]**

W.R. Scowcroft, BioTechnica Canada, Calgary, Alberta T2E 7H7, Canada **[275]**

Dilip Shah, Plant Molecular Biology, Corporate Research Laboratories, Monsanto Company, St. Louis, MO 63198 **[317]**

Raymond D. Shillito, Friedrich Miescher-Institut, CH-4002 Basel, Switzerland **[289]**

David Slade, Research Department, Plant Genetics, Inc., Davis, CA 95616 **[473]**

Cheryl C. Smart, Lehrstuhl für Pflanzenphysiologie, Ruhr-Universität Bochum, D-4630 Bochum, Federal Republic of Germany; present address: Department of Botany, University of Cambridge, Cambridge CB2 3EA, U.K. **[119]**

Hans C. Steinrücken, Lehrstuhl für Pflanzenphysiologie, Ruhr-Universität Bochum, D-4630 Bochum, Federal Republic of Germany; present address: Ciba-Geigy AG, CH-4002 Basel, Switzerland **[119]**

Zora Svab, Advanced Genetic Sciences, Inc., Oakland, CA 94608 **[265]**

Mamoru Tabata, Faculty of Pharmaceutical Science, Kyoto University, Kyoto 606, Japan **[169]**

Steven D. Tanksley, Department of Plant Breeding and Biometry, Cornell University, Ithaca, NY 14853 **[331]**

Terry L. Thomas, Biology Department, Texas A & M University, College Station, TX 77843 **[83]**

Jack Van't Hof, Biology Department, Brookhaven National Laboratory, Upton, NY 11973 **[33]**

Peter Viss, Research Department, Plant Genetics, Inc., Davis, CA 95616 **[473]**

G. Wenzel, Institute for Resistance Genetics, Federal Biological Research Centre for Agriculture and Forestry, D-8059 Grünbach, Federal Republic of Germany **[343]**

Dayton Wilde, Biology Department, Texas A & M University, College Station, TX 77843 **[83]**

Preface

Plant cell and tissue culture has become a major tool in the study of an increasing number of fundamental and applied problems in the plant sciences. Its expanding use to investigate cell and developmental biology, biochemistry, physiology, genetics, and molecular biology is providing new knowledge about fundamental characteristics of plants. Furthermore, it has become an integral part of "plant biotechnology" research as scientists in universities, research institutes and companies pursue the development of improved plants for agriculture, horticulture and forestry.

The 6th International Congress of Plant Tissue and Cell Culture was held in August, 1986, at the University of Minnesota to provide a comprehensive international forum for discussion of the latest developments in this rapidly evolving field. The 1530 participants attending the Congress from 51 countries provided clear evidence that research in this field has expanded rapidly in the four years since the 5th Congress in Tokyo. A total of 355 scientific papers and 472 posters covering virtually all aspects of current research in the field were presented. The numerous developments reported in these presentations and the chapters in this volume clearly indicate that the technology is moving forward on a broad front. Of particular significance has been the development of gene transfer technology and the rapidly expanding investigation of the molecular biology of plants since the 5th Congress in Tokyo.

The content of this volume reflects the organization of the Congress, which was divided into seven broad subject areas: development, biochemistry and physiology, secondary metabolites, genetics, molecular biology, applications to crops, and technology development. These subjects were discussed at length in an All Congress Symposium and 32 Concurrent Symposia. The chapters in this volume were written by leading experts in the field and each was presented as a Plenary Lecture in a symposium. Collectively these chapters will provide the reader with a broad review of recent developments and progress in the field of tissue culture. As is to be expected, each chapter represents the views of the authors, and the editors refer readers desiring further information about research presented at the Congress in specific subject areas to the Congress Abstracts. It is hoped that the wide range of authoritative reviews and views presented by the authors in this volume will provide a useful and exciting stimulus to researchers and students in the field.

The generous financial support given to the 6th Congress by The Rockefeller Foundation, United States Department of Agriculture, and numerous companies interested in plant biotechnology is gratefully acknowledged. This support was especially instrumental in assisting many students and scientists to attend the Congress. Without this financial assistance many would not have been able to participate and contribute to the success of the Congress.

The enthusiastic participation and the devoted and detailed attention to organizing the 6th Congress by many faculty members, graduate students and post-doctoral fellows in the Departments of Agronomy and Plant Genetics, Horticulture, Plant Pathology, and Genetics and Cell Biology, and the staff in the Office of Special Programs at the University of Minnesota are gratefully acknowledged. Without their generous help the 6th Congress would not have been a success.

C.E. Green
August 1986

Section I.
Development

Plant Tissue and Cell Culture, pages 3–15

PLANT CELL BIOLOGY IN THE 21st CENTURY: THE NEEDS OF PLANT CELL AND TISSUE CULTURE.

Edward C. Cocking.

Plant Genetic Manipulation Group, Department of Botany, University of Nottingham, Nottingham, NG7 2RD, U.K.

It is interesting to recall that Philip White in his address to the Seminar on Plant Cell, Tissue and Organ Culture in Delhi in 1967, emphasized that the challenges for the future were still technical ones. One major requirement would be an easy and reliable method of reducing a plant to a single cell without destroying its viability: "ideally we should isolate cells directly from tissues of the plant for there are scores of problems which demand the availability of single cell suspensions without a long previous history of *in vitro* cultivation" (White, 1967). Since 1967, in the intervening nearly thirty years, several techniques have been developed for reducing plants to single cells without destroying their viability, plus numerous other techniques for the dissection of cells at the molecular level. Indeed, emphasis has shifted in many laboratories from the cellular to the molecular aspects: it was H.E. Street, at one time a lecturer in my own University, who foresaw with characteristic pre-science, shortly before his death, the advantages of the *Agrobacterium tumefaciens* plasmid in this respect (Street, 1977).

Perhaps the quotation from Edmond B. Wilson (1925), cited in the Preface to the *Molecular Biology of the Cell* (Alberts *et al*, 1983), is particularly apposite. "Long ago it became evident that the key to every biological problem must finally be sought in the cell, for every living organism is, or at sometime has been, a cell". Alberts *et al*. (1983) also cogently argued that because one is so often aware of cells it is all too easy to underestimate them, and to cease being aware how truly remarkable they really are; and with the dominant role of DNA so clearly established to make the unwarranted assumption that by understanding the nucleic acids one can understand the essence of the living state. It is the incredible subtlety and versatility of cells which serves to highlight the central role of Cell Biology. Chargaff (1971) has also highlighted the fact that we have no real idea of the inside of a living cell, for we lack what could be called a science of compressed spaces. Perhaps the recent nuclear magnetic resonance imaging of a single *Xenopus* ovum will help in this respect (Aguayo *et al.*, 1986).

Molecular Biology, in the words of W.T. Astbury the British Scientist who first coined the name, implies not so much a technique as an approach, an approach from the viewpoint of the so-called basic sciences with the leading idea of searching below large-scale manifestations of classical biology for the corresponding molecular plan (Astbury 1950). Cell Biology also implies not so much a technique as an approach searching below large-scale manifestations of classical biology for the corresponding cellular plan. The unity of these two approaches is to be found in the molecular biology of the cell. The molecular biology of plant cells is now beginning to provide an important new insight into the essence of the living state, within the framework of plant cell and tissue culture. Indeed, the needs of plant cell and tissue culture, as we move into the next century, will increasingly be inextricably linked to molecular aspects of plant cell biology; this is readily confirmed by a glance at a selection of current papers in the leading Plant Science Journals.

Plant cell and tissue culture has, during the past twenty five years, emerged from its purely descriptive and catalogueing phase. Everybody knows, or thinks they know, what a callus is and what somatic embryos are. Very significant progress has been made in the concoction of defined media for the growth of plant cells and tissues from a wide range of species, and these bear witness to the enterprise of a galaxy of plant tissue culturalists. Many of us were not however brought up scientifically as plant tissue culturalists, often being trained in the other disciplines of plant breeding, physiology, genetics and in biochemistry and molecular biology, and sometimes these days even in ecology and population genetics. Much of the new interest and enthusiasm has come from the central role that cell and tissue culture plays in genetic manipulations, ranging from new variation from _in vitro_ culture to transformation, using suitable chimeric plasmids. The desire is to bring greater precision to the cellular aspects of the subject and to utilise where possible molecular approaches. It is no overstatement that our understanding of the fundamental processes involved in growth and development of plants is in urgent need of further research if the activities of plants are to be understood and manipulated in a scientific manner. The challenge for plant cell and tissue culture is to try to improve this understanding. Plant Cell Biology in the 21st Century will have many facets and these will mainly relate to the needs of plant cell and tissue culture. It will be useful in this context to highlight some of the special features of plant cell biology. One of these is the central significance of the cell wall which in plants is closely involved in osmoregulation, reproduction, communication and defence mechanisms. Plant cell culture has a need for a more comprehensive study of cell wall synthesis both at a free plasma membrane surface, as in protoplasts, and during the division of cells. Such studies have often been neglected in the rush to obtain advances in plant genetic manipulations. New biochemical and biophysical analytical procedures, such as reflectance Fourier Transform Infra-red Spectroscopy (Hopkinson _et al._, 1985), are likely to provide further insight into the detailed pattern of wall synthesis. A better understanding of cell wall synthesis could for instance have a profound effect on the ability to obtain single cell suspension cultures, or at second best cell suspensions containing very small cell aggregates. Normally, when plant cells divide in culture the daughter cells remain connected together to varying degrees. Mitotic division of protoplasts can be divorced from cytokinesis and cell wall synthesis by the use of specific inhibitors of cell wall synthesis. With further study it might be possible to affect cell plate formation such that, after mitotic division, the two daughter cells separate from each other with the production in culture of true single cell suspensions. Plant cell cultures have been insufficiently investigated in this respect; the resurgence of interest in the importance of wall matrix cross-links may help greatly (Fry, 1986). The actual cell culture engineering aspects have been largely resolved with little attention paid to the manipulation of the cell biology. In plant cells the plasma membrane functions as a selectively permeable boundary around the cytoplasm, and acts in concert with the cell wall as an interface with the external environment; cryoinjury is primarily a consequence of freeze/thaw-induced destabilization of the plasma membrane. Mersey et al. (1985) have comprehensively surveyed the activity of organelles associated with the inner surface of this plasma membrane; these include endoplasmic reticulum associated with the plasmodesmata, cytoskeletal elements such as cortical microtubules and microfilaments, and smooth and coated vesicles. These studies are important particularly because they may help in our understanding the involvement of coated vesicles in the movement and sorting of plasma membrane components, and defining the role of endocytosis in plant cells. Many of the manipulative activites that are desired to be carried out with cells, whether it involves the attempted "puncturing" of the plasma membrane to facilitate entry of foreign DNA, as in electroporation, or the stimulation of uptake of foreign DNA by endocytosis, all depend for their further development on an improved knowledge of the plasma membrane and the organelles associated with its inner surface. Recently monoclonal antibodies to plant plasma membrane antigens have been prepared (Norman et al. 1986); in addition to use as biochemical markers for fractionation and molecular characterization of plasma membranes, these monoclonal antibodies will provide the basis for new selection tools in plant cell manipulations. Plant Cell Biology has often lacked the precision and manipulation capability inherent in the more molecular aspects. Moreover, Plant Cell Biology has often experimentally lagged behind Animal Cell Biology. Both now, and into the next Century, the pressing needs for greater precision and capability in the manipulation of plant cells will result in an ongoing resurgence of interest in these respects.

Because, in many instances, studies on plant cell and tissue culture are a means to an end, in the sense that the real interest is in various aspects of whole plant improvement, it is to be anticipated that there will be an increasing trend to minimise the length of time cells and tissues are subjected to *in vitro* culture, and to simplify the procedures as much as possible. In this connection the use of single cells in a cell suspension culture, rather than protoplasts, for microinjection is noteworthy. The microinjected carrot cells could divide and differentiate to embryos at a frequency of about 50% (Nomoru and Komamine 1986). The recent simplification of transformation procedures using leaf discs, coupled with regeneration of shoot systems directly from such discs, including those of tomato (Horsch et al. 1985), further highlights this tendency. A better understanding of the role of growth substances and other factors in the induction of division in leaf tissue may well enable this leaf disc approach to be introduced for the transformation of a wide range of species, using whole Agrobacterium as the vector system. This tendency to minimise the extent of in vitro culture in achieving the desired experimental objective is also evident from the recent work on the enzymatic exposure of the plasma membrane at the tips of root hairs of a wide range of crop plants. (Cocking, 1985). Seedlings treated in this manner grew normally when suitably cultured, and the procedure enabled whole seedlings to have the plasma membrane at the tips of their root hairs exposed to foreign DNA, and to Rhizobia and other micro-organisms.

Central to so much of plant cell and tissue culture is the need to be able to regenerate reproducibly whole plants from cultured protoplasts, cells and tissues, or to be able to maintain cells dividing under conditions optimal for the synthesis of particular plant secondary products. In the case of plant secondary product production, an exemplory illustration of the application of a dual approach in this connection has been the elucidation of the physical and chemical factors controlling the biosynthesis of the red naphthoquinone pigment, shikonin, used for medicines, dyes and cosmetics, by Lithospermum erythrohizon cell cultures (Tabata and Fujita, 1985). Stable high-yielding strains were isolated, after repeated selection by the single-cell cloning technique, and for the large-scale production in fermenter systems use was made of an improved growth medium resulting from an examination of all the components of the medium, including inorganic salts, vitamins, sugars and amino acids to determine their optimum concentrations. Computer-aided programmes of this type will unquestionably fulfill an important need in the years ahead. Generally in the field of the use of plant cell cultures for the production of desirable substances insufficient attention has been given to their genetic manipulation to enhance the economics of production. Much could be achieved now in this respect by protoplast fusion and by transformation, particularly using Agrobacterium rhizogenes. Hitherto, for instance, the majority of reports have communicated the use of cultures consisting of either dispersed or immobilised cells. Recently Hamill et al. (1986) established "hairy root" cultures of Beta vulgaris and Nicotiana rustica following infection with Agrobacterium rhizogenes; and the transformed cultures synthesised their characteristic secondary products at levels comparable with those of in vivo roots from the same variety.

The regeneration of plants from cell cultures still has a delightful element of empiricism that for some is an attractive feature. More usefully, however, a scientific insight into the many factors involved is beginning to emerge. Currently there is much interest in the use of biochemical markers that can document progress towards the production of cells that are competent to regenerate; this work is only at an early stage of development, but such markers are already being used to assess the value of the various manipulations being employed to bring about plant regeneration. In the studies of Stuart et al. (1985) on embryo-specific protein expresion in somatic embryos, proteins from somatic embryos were compared with proteins extracted from the seed, and proteins were observed common to both. Some useful correlations were also detected. Stuart et al. (1985) found that, if alfalfa somatic embryos were exposed to too much auxin (2,4-D) during induction, they failed to accumulate such storage proteins and germinated at a lower frequency. Sung (1985) has also isolated a number of biochemical markers for embryogenesis that are being used to evaluate the efficiency of regeneration protocols in carrot. Whilst this work is still at an early stage in its development, it is likely that a combination of the use of mutants, and a biochemical molecular approach using cDNA and antibody probes, will help in the identification of genes responsible for developmental defects (Sung, 1985).

As we progress towards the next century it is perhaps salutory to remember that the initiation and development of embryos from somatic tissues in plant culture was recognized nearly thirty years ago in cultures of carrot tissue derived from the storage taproot (Steward et al., 1958; Reinert 1958) yet, as recently re-emphasised by Ammirato (1983), there has been little effort to use this remarkable tool for basic research - the striking phenomenon of differentiated, or otherwise committed, somatic cells embarking on an embryogenic pathway provides a superb opportunity for exploring some of the fundamental questions of plant growth, differentiation and development, using a combination of new cellular and molecular techniques.

A significant step in this direction has been the recent demonstration that normal, green rice plants can be regenerated efficiently and reproducibly from rice protoplasts via somatic embryogenesis (Abdullah et al., 1986) using an amino acid based medium, and improved culture of rice protoplasts embedded in agarose (Thompson et al., 1986). The recent demonstration of somatic polyembryony in sugar pine, utilising acetocarmine staining, is also significant (Gupta and Durzan 1986); as is the recent report of plant regeneration from protoplasts of Glycine canescens, a wild relative of soybean, (Hammatt et al., 1986).

It is becoming increasingly clear that what plant cell and tissue culture need from plant cell biology is a better basic knowledge of the plant cell itself, and of the relationships between cells. At present plant cell biology lacks a major element of precision. Because of the needs of plant molecular biology, and for the required interplay between plant cell and molecular biology, these deficiencies are being corrected. The application of fluorescence activated cell sorting procedures will unquestionably lead to the ready isolation of cells in large numbers at particular stages of the cell cycle, and to a better understanding of nuclear and cytoplasmic interactions in cell fusion products. The availability of such defined cell types will progressively improve our knowledge of organelle recombinants and segregation in heterokaryons, and the influence of the stages of the cell cycle on the properties of the plasma membrane, including the glycocalyx, and influences on membrane fusion and uptake at the plasma membrane surface. Recent studies on the construction and availability of human chromosome-specific DNA libraries (Van Dilla et al., 1986), using a combination of dual laser flow sorting and molecular cloning procedures, has already resulted in the suggestion that it is now timely to begin the construction of chromosome-specific libraries for higher plants - the plant molecular cloning procedures are already well established; the present deficiency is in the flow karyotyping of plant chromsomes. Here one sees an opportunity for the use of suitably sychronised plant cells, and perhaps protoplasts, for large scale chromosome isolations.

M.S. Swaminathan in his survey of plant research and world agriculture at The First International Congress of Plant Molecular Biology, identified the paramount importance of cell and tissue procedures. Embryo rescue, from sexual hybridisations, protoplast fusion followed by hybrid selection and differentiation, and the use of DNA vectors would accelerate the pace of plant breeding, and also help to generate useful variability. He highlighted the fact that it should not be forgotten that while in unicellular organisms the modification of the cell is the end point of research, in higher plants, the modification of the cell is just the beginning of a long, and often difficult, process of getting a whole plant from the cultured cell and getting the altered gene expressed in the desired manner (Swaminathan 1986). Walbot (1985), in a perceptive comparative survey of the life strategies of plants and animals, has highlighted the fact that in some ways a plant body is more like a population of animals than a single animal, and that considerable physiological data suggests that plants are composed of semiautonomous units. This may further complicate aspects of the control of gene expression. Differences between cellular, including the use of mutants, and molecular approaches to achieve these objectives and the different time scales involved have recently been comprehensively discussed (Pental and Cocking, 1985).

Now that efficient regeneration of plants from rice protoplasts is possible (Abdullah et al., 1986), fusions with wild species of Oryza possessing useful agronomic traits can now be realistically undertaken to begin to transfer genes across sexual barriers (Swaminathan 1986). The recent demonstration (Pirrie and Power 1986) of the production of fertile, interspecific, gametosomatic triploid hybrids of tobacco, following the fusion of somatic leaf protoplasts (diploid) with pollen tetrad (haploid) protoplasts opens up novel possibilities in this respect. Also apart from the present extensive use of haploid plants derived from pollen cultured _in vitro_, this may be a major role in future years for haploid plants. Gene transfer across sexual barriers is also indicated by the recent immunological demonstration of the transfer of the barley nitrate reductase structural gene into tobacco by protoplast fusion, utilising a nitrate reductase deficient recipient tobacco protoplast system (Somers _et al._, 1986). Further introgression of desired genes is usually dependent on the sexual fertility of such hybrids. The finding that sexually fertile interspecic somatic hybrid plants can be produced by protoplast fusions between the cultivated tomato _Lycopersicon esculentum_, and the wild species _Lycoperiscon peruvianum_ (Kinsara _et al._, 1986), augers well for the future. Moreover, the fact that in this study the cultivated tomato was incapable of plant regeneration from protoplasts, will unquestionably simplify the regeneration requirements for somatic hybrid production.

Currently there is a renewal of interest in the application of plant cell and tissue culture for clonal forestry arising from the increasing awareness that clonal approaches to forestry have enormous potential for increased productivity of high quality and more uniform forest products. A major present limitation is the need to use juvenile tissues from young trees, since maturation is usually associated with a declining regenerative capacity. The stage is now set for a comprehensive programme in the development and improvement of methods of regenerating commercially important trees, and also the development of rejuvenation techniques involving, for instance, demethylation of DNA to reactivate inactive genes, and manipulation of the cytoplasm for its rejuvenation using for instance protoplast fusion methodology for cytoplasmic hybrid production (Kumar and Cocking 1986). Whilst in clonal forestry somaclonal variation is usually undesirable, variation arising from cell and tissue culture is often desired in horticultural and crop species.

Nowhere more so than when trying to understand the basis of such somaclonal variation is the need for a better knowledge of the molecular biology of the plant cell more evident. Because regenerated plants are usually derived from single cells in the culture system, mosaics occur at very low frequency, and as a result somaclonal variation is often superior to mutation breeding. It has been suggested (Evans and Sharp, 1986) that plant regeneration from in vitro cultures could act as a sieve that eliminates deleterious genetic changes. Importantly each somaclone differs from the parent by only a single, or small number of genetic changes. As pointed out by Evans and Sharp (1986), this means that plant cell and tissue culture can provide somaclones to identify distinct mRNA correlated with a particular phenotype. In the future these mRNAs could be used to prepare cDNA clones that could be tested for genetic linkage with a specific agronomic character, ultimately resulting in cDNA clones for agronomically useful genes in crops.

In trying to look into the future needs of plant cell and tissue culture from the viewpoint of plant cell biology as it moves into the next century, and trying to forecast what will be significant and worthwhile, I have been struck by the pertinent observation of M.S.Swaminathan at the conclusion of his Presidential Address to the XV International Congress of Genetics held in Delhi in 1983. He recalled that when he visited the village in Czechoslovakia where Mendel was born, the local villagers remembered Mendel for the fire station he had built in the village more than for the laws of heredity he had left behind (Swaminthan 1983). Let us also try to keep in mind what Albert Einstein once said "concern for man himself and his fate must always form the chief interest of all technical endeavours in order that the creation of our minds shall be a blessing and not a curse". In trying to look into the future of Plant Cell Biology in the 21st Century, we should also remember the words of Winston S. Churchill that "The empires of the future are the empires of the mind".

REFERENCES

Abdullah R, Cocking EC, Thompson JA (1986). Efficient plant regeneration from rice protoplasts through somatic embryogenesis. Bio/Technology (in the press).

Aguayo JB, Blackband SJ, Schoeniger J, Mattingly MA,

Hintermann M (1986). Nuclear magnetic resonance imaging of a single cell. Nature 322: 190-191.

Alberts B, Bray D, Lewis J, Raff M, Roberts K, Watson JD (1983). Molecular Biology of the cell. Gartland Publishing Inc, New York and London.

Ammirato PV (1983). Embryogenesis. In Evans DA, Sharp WR, Ammirato PV, Yamada, Y (eds): "Handbook of Plant Cell Culture," Macmillan Publ Co, New York: Vol I, pp 82-123.

Astbury WT (1951). Adventures in Molecular Biology (The Harvey Lecture, 1950) Harvey Soc Series 46.

Chargaff E (1971). Preface to a grammar of Biology: A hundred years of nucleic acid research. Science 172: 637-642.

Cocking EC (1985). Protoplasts from root hairs of crop plants. Biotechnology 3: 1104-1106.

Evans DA, Sharp WR (1986). Applications of somaclonal variation. Biotechnology 4: 528-532.

Fry SC (1986). Cross-linking of matrix polymers in the growing cell walls of angiosperms. Ann Rev Plant Physiol 37: 165-186.

Gupta PK, Durzan DJ (1986). Somatic polyembryogenesis from callus of mature sugar pine embryos. Biotechnology 4: 643-645.

Hamill JD, Parr AJ, Robins RJ, Rhodes MJC (1986). Secondary product formation by cultures of Beta vulgaris and Nicotiana tabacum transformed with Agrobacterium rhizogenes. Plant Cell Reports 5: 111-114.

Hammatt N, Kim HI, Davey MR, Nelson RS, Cocking EC (1986). Plant regeneration from cotyledon protoplasts of wild Glycine species sexually incompatible with the soybean (G. max L. Merr). (Submitted).

Hopkinson J, Moustou CM, Charlwood IKA, Newberry JE, Charlwood BV (1985). Fourier transform infra-red spectra analysis of plant cells and protoplasts. Plant Cell Reports 4: 321-324.

Horsch RB, Fry JE, Hoffmann NL, Eichholtz D, Rogers SG, Fraley RT (1985). A simple and general method for transferring genes into plants. Science 227: 1229-1231.

Kinsara AM, Patnaik SN, Cocking EC, Power JB (1986). Somatic hybrid plants of Lycopersicon esculentum Mill. and Lycopersicon peruvianum Mill. Journal of Plant Physiology (in the press).

Kumar A, Cocking EC (1986). Protoplast fusion: a novel approach to organelle genetics in higher plants. Amer J Bot (in the press).

Mersey BG, Griffing LR, Fowke LC (1985). The use of protoplasts in the study of coated vesicles. In Pilet PE (ed): "The Physiological Properties of Plant Protoplasts," Springer-Verlag, Heidelberg, pp 45-53.

Nomoru K, Komamine A (1986). Embryogenesis from microinjected single cells in a carrot cell suspension culture. Plant Science 44: 53-58.

Norman PM, Wingate VPM, Fitter MS, Lamb CJ (1986). Monoclonal antibodies to plant plasma-membrane antigens. Planta 167: 452-459.

Pental D, Cocking EC (1985). Some theoretical and practical possibilities of plant genetic manipulation using protoplasts. Hereditas (suppl) 3: 83-93.

Pirrie A, Power JB (1986). The production of fertile triploid somatic hybrid plants Nicotiana glutinosa (n) + N. tabacum (2n) via gametic:somatic protoplast fusion. Theor Appl Genet 72: 48-52.

Reinert J (1958). Morphogenese und ihre kontrolle an gewebekulturum aus carotten. Naturwissenschaften 45: 344-345.

Somers DA, Narayanan KR, Kleinhofs A, Cooper-Bland S, Cocking EC (1986). Immunological evidence for transfer of the barley nitrate reductase structural gene to Nicotiana tabacum by protoplast fusion. Mol Gen Genet (in the press).

Steward FC, Mapes MO, Mears K (1958). Growth and organized development of cultured cells II. Organization in cultures grown from freely suspended cells. Am J Bot 45: 705-708.

Street HE (1977). Engineering with plant cells. Interdisciplinary Science Rev 2: 62-74.

Stuart DA, Nelsen J, McCall CM, Strickland SG, Walker KA (1985). Physiology of the development of somatic embryos in cell cultures of alfalfa and celery. In Zaitlin M, Day P, Hollaender A (eds): "Biotechnology in Plant Science," Academic Press, pp 35-47.

Sung ZR (1985). Developmental Biology of plant embryogenesis. In Freeling M (ed): "Plant Genetics," New York: Alan R Liss, 35, pp 115-128.

Swaminathan MS (1983). Genetic conservation: microbes to man. In Chopra VL, Josh BC, Sharma RP, Bansal HC (eds): "Genetics: New Frontiers, (Proceedings of the XV Int Congress of Genetics)": Bowker, Vol 1.

Swaminathan MS (1986). Plant research and world agriculture. Plant Molecular Biology Reporter 4: 1-17.

Tabata M, Fujita Y (1985). Production of shikonin by

plant cell cultures. In Zaitlin M, Day P, Hollaender A (eds): "Biotechnology in Plant Science," Academic Press, pp 207-218.

Thompson JA, Abdullah R, Cocking EC (1986). Protoplast culture of rice (Oryza sativa L.) using media solidified with agarose. Plant Science (in the press).

Van Dilla MA, Deaven LL, Albright KL, Allen NA, Aubuchon MR, Bartholdi MF, Brown NC, Campbell EW, Carrano AV, Clark LM, Cram LS, Crawford BD, Fuscoe JC, Gray JW, Hildebrand CE, Jackson PJ, Jett JH, Longmire JL, Lozes CR, Luedemann ML, Martin JC, McNinch JS, Meincke LJ, Mendelsohn ML, Meyne J, Moyzis RK, Munk AC, Perlman J, Peters DC, Silva AJ, Trask BJ. Human Chromosome-specific DNA libraries: construction and availability. Biotechnology 4: 537-552.

Plant Tissue and Cell Culture, pages 17–31

INVESTIGATIONS OF CELL STRUCTURE USING CULTURED CELLS AND PROTOPLASTS

Larry C. Fowke

Department of Biology, University of Saskatchewan, Saskatoon, Saskatchewan, Canada S7N OWO

INTRODUCTION

Cultured plant cells and protoplasts provide excellent experimental material for studies of the structure and function of plant cells and their organelles. Current research with such tissue culture systems is making a significant contribution to the field of plant Cell Biology. The major purpose of this chapter is to illustrate this point by providing specific examples of the type of research which is in progress. Three examples have been chosen: the discovery and exploration of a pathway for endocytosis in plants, the isolation of plant cell organelles for studies of their structure and function and the examination of the distribution and role of the plant cytoskeleton. The use of tissue culture systems has greatly facilitated research in each of these areas.

ENDOCYTOSIS BY PLANT PROTOPLASTS

Research over the past few years has revealed a very specific pathway (receptor mediated endocytosis) for the uptake of proteins into animal cells (see reviews by Pearce and Bretscher 1981, Fine and Ockleford 1984, Willingham and Pastan 1984, Goldstein et al. 1985). This pathway involves tiny cytoplasmic vesicles, known as coated vesicles, which consist of a membranous sac enclosed within a characteristic cage made primarily of the protein clathrin. The first step of the internalization process involves binding of proteins (ligands) to specific receptors on clathrin coated regions of the plasma membrane. Coated pits formed by invaginations of the plasma membrane pinch off to form coated vesicles. These deliver the ligands first to the endosomal compartment where they are dissociated from their receptors and finally to the lysosome for degradation. A number of proteins (eg. low density lipoprotein, transferrin, insulin) are endocytosed by animal cells using this pathway or a variation of it.

Although coated vesicles are widespread in plants (Newcomb 1980), our understanding of their function in plant cells is rather limited. Indirect evidence has suggested that they are exocytic and probably involved in the secretion of cell wall components (see reviews by Newcomb 1980, Fowke et al. 1983). Some authors have also suggested that coated vesicles may be endocytic.

Research with plant protoplasts has clearly demonstrated that coated vesicles are involved in endocytosis via a pathway very similar to that found in animal cells (Tanchak et al. 1984, Joachim and Robinson 1984). Protoplasts will endocytose cationized ferritin (CF) by coated pits which invaginate to form coated vesicles (Fig. 1). The vesicles uncoat and deliver the CF to dictyosomes, multivesicular bodies and a very interesting tubular membrane system bearing clathrin coated regions, the partially coated reticulum. The latter resembles in many ways the endosomal component of animal cells. Recent research indicates that with prolonged incubation, the CF is delivered to the vacuole of the protoplast (Tanchak, personal communication). This pathway is not restricted to protoplasts but apparently also operates during the uptake of heavy metal salts by root cap cells (Hubner et al. 1985).

The existence of an uptake pathway involving coated vesicles in plant cells is exciting. Perhaps plant cells are able to select particular molecules for endocytosis as do animal cells. Possible candidates for such a specific mechanism would include plant hormones, metal ions, fungal elicitors and toxins. The ability of plant cells to regulate the entry of such molecules could be very important. Further work is necessary to establish whether in fact these molecules enter cells via coated vesicles and to establish the nature of plasma membrane receptors if they are involved. A second possible role for this pathway might be in the process of detoxification. The fate of endocytosed CF in protoplasts is the central vacuole, the lysosomal compartment

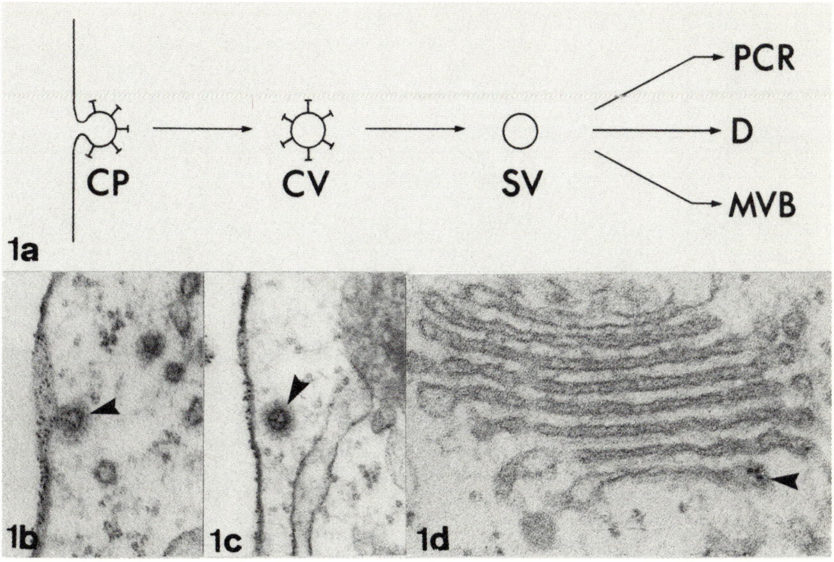

Figure 1. (a) Diagrammatic representation of the pathway for uptake of cationized ferritin (CF) by soybean protoplasts. Based on the data of Tanchak et al. 1984. CP - coated pit, CV - coated vesicle, SV - smooth vesicle, PCR - partially coated reticulum, D - dictyosome, MVB - multivesicular body. (b) electron micrograph showing CF in coated pit (arrow) of soybean protoplast after 10 seconds labelling. (c) electron micrograph showing CF in coated vesicle (arrow) of soybean protoplast after 10 seconds labelling. (d) electron micrograph showing CF (arrow) in dictyosome of soybean protoplast after 12 minutes labelling. 1b and 1c = X 87,000; 1d = X 110,000. From Tanchak et al. 1984.

in plants (Marty et al. 1980). Toxic materials could thus be directed by coated vesicles to the vacuole of the cell for degradation. Finally, coated vesicles may provide a mechanism to recycle the plasma membrane. The secretion of large amounts of matrix materials via smooth dictyosome vesicles during active cell wall formation results in an increase in the surface area of the plasma membrane. Coated vesicles may be responsible for retrieving the excess plasma membrane.

Coated vesicles likely have other important functions in plant cells. Evidence to date suggests that they are involved in enzyme sorting (Griffing and Fowke 1985) and possibly exocytosis of materials to the developing cell wall (Griffing et al. 1986).

ISOLATION OF CELL ORGANELLES

Most cell organelles have been isolated from either cultured plant cells or protoplasts (see Table 1 for list of organelles and selected references). Cell suspensions are suitable for isolation experiments because they are easily cultured and consist of a relatively homogeneous population of rapidly growing thin walled cells. In some cases it is even possible to achieve a certain level of synchronization of the cell population. Protoplasts, however, are the system of choice for organelle isolation. The absence of a cell wall makes them very fragile and easily lysed by rather gentle physical methods or simple changes in the osmoticum of the medium. The harsh methods used to disrupt cells are not necessary. The use of protoplasts is not always possible since not all cell types are amenable to protoplasting. The most common sources of protoplasts are suspension cultured cells and leaf mesophyll.

Cultured cells have proved particularly useful for isolating very small organelles such as coated vesicles or the tubulin subunits of microtubules (Table 1 and Fig. 2a). The rather rough methods necessary to break plant cells apparently do not damage these organelles. Microtubule subunits isolated from cell suspensions have been used to repolymerize intact microtubules (Morejohn and Fosket 1982, Mizuno 1985).

Successful isolation of larger organelles (e.g. nuclei, plastids, vacuoles) has been achieved from protoplasts rather than from cells. Structural and biochemical studies of such organelles have provided important information regarding the role of these organelles in plant cells. Our understanding of vacuole function, for example, has benefitted directly from studies of vacuoles isolated from plant protoplasts (Marty et al. 1980). The demonstration of many hydrolytic enzymes in isolated vacuoles provides strong support for the suggestion that the vacuole serves a lysosomal function in plant cells. The endocytosis of CF and its delivery to the central vacuole of protoplasts is consistent with this role (see above). Isolated vacuoles also provide an excellent tool for studying the storage and transport properties of the vacuole (see review by Leonard and Rayder 1985).

TABLE 1. Isolation of Cell Organelles

Organelle	Source	Selected References
Nuclei	P	Dunham and Bryant 1983 (Review)
	P	Saxena et al. 1985
Chromosomes	P	Hadlaczky et al. 1983
Vacuoles	P	Leonard and Rayder 1985 (Review)
Plasma Membrane	P	Perlin and Spanswick 1980
	P	Polonenko and MacLachlan 1984
Mitochondria	P	Nishimura et al. 1982
Plastids	P	Kobza and Edwards 1984
	P	Echeverria et al. 1985
Microbodies	P	Robinson 1985 (Review)
Coated Vesicles	P	Mersey et al. 1985
	C	Depta and Robinson 1986
Microtubule Subunits	C	Morejohn and Fosket 1982
	C	Mizuno 1985

P = Protoplasts
C = Suspension cultured cells

Organelles isolated from plant protoplasts have also been used for uptake experiments. Techniques are available for inducing the uptake of different organelles (e.g. chloroplasts, nuclei, chromosomes) into protoplasts (see reviews

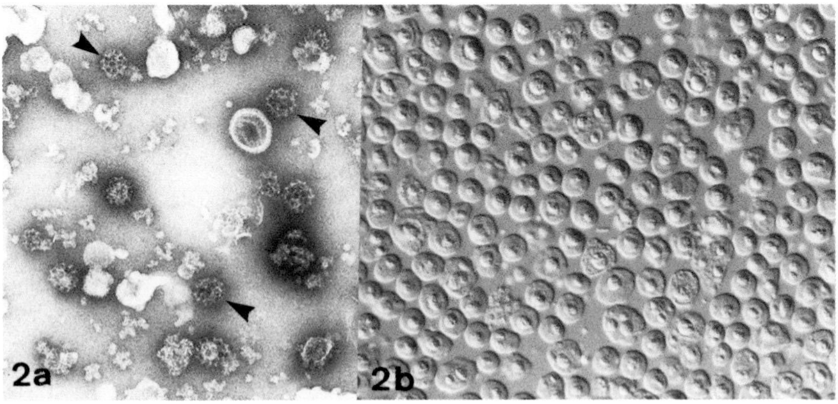

Figure 2. (a) electron micrograph of negatively stained coated vesicles (arrows) isolated from cultured tobacco cells. (b) light micrograph (Nomarski optics) showing nuclei isolated from protoplasts of *Brassica nigra*. 2a = X 45,000, 2b = X 500. Figure 2b kindly supplied by P.K. Saxena.

by Lorz 1985, Fowke 1986). The primary goal of this research has been to demonstrate the uptake and expression of foreign DNA by the host protoplasts. Uptake of both chloroplasts and chromosomes has been demonstrated but expression of the transferred DNA has not. Successful uptake of nuclei into protoplasts was reported in the 1970s (e.g. Potrykus and Hoffman 1973); however, in these early experiments genetic transformation was not possible. Recently, successful nuclear uptake and expression of foreign DNA has been achieved (Saxena et al. 1986). A technique was devised for the rapid and efficient isolation of nuclei from protoplasts (Saxena et al. 1985 and Fig. 2b). Uptake of *Vicia hajastana* nuclei isolated in this manner, into protoplasts of an auxotrophic cell line of *Datura innoxia* resulted in the recovery of prototrophic clones. The presence of *Vicia* genomic DNA in some prototrophic clones was confirmed by dot-blot hybridization. The successful transformation of plant cells with nuclei certainly provides encouragement for further uptake studies using isolated chromosomes and possibly other DNA containing organelles such as mitochondria and chloroplasts.

STUDIES OF THE CYTOSKELETON

The plant cytoskeleton consists of different types of protein filaments, the best characterized being microtubules and actin microfilaments. Microtubules play a central role in cell division and cell wall formation while microfilaments are believed to interact with myosin molecules to produce cytoplasmic streaming (see review by Hepler 1985).

Microtubules

Most cylindrical plant cells contain a network of microtubules in the cortical cytoplasm. These cortical microtubules are arranged in a transverse orientation around the cell and are believed to orient the deposition of cellulose microfibrils. Recent research suggests that the cortical microtubules are arranged in helicies of varying pitch and that the complete array of microtubules is capable of rapidly shifting orientation (Lloyd 1984).

Current studies of cultured plant cells and protoplasts are providing exciting information regarding the distribution of cortical microtubules and their role in wall deposition and cell shape determination.

One cell culture system which is particularly interesting was developed by Kohlenbach and Schmidt (1975). They defined conditions under which *Zinnia* leaf mesophyll cells would differentiate into mature xylem elements in culture. Since initially described, this system has been examined biochemically (e.g. Fukuda and Komamine 1982) and cytologically (e.g. Burgess and Linstead 1984). Differentiating *Zinnia* cells have been explored recently using antitubulin and immunofluorescence techniques (e.g. Falconer and Seagull 1985, Roberts et al. 1985). During early differentiation microtubules which are initially aligned longitudinally in the leaf cells, reorient to a more transverse orientation, group into bands and underlie newly forming secondary wall bands (Fig. 3). In the presence of Taxol, a microtubule stabilizer, the microtubules fail to reorient. However, they do aggregate into bundles which predict the position of new secondary cell wall bands. These bands are oriented longitudinally rather than transversely as observed during normal differentiation (Falconer and Seagull 1985 and Fig. 3). If cells are treated with the herbicide amiprophosmethyl, a

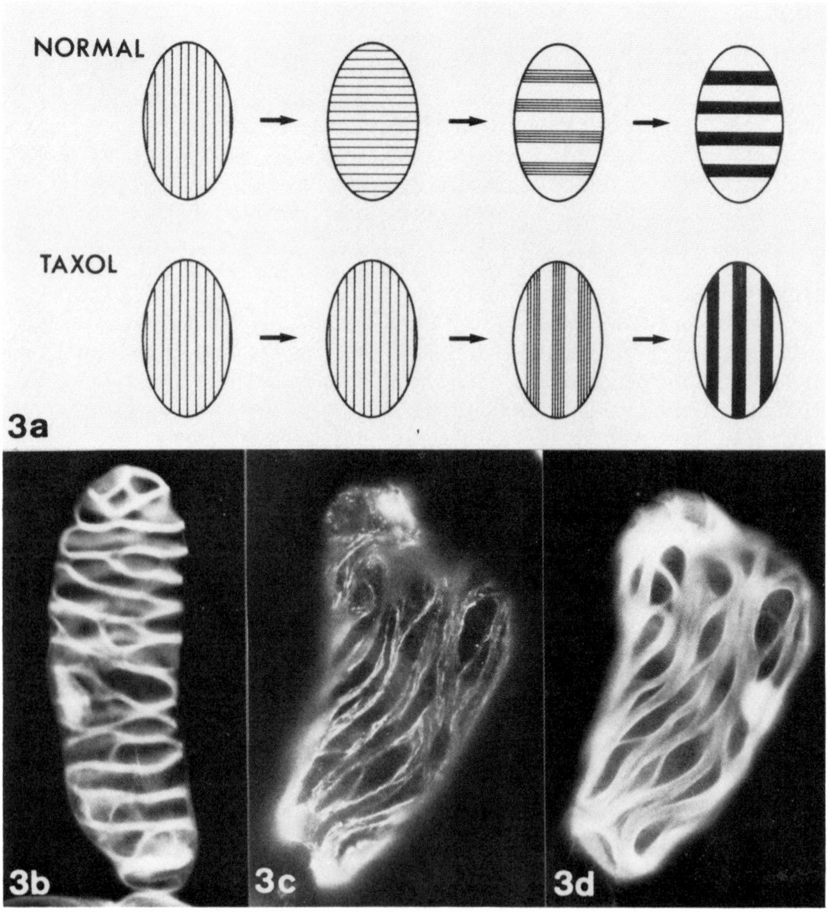

Figure 3. (a) Diagram showing effect of Taxol on the distribution of microtubules (thin lines) and subsequent formation of secondary cell walls (heavy lines) in cultured Zinnia leaf cells. Based on data of Falconer and Seagull (1985). (b) Normal Calcofluor stained tracheary element derived from cultured leaf cell showing secondary wall bands. (c) Microtubule pattern in tracheary element which differentiated in the presence of Taxol. (d) Same cell as in 3c, stained with Calcofluor to show secondary cell wall bands. 3b = X 400; 3c and 3d = X 550. Micrographs kindly supplied by R.W. Seagull (3c and 3d from Falconer and Seagull 1985).

microtubule depolymerizer, during the early stages of culture, microtubules are removed and the cells become spherical (Seagull, personal communication). During a subsequent wash period random arrays of microtubules are reestablished and secondary wall formation commences. The new cell wall is established as a reticulate network, a pattern which seems to reflect the distribution of cortical microtubules. These results lend weight to the proposal that cortical microtubules orient cellulose microfibrils during cell wall formation and are involved in the determination of cell shape. It has been suggested that during the differentiation of normal xylem vessel elements, the aggregation of microtubules into bundles prior to secondary wall deposition may in fact collect cellulose synthases together to permit the formation of discrete secondary cell wall bands (Herth 1985).

Recent studies with cultured cotton fibers have provided strong supporting evidence for the idea that cortical microtubules exist in dynamic helices in plant cells (Seagull 1986). Microtubule helices are clearly visible in developing cotton fibers. During the transition from primary to secondary cell wall deposition the microtubule arrangement shifts from a flat helix to a steep helix. It is not clear how this change in helical pattern occurs but it is consistent with the idea that complete arrays of microtubules are capable of shifting orientation. The pattern of cellulose microfibrils parallels the microtubules during cell growth as would be expected if microtubules do indeed orient the deposition of cellulose microfibrils.

The application of immunofluorescence techniques and different inhibitors to both the *Zinnia* system and cultured cotton fibers is providing important information regarding the relationship of microtubules to plant cell development. Future research will contribute to our basic understanding of the role of microtubules in plant cells.

MICROFILAMENTS

Actin microfilaments act in conjunction with myosin to facilitate contractile processes in non-muscle cells of animals. Actin is also present in plant cells and the elegant work with Characean algae indicates that they are also involved in plant cytoplasmic streaming (Jackson 1982, Sheetz and Spudich 1983).

The availability of fluorescent fungal metabolites (e.g. phalloidin) which bind specifically to F-actin makes it possible to examine the distribution of actin microfilaments in whole cells of plants. This technique has been used to study actin distribution in a wide variety of plant cells (e.g. Parthasarathy et al. 1985). Such studies indicate that actin microfilaments are often arranged along the longitudinal axis of cells parallel to the direction of cytoplasmic streaming and sometimes form quite distinct cables.

These fluorescence techniques have also been applied to cultured plant cells (Fig. 4). A recent study of cultured alfalfa and *Vicia* cells using rhodamine labelled phalloidin has revealed new and important information regarding the dynamic nature of actin microfilaments during the cell cycle (Seagull - personal communication). Cultured plant cells contain different arrays of microfilaments which change in a predictable manner throughout the cell cycle. Major reorganization occurs in relation to the processes of cell expansion and mitosis. The observation of actin microfilaments in the spindle region at metaphase and anaphase is of particular interest in light of the current controversy regarding the mechanism of chromosome movement to the poles at anaphase (Forer 1985). It is also interesting that microfilaments seem to be initiated at the nuclear envelope surface in a manner reminiscent of microtubule initiation. This research clearly demonstrates the value of using cultured plant cells for investigations of the plant cytoskeleton. Further studies of this nature should provide critical information regarding the function of actin microfilaments (e.g. in cell division and cytoplasmic streaming) as well as the nature of the factors which control the dynamic microfilament arrays in plant cells.

SUMMARY

Current research with suspension cultured cells and protoplasts is providing new and exciting insights into the structure and function of the plant cell and its organelles. Studies of freshly isolated protoplasts resulted in the discovery of a coated vesicle-mediated endocytic pathway in plants. Further work with plant protoplasts should determine whether this pathway plays an important role in plants, perhaps in regulating molecular uptake, detoxification

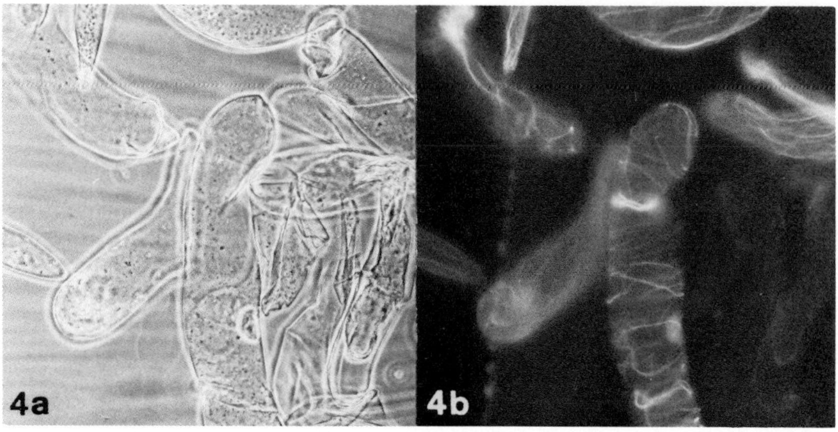

Figure 4. (a) phase contrast light micrograph showing cultured cells of white spruce. (b) same cells stained with rhodamine labelled phalloidin and examined in a fluorescence microscope. The actin microfilaments are arranged transversely in some cells and longitudinally in others. Both micrographs X 200. Kindly supplied by I.C. Hakman.

reactions or plasma membrane recycling. Cultured cells and protoplasts are particularly well suited for the isolation of plant cell organelles. Because protoplasts are more easily ruptured, they are used more extensively for the isolation of larger membrane bounded organelles. Organelles isolated from cultured cells and protoplasts have been used for studies of organelle structure and function as well as in research directed towards genetic transformation. Finally, studies of cultured cells and protoplasts are providing new and exciting information regarding the cytoskeleton of plant cells. The application of immunofluorescence techniques using antitubulin to studies of cytodifferentiation with _Zinnia_ leaf cells and cotton fibers has contributed to our understanding of the distribution of microtubules and their role in cell wall formation and cell shaping. Studies of cultured cells with fluorescent labelled phalloidin has also expanded our knowledge of microfilaments and their role in dividing and elongating cells.

ACKNOWLEDGEMENTS

I wish to express my appreciation to P. Rennie, D. Dyck and E. Peters for their help in preparing this manuscript. A special thanks to R. Seagull for generously supplying both published and unpublished data.

REFERENCES

Burgess J, Linstead P (1984). In vitro tracheary element formation: structural studies and the effect of tri-iodobenzoic acid. Planta 160:481-489.

Depta H, Robinson DG (1986). The isolation and enrichment of coated vesicles from suspension-cultured carrot cells. Protoplasma 130:162-170.

Dunham VL, Bryant JA (1983). Nuclei. In Hall JL, Moore AL (eds): "Isolation of Membranes and Organelles from Plant Cells," New York, Academic Press, pp 237-275.

Echeverria E, Boyer C, Liu K-C, Shannon J (1985). Isolation of amyloplasts from developing maize endosperm. Plant Physiol 77:513-519.

Falconer MM, Seagull RW (1985). Xylogenesis in tissue culture: taxol effects on microtubule reorientation and lateral association in differentiating cells. Protoplasma 128:157-166.

Fine RE, Ockleford CD (1984). Supramolecular cytology of coated vesicles. Int Rev Cytol 91:1-43.

Forer A (1985). Does actin produce the force that moves a chromosome to the pole during anaphase? Can J Biochem Cell Biol 63:585-598.

Fowke LC (1986). Ultrastructural cytology of cultured plant tissues, cells and protoplasts. In Vasil IK (ed): "Cell Culture and Somatic Cell Genetics of Plants," Vol. 3, "Regeneration and Variability", Orlando, Academic Press (in press).

Fowke LC, Griffing LR, Mersey BG, Van der Valk P (1983). Protoplasts for studies of the plasma membrane and associated cell organelles. In Potrykus I, Harms CT, Hinnen A, Hutter R, King PJ, Shillito RD (eds): "Protoplasts 1983, Lecture Proceedings", Basel, Birkhauser Verlag, pp 101-110.

Fukuda H, Komamine A (1982). Lignin synthesis and its related enzymes as markers of tracheary-element differentiation in single cells isolated from the mesophyll of _Zinnia elegans_. Planta 155:423-430.

Goldstein JL, Brown MS, Anderson RGW, Russel DW, Schneider WJ (1985). Receptor-mediated endocytosis: concepts emerging from the LDL receptor system. Ann Rev Cell Biol 1:1-39.

Griffing LR, Fowke LC (1985). Cytochemical localization of peroxidase in soybean suspension culture cells and protoplasts: intracellular vacuole differentiation and presence of peroxidase in coated vesicles and multivesicular bodies. Protoplasma 128:22-30.

Griffing LR, Mersey BG, Fowke LC (1986). Cell fractionation analysis of glucan synthase I and II distribution and polysaccharide secretion in soybean protoplasts. Evidence for the involvement of coated vesicles in wall biogenesis. Planta 167:175-182.

Hadlaczky G, Bisztray G, Praznovszky T, Dudits D (1983). Mass isolation of plant chromosomes and nuclei. Planta 157:278-285.

Hepler PK (1985). The plant cytoskeleton. In Robards AW (ed): "Botanical Microscopy 1985", Oxford, Oxford University Press, pp 233-262.

Herth W (1985). Plasma-membrane rosettes involved in localized wall thickening during xylem vessel formation of Lepidium sativum L. Planta 164:12-21.

Hubner R, Depta H, Robinson DG (1985). Endocytosis in maize root cap cells. Evidence obtained using heavy metal salt solutions. Protoplasma 129:214-222.

Jackson WT (1982). Actomyosin. In Lloyd CW (ed): "The Cytoskeleton in Plant Growth and Development," London, Academic Press, pp 3-29.

Joachim SA, Robinson DG (1984). Endocytosis of cationic ferritin by bean leaf protoplasts. Eur J Cell Biol 34: 212-216.

Kobza J, Edwards GE (1984). Isolation of organelles: chloroplasts. In Vasil IK (ed): "Cell Culture and Somatic Cell Genetics", Vol 1, "Laboratory Procedures and Their Applications", Orlando, Academic Press, pp 471-482.

Kohlenbach HW, Schmidt B (1975). Cytodifferenzierung in Form einer direkten Umwandlung isolierter Mesophyllzellen zu Tracheiden. Z Pflanzenphysiol 75:369-374.

Leonard RT, Rayder L (1985). The use of protoplasts for studies on membrane transport in plants. In Fowke LC, Constabel F (eds): "Plant Protoplasts", Boca Raton, CRC Press, pp 105-118.

Lloyd CW (1984). Toward a dynamic helical model for the influence of microtubules on wall patterns in plants. Int Rev Cytol 86:1-51.

Lorz H (1985). Isolated cell organelles and subprotoplasts - their roles in somatic cell genetics. In Dodds JH (ed): "Plant Genetic Engineering", Cambridge, Cambridge University Press, pp 27-59.

Marty F, Branton D, Leigh RA (1980). Plant vacuoles. In Tolbert NE (ed): "The Biochemistry of Plants", Vol. 1, "The Plant Cell", New York, Academic Press, pp 625-658.

Mersey BG, Griffing LR, Rennie PJ, Fowke LC (1985). The isolation of coated vesicles from protoplasts of soybean. Planta 163:317-327.

Mizuno K (1985). In vitro assembly of microtubules from tubulins of several higher plants. Cell Biol Int Rep 9:13-21.

Morejohn LC, Fosket DE (1982). Higher plant tubulin identified by self-assembly into microtubules in vitro. Nature 297:426-428.

Newcomb EH (1980). Coated vesicles: their occurrence in different plant cell types. In Ockleford CJ, Whyte A (eds): "Coated Vesicles", Cambridge, Cambridge University Press, pp 55-68.

Nishimura M, Douce R, Akazawa T (1982). Isolation and characterization of metabolically competent mitochondria from spinach leaf protoplasts. Plant Physiol 69:916-920.

Parthasarathy MV, Perdue TD, Witztum A, Alvernaz (1985). Actin network as a normal component of the cytoskeleton in many vascular plant cells. Amer J Bot 72:1318-1323.

Pearse BMF, Bretscher MS (1981). Membrane recycling by coated vesicles. Ann Rev Biochem 50:85-101.

Perlin DS, Spanswick RM (1980). Labelling and isolation of plasma membranes from corn leaf protoplasts. Plant Physiol 65:1053-1057.

Polonenko DR, Maclachlan GA (1984). Plasma-membrane sheets from pea protoplasts. J Exp Bot 35:1342-1349.

Potrykus I, Hoffmann F (1973). Transplantation of nuclei into protoplasts of higher plants. Z Pflanzenphysiol 69: 287-289.

Roberts K, Burgess J, Roberts I, Linstead P (1985). Microtubule rearrangement during plant cell growth and development: an immunofluorescence study. In Robards AW (ed): "Botanical Microscopy 1985", Oxford, Oxford University Press, pp 263-283.

Robinson DG (1985). "Plant Membranes: Endo- and Plasma Membranes of Plant Cells". New York, John Wiley and Sons.

Saxena PK, Fowke LC, King J (1985). An efficient procedure for isolation of nuclei from plant protoplasts. Protoplasma 128:184-189.

Saxena PK, Mii M, Crosby WL, Fowke LC, King J (1986). Transplantation of isolated nuclei into plant protoplasts: a novel technique for introducing foreign DNA into plant cells. Planta (in press).

Seagull RW (1986). Changes in microtubule organization and wall microfibril orientation during in vitro cotton fiber development: an immunofluorescence study. Can J Bot (in press).

Sheetz MP, Spudich JA (1983). Movement of myosin-coated beads on actin cables in vitro. Nature 303:31-35.

Tanchak MA, Griffing LR, Mersey BG, Fowke LC (1984). Endocytosis of cationized ferritin by coated vesicles of soybean protoplasts. Planta 162:481-486.

Willingham MC, Pastan I (1984). Endocytosis and exocytosis: current concepts of vesicle traffic in animal cells. Int Rev Cytol 92:51-92.

Plant Tissue and Cell Culture, pages 33–44

EXTRACHROMOSOMAL DNA AND CELL DIFFERENTIATION IN CULTURED PEA ROOTS (*PISUM SATIVUM*)

Jack Van't Hof

Biology Department, Brookhaven National Laboratory, Upton, New York 11973

INTRODUCTION

Locked within the root meristem are the secrets of cell division, cell differentiation and organ formation. Any median longitudinal section of a growing root provides visible evidence of these biological phenomena. Cell division is seen near the apex, enlarging differentiating cells are seen in the elongation zone, and a few mm from the apex the tissue is filled with a variety of mature cells, some with distinctive secondary walls, some with root hairs. The fixed pattern of cell files in the root tells us that the dividing cells are the progenitors of those that are differentiated but how a cell is transformed from a meristematic progenitor to a xylem element remains a mystery. It is safe to assume that certain molecular and metabolic changes occur in precursor cells before they are committed to a particular path of differentiation and our task is to determine the nature of these initial changes. The work described here addresses this task. It is focused specifically on the transition of cells from the meristematic to the differentiated state and it is guided by the idea that this transition requires changes in genetic information reflected at the level of chromosomal DNA replication.

The task requires four essentials. First, the biological material must have identifiable meristematic cells capable of unambiguous differentiation. Second, the activity of the meristematic cells and their ultimate fate must be experimentally controllable. Third, the DNA of the cells must lend itself to selective labeling and isolation procedures, and fourth, the DNA must be amenable to biophysical and molecular analyses.

These four essentials are provided by cultured excised pea roots and the cells therein. The results presented below were obtained with this manipulatable system.

There are two parts to the discussion. The first part concerns the cytological and histological aspects of the system. It details the distribution of meristematic cells within the cell cycle, how this distribution is changed by altering the carbohydrate concentration in the culture medium, and the selective labeling of precursors that differentiate into specific cell types in mature tissue of the root. The second part discusses chromosomal DNA replication and the excision of certain DNA molecules from replicating chromosomal DNA prior to cell differentiation. These free molecules are called extrachromosomal DNA. They are of nuclear origin and it is proposed that their production is an initial step in the transition from a meristematic cell to one that will form vascular parenchyma or a part of the root cap. A step that involves the elimination and/or rearrangement of chromosomal DNA.

RESULTS

Carbohydrate Starvation Allows Selective Labeling of Meristematic Precursor Cells with a Latent Distinctive Nuclear Phenotype

In the meristem of a growing pea root approximately 88% of the cells are meristematic. At any given moment 35% of these cells are in G_1 phase, 22% are in S phase, 34% are in G2 phase, and 9% are in mitosis (Evans and Van't Hof, 1975). This distribution is altered significantly if the cells are deprived of carbohydrate. When primary roots are excised and cultured for 24 hr or longer in White's medium without sucrose, the meristematic cells stop dividing and accumulate in the G_1 and G_2 phases in approximately equal numbers (Van't Hof, 1966; Webster and Van't Hof, 1970; Evans and Van't Hof, 1974; Van't Hof, 1976).

Cultured carbohydrate-starved meristems have no division figures and very few cells in S phase. During the first 24 hr of starvation the number of cells in S phase, those that are replicating chromosomal DNA, decreases rapidly from 22% to 1.3%. Later, after 24 hr, the number further decreases at a slow constant rate reaching less than 0.07% by 100 hr (Van't Hof et al, 1983). The cells remaining in S phase accumulate temporarily in late S phase after replicating about 80% of their genome (Van't Hof et al, 1983, 1986). These kinetic data indicate that approximately 2.2% of the cells in a 2 mm meristematic root tip are subject to accumulation in late S phase and display a retarded DNA replication phenotype.

The retardation of certain cells in late S phase while all others are arrested in either the G_1 or G_2 phases makes it possible to selectively label these cells with [^{3}H]-thymidine while they are still completing DNA replication. Their replication rate is slow but a high concentration of isotope (10 to 25 μCi/ml) and an hour pulse label the nuclei well. Autoradiograms of nuclei isolated from the meristematic tip show that 90% of the labeled nuclei are spherical or round (Figure 1a). These nuclei remain

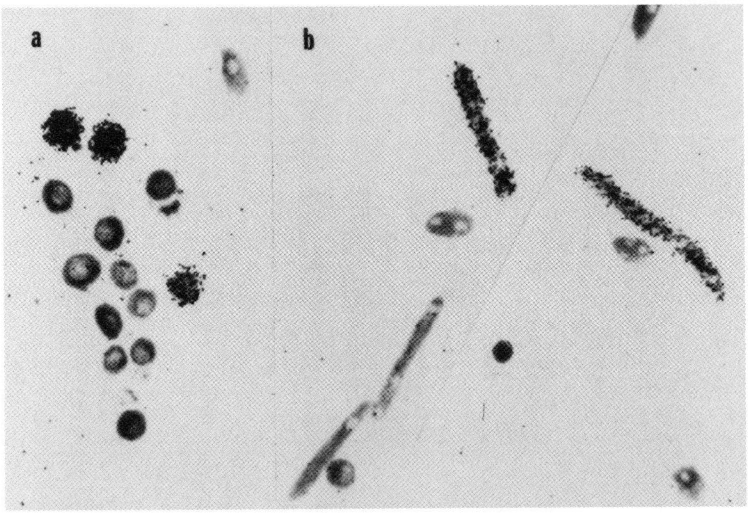

Figure 1. Autoradiograms of isolated nuclei from cultured pea roots. 1a, Round labeled nuclei from tips starved of sucrose for 48 hr. 1b, Elongated labeled nuclei from starved roots 72 hr after being labeled then fed 2% sucrose. (from Van't Hof et al, 1986).

spherical as long as the root tips are not fed carbohydrate. On the other hand, if fed sucrose the labeled nuclei undergo a change. Within 72 hr 90% become elongated having a distinctive oblong shape (Figure 1b). The transformation from a spherical to an elongated shape upon provision of sucrose is accompanied by the resumption of cell division, DNA replication and root growth, but the cells with labeled nuclei do not participate in these functions. Instead, the cells elongate and differentiate from the G2 phase to form vascular parenchyma and a portion of the root cap (Van't Hof et al, 1986). As seen in the autoradiograms shown in Figures 2a and 2b, the anatomical location of most of the selectively labeled cells in a histologically prepared root is unmistakably peristelar, the position of vascular parenchyma.

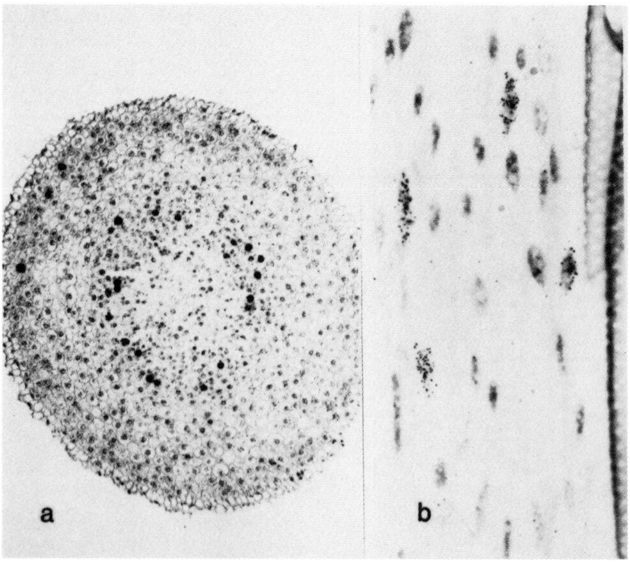

Figure 2. Autoradiograms of sections of cultured pea roots. 2a, A cross-section of a root starved for 48 hr and pulsed with tritiated thymidine for 1 hr. Note the ring of labeled nuclei in cells surrounding the presumptive stelar cylinder. 2b, A tangential section of a root grown for 72 hr after labeling. Note the labeled oblong nuclei and their position relative to the xylem vessels with secondary walls. The position of the labeled cells is that of vascular parenchyma. (from Van't Hof et al, 1986).

Extrachromosomal DNA Molecules Fail to Mature to Chromosomal Size.

The average length of a completely replicated chromosomal DNA molecule in pea is 20 cm, a length equivalent to about 4 x 10^{11} daltons or 6 x 10^5 kb. This lengthy molecule is replicated in a stepwise manner by 7.7 x 10^4 replicons (Van't Hof and Bjerknes, 1977). First replicated are Okazaki fragments of about 300 bases and these are joined to form replicon sized nascent chains of about 20 x 10^6 daltons. The nascent chains of tandem replicons are eventually joined to form intermediate sized molecules of 7 to 14 x 10^7 daltons. Finally, these intermediates are attached to give molecules of 8 x 10^8 daltons and larger (Schvartzman et al, 1981, 1982). The entire process is termed chromosomal DNA maturation and the end product is a lengthy mature double-stranded chromosomal molecule. Under normal circumstances maturation requires about 8 hr, the duration of S phase in cultured roots. To avoid breakage and achieve accurate distribution of genetic information to daughter cells, all chromosomal molecules must be mature at the time of mitosis. In pea roots, however, not all nascent DNA chains reach chromosomal size. Velocity sedimentation in an alkaline sucrose gradient shows that DNA from nuclei of growing roots has some nascent molecules that fail to mature even when given 48 hr to do so (Van't Hof and Bjerknes, 1982). These molecules have a size near that of pea genomic replicons (54 to 73 kb). Initially, the molecules are seen only in replicating DNA from nuclei of cells in the meristematic tip but later on, after the cells of the tip are displaced to the elongation zone, the molecules also are found in DNA isolated from nuclei of differentiating cells. These DNA molecules, produced by cells in the meristem, that are carried within nuclei as cells differentiate in the elongation zone and yet do not mature to chromosomal size, are called extrachromosomal DNA.

Sources of the extrachromosomal molecules are the precursors of vascular parenchyma and a portion of the root cap. Only these cells are delayed in S phase after 24 hr of starvation and it can be shown by velocity sedimentation in an alkaline sucrose gradient that they generate extrachromosomal molecules while completing DNA replication (Van't Hof et al, 1983). Most asynchronously dividing cells of the meristem replicate nascent DNA chains of replicon size in 20 to 30 min but it takes the precursor cells 40 to 70 min to produce chains of comparable length. The lower rate of replication by precursor cells is responsible, at least partially, for the delay in S phase. The lower replication rate also is accompanied by the lack of subsequent maturation of all the nascent chains. Maturation proceeds to sizes of 18 x 10^6 daltons or larger but then many of the chains remain at 18 x 10^6 daltons or are reduced to this size or smaller. The change in size from small to large and back to small suggests that extrachromosomal DNA is excised from chromosomal DNA after first being replicated as an integral part of the chromosome.

A closer look at extrachromosomal DNA selectively extracted from isolated nuclei indicates that after or concomitant with excision it further replicates two nascent chains, one of 3 and another of 7 x 10^6 daltons (Van't Hof et al, 1983). The size of these chains corresponds well with that of free single-stranded molecules and that of single-stranded branches on linear duplex molecules seen by electron microscopy (Krimer and Van't Hof, 1983). Moreover, the electron micrographs of selectively isolated extrachromosomal DNA indicate that the single-stranded DNA is most likely produced by displacing a portion of one of the two strands in duplex extrachromosomal DNA.

Extrachromosomal DNA Has Repeated Sequences And Ribosomal Genes.

The paragraphs above describe the cellular source, the fate and some of the biophysical aspects of extrachromosomal DNA. Though necessary, these observations tell us little about possible genetic information contained in the molecules. To obtain this information other analyses are needed. First and foremost it is necessary to prove that extrachromosomal DNA is homologous to nuclear DNA. The velocity sedimentation data indicated that the source of extrachromosomal molecules was nuclear but unequivocal proof awaited results from Southern blotting and hybridization procedures.

First on the agenda was the demonstration that extrachromosomal molecules had little to no homology with either plastid or mitochondrial DNAs. This demonstration was needed because the procedure used to selectively extract extrachromosomal DNA from isolated nuclei is subject to organelle contamination. The presence of organelle DNAs, however, is readily detected by certain restriction endonucleases that are sensitive to methylated bases in the DNA substrate. Nuclear DNA is partially refractory to such enzymes because it has a relatively high percentage of methylated bases as compared with DNA from organelles which has few methylated bases. Further, hybridization procedures using labeled organelle DNAs allow the identification of organelle DNAs in extrachromosomal DNA extracts digested by methyl-sensitive restriction enzymes. Hybridization of organelle DNAs to extrachromosomal DNA extracts subjected to Hpa II digestion, a methyl-sensitive enzyme, shows that more than 90% of the molecules in the extract are highly methylated and none of these are homologous to either mitochondrial or plastid DNAs (Kraszewska et al, 1986). These results demonstrate that extrachromosomal DNA is of nuclear origin.

Given the nuclear origin of extrachromosomal DNA, the molecules were hybridized with total pea RNA to determine what, if any, genetic information is contained in the molecules. This work showed that extrachromosomal DNA molecules have rRNA genes but no other readily detectable genes. The molecules are, however, abundant in nontranscribed repeated sequences and some of these sequences differ quantitatively in copy number, as do the rRNA genes, from that of total genomic DNA (Kraszewska et al, 1986).

The electron micrographs of selectively extracted extrachromosomal DNA indicated that about 74% of the duplex molecules have single-stranded branches and that the extract is rich in single-stranded molecules comparable in size to the branches (Krimer and Van't Hof, 1983). These findings suggested that extrachromosomal DNA could be separated from other genomic DNA by conventional procedures of molecular biology, i. e. from DNA extracted by conventional methods not involving the need to isolate nuclei. To test this possibility, we used four procedures and applied them sequentially to DNA extracted from the 3 mm tip of roots that were starved of carbohydrate for 48 hr. In these tips only the precursors of vascular parenchyma and certain root cap cells are replicating DNA in late S phase. The DNA was extracted by the method of Burr and Burr (1982) and digested with the restriction enzyme, Hind III, to generate fragments of rDNA of known size. A partial separation (85% or more) of single-stranded and branched double-stranded molecules from double-stranded molecules was obtained by benzoylated naphthoylated DEAE column chromatography (Gamper et al, 1985). A portion of the fractionated DNAs was treated with S_1 nuclease to digest the single-stranded molecules and the branches on the duplex molecules. The treated and untreated fragments were separated by agarose gel electrophoresis to distinguish further between branched and unbranched molecules by differential migration rates. Branched molecules migrate more slowly than their double-stranded counterparts (Bell and Byers, 1983). Finally, the separated fragments were blotted to Zeta-probe and hybridized to a cloned rDNA probe.

If successful, the results of this experiment are predictable. First, we must be mindful that rDNA from var. Alaska peas is located on four Hind III fragments (Ellis et al, 1984). Most rDNA is on fragments of 8.6 and 9 kb and on their rare dimers of about 17 and 18 kb. Second, the autoradiogram of DNA fragments with branches should have rDNA fragments that are digestable by S_1 nuclease and also migrate slowly at rates corresponding to fragments larger than 18 kb and third, the fragments of unbranched duplex DNA should have no slowly migrating rDNA fragments larger than 18 kb that are sensitive to S_1 nuclease.

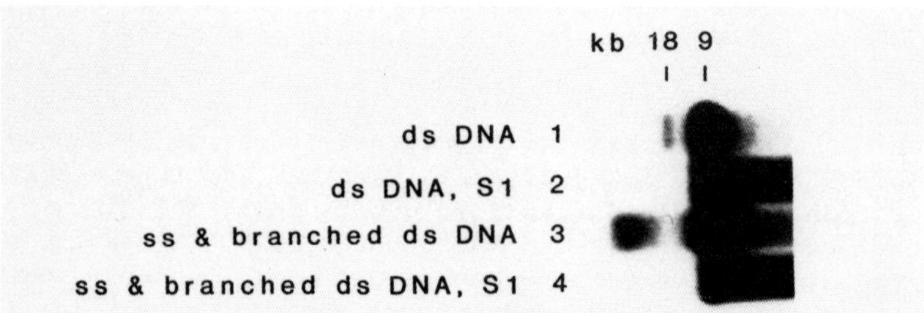

Figure 3. An autoradiogram of cloned rDNA hybridized to Hind III fragments of pea DNA isolated from 3 mm tips of roots starved of carbohydrate for 48 hr. Before hybridization to the membrane-blotted fragments the DNA was digested with Hind III, separated on a benzoylated naphthoylated DEAE column, treated with S_1 nuclease and separated by electrophoresis. ds, double-stranded Hind III fragments; ss, single-stranded DNA and branched double-stranded fragments; S 1, S_1 nuclease treated.

The autoradiogram of the blotted DNAs confirm these predictions (Figure 3). The autoradiogram shows that only the branched and single-stranded DNA (lane 3) has rDNA fragments that migrate at a rate slower than the 18 kb dimer seen in the double-stranded DNA (lane 1) and that the retarded fragments are also sensitive to S_1 nuclease digestion (lane 4). These results, using conventional molecular biological procedures, give further evidence of extrachromosomal DNA in cultured pea root cells.

SUMMARY AND GENERAL DISCUSSION

Histological, cytological and molecular analyses of DNA replication and differentiation of meristematic precursors of vascular parenchyma in cultured pea roots show that the initial steps in the transition from a dividing to a differentiated cell involve the retardation of DNA replication in late S phase and the production of extrachromosomal molecules by excision from late replicating chromosomal DNA. Portions of the extrachromosomal DNA are displaced by a strand displacement mechanism thereby producing free single-stranded molecules.

That extrachromosomal molecules have rDNA reflects the fact that in pea roots replication of the ribosomal genes is maximal during late S phase (unpublished results). It is likely, therefore, that the excised late replicating DNA contains some rDNA sequences.

Why extrachromosomal DNA is produced is an unanswered question but it is not a phenomenon exclusive to cells of pea roots. Extrachromosomal rDNA molecules are common in lower organisms such as slime molds (Zellweger et al, 1972; Vogt and Braun, 1976; Cockburn et al, 1978), *Tetrahymena* (Engberg et al, 1976; Pan and Blackburn, 1981), *Paramecium* (Findly and Gall, 1978), and in yeast (Larionov et al, 1980). In slime molds and in *Tetrahymena* (Pan and Blackburn, 1981) it is the means of rDNA amplification and in *Tetrahymena* it is associated with cell development as well (Yao, 1981). The classical example of extrachromosomal rDNA is that found in amphibian oocytes (Long and Dawid, 1980). The molecules found in pea cells are more analogous to L1 sequences found in mammalian cells (Jones and Potter, 1985; Schindler and Rush, 1985) or to the products of reinitiation events by impeded chromosomal replicons

(Varshavsky, 1981; Schimke et al, 1986). Only in peas, however, is the impediment traceable to certain cells that differentiate into specific tissues. The retarded DNA replication, the production of extrachromosomal DNA, and the change in nuclear phenotype of these specific cells argue strongly for the idea that chromosomes are modified either by breakage and/or rejoining or by DNA elimination preceding cell differentiation.

Research supported by the U.S. Department of Energy, Office of Health and Environmental Research.

REFERENCES

Bell, LR, Byers, B (1983). Homologous association of chromosomal DNA during yeast meiosis. Cold Spring Harbor Symp 47:829-840.

Burr, B, Burr F (1981). Controlling element events at the shrunken locus in maize. Genetics 98:143-156.

Cockburn, AF, Taylor, WC, Firtel, RA (1978). Dictyostellium rDNA consists of non-chromosomal palindromic dimers containing 5S and 35S coding regions. Chromosoma 70:19-29.

Ellis, THN, Davies, DR, Castleton, JA, Bedford, ID (1984). The organization and genetics of rDNA length variants in pea. Chromosoma 91:74-81.

Engberg, J, Andersson, P, Leick, V, Collins, J (1976). Free ribosomal DNA molecules from Tetrahymena pyriformis GL are giant palindromes. J Mol Biol 104:455-470.

Evans, LS, Van't Hof, J (1974). Is the nuclear DNA content of mature root cells prescribed in the root meristem?. Amer J Bot 61:1101-1111.

Evans, LS, Van't Hof, J (1975). The age-distribution of cell cycle populations in plant root meristems. Exptl Cell Res 90:401-410.

Findly, RC, Gall, JG (1978). Free ribosomal RNA genes in Paramecium are tandemly repeated. Proc Natl Acad Sci U.S.A. 75:3312-3316.

Gamper, H, Lehman, N, Piette, J, Hearst, JE (1985). Purification of circular DNA using benzoylated naphthoylated DEAE-cellulose. DNA 4:157-164.

Jones, RS, Potter, SS (1985). L1 sequences in HeLa extrachromosomal circular DNA: Evidence for circularization by homologous recombination. Proc Natl Acad Sci U.S.A. 83:1989-1993.

Kraszewska, EK, Bjerknes, CA, Lamm, SS, Van't Hof, J (1985). Extrachromosomal DNA of pea-root (Pisum sativum) has repeated sequences and ribosomal genes. Plant Mol Biol 5: 353-361.

Krimer, DB, Van't Hof, J (1983). Extrachromosomal DNA of pea (Pisum sativum) root-tip cells replicates by strand displacement. Proc Natl Acad Sci U.S.A. 80:1933-1937.

Larionov, VL, Grishin, AV, Smirnov, MN (1980). 3 µm DNA-an extrachromosomal ribosomal DNA in the yeast Saccharomyces cerevisiae. Gene 12:41-49.

Long,EO, Dawid, IB (1980). Repeated genes in eukaryotes. Ann Rev Biochem 49:727-764.

Pan, WC, Blackburn, EH (1981). Single extrachromosomal ribosomal RNA gene copies are synthesized during amplification of the rDNA in Tetrahymena. Cell 23:459-466.

Schimke, RT, Sherwood, SW, Hill, AB, Johnston, RN (1986). Overreplication and recombination of DNA in higher eukaryotes: Potential consequences and biological implications. Proc Natl Acad Sci U.S.A. 83:2157-2161.

Schindler, CW, Rush, MG (1985). Discrete size classes of monkey extrachromosomal circular DNA containing the L1 family of long interspersed nucleotide sequences are produced by a general non-sequence specific mechanism. Nuc Acid Res 13:8247-8258.

Schvartzman, JB, Chenet, B, Bjerknes, CA, Van't Hof, J (1981). Nascent replicons are synchronously joined at the end of S phase or during G_2 phase in peas. Biochim et Biophys Acta 653:185-192.

Schvartzman, JB, Van't Hof, J (1982). In the higher plant Pisum sativum maturation of nascent DNA is blocked by cycloheximide, but only after 4-8 replicons are joined. Nuc Acid Res 10:6191-6205.

Van't Hof, J (1966). Experimental control of DNA synthesizing and dividing cells in excised root tips of Pisum. Amer J Bot 53:970-976.

Van't Hof, J (1976). Replicon size and rate of fork movement in early S of higher plant cells (Pisum sativum). Exptl Cell Res 103:395-403.

Van't Hof, J, Bjerknes, CA (1977). 18 um replication units of chromosomal DNA fibers of differentiated cell of pea (Pisum sativum). Chromosoma 64:287-294.

Van't Hof, J, Bjerknes, CA (1982). Cells of pea (Pisum sativum) that differentiate from G_2 phase have extrachromosomal DNA. Mol Cell Biol 2:339-345.

Van't Hof, J, Bjerknes, CA, Delihas, NC (1983). Excision and replication of extrachromosomal DNA of pea (*Pisum sativum*). Mol Cell Biol 3:172-181.

Van't Hof, J, Bjerknes, CA, Lamm, SS (1986). Meristematic precursors of vascular parenchyma differentiate from G_2 phase after replicating DNA discontinuously. Amer J Bot 73:87-95.

Varshavsky, A (1981). On the possibility of metabolic control of replicon 'misfiring': Relationship to emergence of malignant phenotypes in mammalian cell lineages. Proc Natl Acad Sci U.S.A. 78:3673-3677.

Vogt, VM, Braun, R (1976). Structure of ribosomal DNA in *Physarum polycephalum*. J Mol Biol 106:567-587.

Webster, PL, Van't Hof, J (1970). Recovery of G_2 cells in pea root meristems: survival and mitotic delay following irradiation. Rad Bot 10:145-154.

Yao, MC, (1981). Ribosomal RNA gene amplification in *Tetrahymena* may be associated with chromosomal breakage and DNA elimination. Cell 24:765-774.

Zellweger, A, Ryser, U, Braun, R (1972). Ribosomal genes of *Physarum*: Their isolation and replication in the mitotic cycle. J Mol Biol 64:681-691.

Plant Tissue and Cell Culture, pages 45–55

Causal Events In Morphogenesis

Michael L. Christianson

Zoecon Research Institute, Sandoz Crop Protection Corp., Palo Alto, California 94304

The major problem in research on morphogenesis from plant cells or tissues cultured *in vitro* is the fact that just about every species has now been regenerated. Industrious, creative, and yes, lucky, scientists have successfully pushed and extended Skoog and Miller's (1957) observations on the control of organogenesis by the relative amounts of auxin and cytokinin. Admittedly, there are no reports of shoot organogenesis for some species; this seems a moot point in light of reports of somatic embryogenesis in those species. A substantial amount of work remains to turn success with primary explants of only one cultivar of a species into regular success with explants or short-to-medium term callus cultures of any cultivar and to extend that to regeneration from protoplasts from such material. Past these points, research becomes a kind of "bio-atheletics" -- faster, more simple, better, quicker, higher, longer regeneration from increasingly difficult or obscure starting materials.

The formation of organs from explants or callus cultured *in vitro* is not only a remarkable technical feat, it is also a complex piece of plant morphogenesis. As such, organogenesis *in vitro*, like *in vitro* systems of cellular differentiation or somatic embryogenesis, can serve as model systems for the *in vivo* processes of differentiation and growth. Advancing our knowledge of the mechanisms behind tissue and organ differentiation not only adds to our store of "pure biology" but would also seem to be necessary to all but trivial successes in the genetic engineering of higher plants.

As cellular differentiation and aspects of somatic embryogenesis _in vitro_ are covered elsewhere in this volume, this paper will be restricted to a consideration of organogenesis _in vitro_. It is fairly well established that organogenesis _in vitro_ does not merely represent an aberrant form of somatic embryogenesis; in cultures of _Convolvulus_, for example, capable of making shoots or roots, the two types of organ arise from distinct parts of the callus, and not as if the organs represented stimulation of either the root or shoot pole of an aberrant embryo (Christianson and Warnick, 1985).

The distinct processes of organogenesis and somatic embryogenesis _in vitro_ have parallels in the biology of the plant _in vivo_. Plants make embryos, even if these embryos sometimes are not from zygotes. Plants also make shoot apices in the axil of each leaf, and make additional root apices during the growth and ramification of their root system. Most whole plants have, in addition, the ability to form adventive roots or shoot buds. Organogenic abilities are distinct from embryogenic abilities _in vivo_ just as they are _in vitro_.

That is not to say that aberrant forms of somatic embyogenesis that appear to be shoot or root organogenesis do not exist. The roots from certain alfalfa cultures will become somatic embryos with improvements in the medium (c.f., Walker _et al._, 1979; Stuart _et al._, 1985); such systems do not represent organogenesis as restricted in this paper.

A number of reviews document the increasing success in achieving shoot and root organogenesis from a number of species, examine the chemical factors behind those successes and consider the diverse morphogenetic procedures. Few reviews focus on the developmental biology of morphogenesis _in vitro_ (e.g., Halperin, 1969; Hicks, 1980). Hicks (1980) is correct in his conclusion that "there is a need in plant development for greater understanding of the _early determinative events_ in organ development _in vitro_" (emphasis mine); some small progress toward that end has been made.

Many authors refer to the "induction" of shoots. Induction, in its widest developmental sense, is a process seen by its end result, a change in the fate or destiny of a cell or group of cells. This change in fate is cellular determination. More explicitly, determination is exactly that event "which initiates a specific pathway of development by singling it out from among various possibilities for which a cellular system is competent (Roux, 1905)" (Rieger _et al._, 1968). Determinative events are, by definition, the causative events in development. While the morphological differentiation and development of shoots or roots _in vitro_ are exciting and interesting processes, the crucial point in the recovery of organs from _in vitro_ culture is the developmental decision to do growth and development.

A Close Look at Organogenesis

This emphasis on the very early events of organogenesis _in vitro_ results in a very simple picture of the process. Tissue first becomes competent to respond to the organogenic properties of the medium. This developmental state, competence, is much like the notion of the dedifferentiation of the explant (Gautheret, 1966). Unlike dedifferentiation, competence does not imply proliferation of seemingly unorganized callus tissue. Once competent, tissue responds to the phytohormone balance in the medium and undergoes induction. The end of the induction process, however, is not the time when shoots can first be seen, but the time when cells or groups of cells are committed to the differentiation and growth of shoots. Waddington (1966) makes the observation that determination, the decision to do a piece of morphogenesis, necessarily precedes the actual morphogenesis. States of competence and determination cannot be measured by direct inspection of the tissue, but only demonstrated by means of experimental manipulations (Meins and Binns, 1979). Fortunately, the artificial nature of tissue culture and the dependence on exogenous hormones and nutrients make cell and tissue culture phenomena remarkably amenable to such demonstration.

If determination is the end result of an induction process controlled by a particular phytohormone balance in the culture medium, "determination" can be assessed operationally: determined tissue will go on to produce shoots, for example, if explants are removed from shoot-inducing medium, SIM, and placed on basal medium -- salts, vitamins and carbon source but no phytohormones. In the experimental protocol leaf explants are placed on shoot-inducing medium for various lengths of time. When the explants are removed to basal medium, only the determined explants will go on to produce shoots. Such experiments show that phytohormone balance does, indeed, induce shoot formation, producing cells or groups of cells fated or determined for shoot formation (Christianson and Warnick, 1983).

The demonstration of competence involves the same kind of simple transfers. If tissue first becomes competent for induction, and is then specifically induced by SIM, that tissue could just as well be cultured on root-inducing medium, RIM, or callus-inducing medium, CIM, to produce competent tissue. Preculture of explants on RIM or CIM can substitute for the time the tissue spends on SIM simply becoming competent for the specific shoot-inducing properties of SIM. In the experimental protocol, explants are placed on RIM or CIM for known amounts of time, transferred to SIM for additional lengths of culture, and, finally, transferred to basal medium, BM. Such experiments show that explants spend a discrete amount of time in culture becoming competent for induction (Christianson and Warnick, 1983).

Beyond Competence and Determination

Henshaw _et al_. (1982) observe that there is little doubt that competence and determination exist in plants, but it is not at all known whether these are properties of individual cells or of groups of cells. This uncertainly is a different statement of the long-standing question about the single or multi-cellular origin of shoots formed _in vitro_. Certainly, shoots formed _in vitro_ have, as their ultimate source, a single cell -- the zygote that gave rise to the plant from which the explants were taken. Recently, a very nice set of papers have shown, as Carlson and Chaleff (1974) suggest by omission in their original report, that shoots formed _in vitro_ can arise from more than one cell in the explant, but usually do not

(Marcotrigiano and Gouin, 1984a,b). (This should reassure molecular biologists who use leaf disc transformation schemes and worry about chimerism.) The demonstration that culture *in vitro* results in cells or groups of cells fated or determined for shoot formation leads to a precise formulation of the question about the origin of shoots: at the time when culture on SIM results in determination in the cultured tissue, is the precurser of the shoot a cell or a group of cells?

Acute irradiation at the time of determination of cultured leaf explants from a plant heterozygous for a chlorophyll mutation produces rare chimeral shoots (Christianson, 1984). An analysis based on 23 such chimeras reveals that at the time of determination, a number of cells contribute cell lineages to the nascent shoot. The lower 3-5 nodes on the shoot each come from a distinct developmental "compartment" comprised of 2-4 cells; the upper nodes of the new shoot come from a single cell.

Please note that while this analysis is a fairly precise description of the cellular origins of the shoots formed *in vitro*, the results are compatible with the determinative event compassing the whole field of cells or only the single cell that gives rise to the apex of the new shoot. This second interpretation only requires that the single, determined, cell acquires the physiology and organising properties of the apical initials soon enough to "capture" or "recruit" adjacent cells into the forming shoot apex.

Proving Causality

A number of chemicals are known to inhibit or block the process of shoot organogenesis. Some of these chemicals can be shown to be stage-specific inhibitors (Christianson and Warnick, 1984; Tanimoto and Harada, 1982; Thorpe and Meier, 1973). The use of a schedule of transfers between SIM with and without the inhibitor can demonstrate exactly when in the organogenic process a particular "transient sensitivity" occurs (Christianson and Warnick, 1984). Such sensitivities, localized to developmental times prior to that time when cells in the explant become determined for shoot formation, are likely

Table 1. Mean numbers of shoots from leaf explants of *Convolvulus arvensis* exposed to SIM plus ribose or sorbitol for various lengths of time: Genotype C30, n=15, after 3 weeks in culture, ribose at 10 gm/l, sorbitol at 20 gm/l.

Transfer sequence	Day of Transfer														
	0	3	5	6	7	8	9	10	11	12	13	15	17	19	21
SIM+s --> SIM+r	0.00*	0.00*	0.00*	0.33	0.00*	0.50	0.20	0.00*	0.00	0.00	----	0.00	0.00*	0.00	0.00
SIM+r --> SIM+s	0.00	0.00*	0.00*	0.00*	0.00	----	0.00	0.00	0.00	0.00	0.00*	0.00	0.00	0.00	----
SIM+s --> SIM	(3.90)	1.50	1.80	0.29	1.46	2.90	0.50	----	----	0.50	0.00	----	0.00	0.00	0.00
SIM --> SIM+s	0.00	0.00	0.00	0.00	0.00	0.11	0.00*	0.20	0.00	0.80	2.07	0.86	1.46	4.50	2.50
SIM+r --> SIM	3.90	1.75	0.00	0.00	0.00	0.00	0.00	0.00	0.00	0.00	0.00	0.00	0.00	0.00	0.00
SIM --> SIM+r	0.00*	0.00	0.00	0.47	0.67	2.08	1.08	1.07	1.30	1.86	3.38	1.69	4.20	4.53	1.29

* = explants did not produce significant amounts of callus tissue.

to be the "causal events" behind regeneration. We would like to know the number of such "causal events" and the identity of each. It is necessary to demonstrate the distinctness of each of a number of stage-specific inhibitions, to suggest an underlying mechanism for an inhibition, and to confirm that mechanism by an independent means.

In many cases, the distinctness of two transient inhibitions is suggested by the different appearance of the inhibited explants and by the fact that several days separate the times at which the inhibitions occur. This indirect evidence can be confirmed with direct evidence from experiments conceptually analogous to experiments used to order the sequence of gene action in the cell cycle (Jarvik and Botstein, 1973). Both cold-sensitive, CS, and heat-sensitive, HS, alleles of cell cycle mutations will function at normal temperatures; CS alleles are inactive at cool temperatures, HS alleles are inactive at warm temperatures. Combining a HS allele at one locus with a CS allele at a second cell cycle locus, submitting such double mutants to shifts from one non-permissive temperature to the other non-permissive temperature and observing the results leads to the unambiguous ordering of the two genes in the developmental pathway: a linear order, independent action, coincident action (Jarvik and Botstein, 1973). Experiments shifting explants from one non-permissive medium to the other non-permissive medium lead to the same kind of results.

Two carbohydrates, ribose and sorbitol, both inhibit shoot organogenesis from leaf explants of *Convolvulus arvensis*. Shifting explants to and from SIM and SIM plus ribose shows that only those transfer schedules where explants "see" ribose on day 5 of culture are inhibited (Table 1). Similar transfers place the time of sorbitol inhibition to day 11 of culture (Table 1). Transfers from SIM with one inhibitor to SIM with the other inhibitor do not give vigorous explants; the inhibitory level of ribose, for example, reduces callus growth some 33% (Christianson and Warnick, 1984). Transfers from SIM plus ribose to SIM plus sorbitol lead to no shoots at all transfer times. Transfers from SIM plus sorbitol to SIM plus ribose result in explants making a few shoots but only if the transfer is done on day 6, 7, 8 or 9 of culture (Table 1). Explants on SIM plus sorbitol on day 5 can do whatever process ribose inhibits; if they are moved to SIM plus ribose before day 10, they will be able, on SIM plus ribose, to do whatever process sorbitol inhibits on day 11. This result confirms the fact that the ribose-inhibition occurs some 5 days previous to the sorbitol inhibition and that they must represent two distinct events.

Attaching Mechanisms to Inhibitors

The most desirable and direct way of attaching a mechanism to a particular transient sensitivity would be through the isolation of mutants resistant to the inhibitor; these mutants would be relatively easy to recover as they would be those shoots that form in the presence of the inhibitor. (I thank Dr. R.D. Locy, Native Plants, Inc., for pointing this out to me.) Recent advances in molecular biology make it possible to think about generating these mutants with a known transposon so that these "genes of regeneration" would be directly accessible to gene cloning.

Attaching a function to an unknown nucleotide sequence in the absence of other information is as chancy as attaching a function to "an inhibition by ribose". It is possible to gather additional information. A series of papers by Thorpe's group have indicated the importance of the hexose monophosphate shunt in the regeneration process (Thorpe and Laishley, 1973). This pathway could supply carbon skeletons or reducing equivalents, or both. Both ribose and erythrose are intermediates in this pathway, and both inhibit shoot regeneration. Since both compounds (as phosphates) are products of reactions that also produce NADPH, excess ribose or erythrose should block the production of NADPH. It is possible to test this proposition by using some other way to use up NADPH. The dye, methylene blue, is known to mediate oxidation of NADPH (Barron and Harrop, 1928); it also inhibits shoot regeneration from explants of <u>Convolvulus</u> at 10^{-7} to 10^{-5} M. These data support a conclusion that the hexose monophosphate shunt supplies reducing equivalents, and that the inhibitory effects of ribose result from a direct effect on NADPH production. This conclusion would be strengthened by the demonstration that the inhibitions by ribose, erythrose and methylene blue represent a single event; extensive transfer schedules from one of these inhibitors to another never result in explants making shoots, in sharp contrast to the comparison of the ribose and sorbitol inhibitions in Table 1.

SUMMARY

Morphogenesis *in vitro* or *in vivo* are complex developmental processes. Morphogenesis involves first the fating of cells or groups of cells to undergo morphological change and then the carrying-out of that fating, the processes of morphological differentiation and growth. The study of such determinative events is the study of the causal events behind morphogenesis.

The fact that organogenesis *in vitro* occurs mediated by exogenously supplied phytohormone and nutrients makes it possible to experimentally dissect the process. Explants placed into culture acquire competence, the ability to respond to the inductive nature of the phytohormone balance in the medium. Culture of competent tissue results in organogenic induction; this induction process, although controlled by a single medium, is comprised of distinct, identifiable steps. That these transient sensitivities to inhibitors represent distinct events can be shown by simple schemes of transfers of explants to and from the inhibitors. Hypotheses about the nature or mechanism of any one inhibitor can be put to critical tests; alternate substances known to act in the proposed way should also inhibit regeneration, and will identify the same transient sensitivity in media exchange experiments. Such stage-specific inhibitors also present the opportunity to recover mutants resistant to the inhibitors; while some of the mutants will be defects in uptake of the inhibitor, some will represent an altered gene used in the process of regeneration. Although exogenous phytohormones control the process of organogenesis, our focus must remain on the genes and physiological activity that are the organogenic process. More specifically, we must focus on the events that precede the actual differentiation and growth of the new organ: competence for induction, induction, and the nature of the determinative event that is the cause of organogenesis *in vitro*.

REFERENCES

Barron ESG, Harrop GA, jr (1928) Studies on blood cell metabolism II. The effect of methylene blue and other dyes upon the glycolysis and lactic acid formation of mammalian and avian erythrocytes. J Biol Chem 89:65-87.

Carlson PS, Chaleff RS (1974). Heterogeneous associations of cells formed in vitro. In Ledoux L (ed): "Genetic manipulations with plant material," New York: Plenum, pp245-261.

Christianson ML (1984). Clonal analysis of organizing apices. Am J Bot 71:22.

Christianson ML, Warnick DA (1983). Competence and determination in the process of in vitro shoot organogenesis. Dev Biol 95:288-293.

Christianson ML, Warnick DA (1984). Phenocritical times in the process of in vitro shoot regeneration. Dev Biol 101:382-390.

Christianson ML, Warnick DA (1985). Temporal requirement for phytohormone balance in the control of organogenesis in vitro. Dev Biol 112:494-497.

Gautheret RJ (1966). Factors affecting differentiation of plant tissues grown in vitro. In Beermann W (ed): "Cell differentiation and morphogenesis," Amsterdam: North Holland Publishers, pp55-95.

Jarvick J, Botstein D (1973). A genetic method for determining the order of events in a biological pathway. Proc Nat Acad Sci (USA) 70:2046-2050.

Halperin W (1969). Morphogenesis in cell cultures. An Rev Plant Physiol 20:395-418.

Henshaw GG, O'Hara JF, Webb KJ (1982). Morphogenetic studies in plant tissue cultures. In Yeoman MM, Truman DES (eds): Differentiation in vitro," New York: Cambridge Press, pp231-251.

Hicks GS (1980). Patterns of organ development in plant tissue culture and the problem of organ determination. Bot Rev 46:1-23.

Marcotrigiano M, Gouin FR (1984a). Experimentally synthesized plant chimeras 1. In vitro recovery of Nicotiana tabacum L. chimeras from mixed callus cultures. Ann Bot 54:503-511.

Marcotrigiano M, Gouin FR (1984b). Experimentally synthesized plant chimeras 2. A comparison of in vitro and in vivo techniques for the production of interspecific Nicotiana chimeras. Ann Bot 54:513-521.

Meins F, Binns AN (1979). Cell determination in plant

development. Bioscience 29:221-225.

Reiger R, Michaelis A, Green MM (1968). "A Glossary of Genetics and Cytogenetics," New York: Springer-Verlag.

Roux, W (1905). Die Entwicklungsmechanik, ein neuer Zwieg der biologischen Wissenschaft. In: Roux, W (ed): "Vortrage and Aufsatze uber Entwicklungsmechanik der Organismen", Leipzig: Engelman.

Skoog F, Miller CO (1957). Chemical regulation of growth and organ formation in plant tissues cultured _in vitro_. Symp Soc Exp Biol 11:118-140.

Stuart DA, Nelson J, Strickland SG, Nichol JW (1985). Factors affecting developmental processes in alfalfa cell cultures. In Henke RR, Hughes KW, Constantin MP, Hollaender A (eds): "Tissue Culture in Forestry and Agriculture," New York: Plenum, pp59-73.

Tanimoto S, Harada H (1982). Studies on the initial process of adventitious bud differentiation in _Torenia_ stem segments cultured _in vitro_ I. Effects of cytokinin. Biochem Physiol Pflanzen 177:222-228.

Thorpe TA, Laishley EJ (1973). Glucose oxidation during shoot initiation in tobacco callus cultures. J Exp Bot 24:1082-1089.

Thorpe TA, Meier DP (1973). Effects of gibberellic acid and abscisic acid on shoot formation in tobacco callus cultures. Physiol Plant 29:121-124.

Waddington CH (1966). "Principles of Development and Differentiation." New York: MacMillan.

Walker KA, Wendeln ML, Jaworski EG (1979). Organogenesis in callus tissue of _Medicago sativa_. The temporal separation of induction processes from differentiation processes. Plant Sci Lett 16:23-30.

Plant Tissue and Cell Culture, pages 57–81

ORGANIZATIONAL EVENTS DURING SOMATIC EMBRYOGENESIS

Philip V. Ammirato

Department of Biological Sciences, Barnard College, Columbia University, New York, NY 10027 and DNA Plant Technology Corporation, Cinnaminson, NJ 08077

INTRODUCTION

The list of plants that can successfully regenerate somatic embryos in culture (Ammirato, 1983b) continues to grow, with the recent addition of a number of hitherto recalcitrant crops including, among the large seeded legumes, soybean (Ranch et al., 1985) and, among the conifers, sugar pine (Gupta and Durzan, 1986).

One of the most promising applications for somatic embryogenesis is large scale propagation (Ammirato and Styer, 1985). Somatic embryogenesis provides many advantages: large numbers (e.g., 60,000 embryos per liter of medium); the presence of both root and shoot meristems in the same unit; easy scale-up, transfers and other manipulations since embryos can be grown individually and freely floating in liquid medium; the promise of dormancy induction and long term storage; the possibilities of encapsulation (Lutz et al., 1985) and other methods of packaging and of direct delivery systems, whether as naked embryos or young plantlets.

Populations of somatic embryos also provide an important resource for the analysis of molecular and biochemical events that occur during induction and maturation. Work has been proceeding along these lines, for example, the formation of developmental (temperature-sensitive) mutants in carrot cultures (Breton and Sung, 1982; Terzi et al., 1982), and the isolation of embryo-specific proteins (Sung and Okimoto, 1981; 1983) even

stage-specific storage proteins (Crouch, 1982).

For both practical applications and basic molecular and biochemical studies, it is important to understand the organizational events that occur during somatic embryogenesis and in particular to compare development in culture with the similar events that occur during zygotic embryogenesis. Are the stages of somatic embryogeny accurate representations of their zygotic counterparts? Will somatic embryos develop in such a manner than they can be mechanically delivered? Will they germinate or convert into plants? Deviations from the normal course of events may affect the ability of somatic embryos to perform as embryos in practical applications. They may also alter molecular or biochemical profiles seen when the somatic embryos are analyzed.

PROBLEMS IN SOMATIC EMBRYOGENY

Although one can examine a population of somatic embryos and pick out individuals that appear identical to zygotic embryos at all the traditional stages (globular, heart, torpedo, etc.), it has been recognized for some time that many abnormal forms can be found (Halperin, 1966; Ammirato and Steward, 1971; Konar et al., 1972b; Ammirato, 1985). (1) Embryogenesis in cultured cells is highly repetitive and somatic embryos can bear additional and smaller embryos along the axis, sometimes restricted to certain loci, such as the radicle-hypocotyl junction or at the base of the radicle. In other cases, accessory or secondary embryos can occur along the whole axis. This is one of the major causes of the absence of synchrony in populations of somatic embryos. (2) In addition to single embryos, twins, triplets and highly multiple clusters of embryos can be found. This generate multiple shoots when plantlets form. (3) Cotyledon development may be affected both in terms of number and morphology. There may be one, two, three or even more cotyledons on a single axis. Their development may be exceptionally poor or they may be fused either partially or completely. (4) The shoot apex may be affected. Embryos may look normal in terms of their external morphology but fail to produce a shoot with leaves and stem, as when the embryo in the seed germinates. (5) As embryos mature, they may proceed directly into plantlet formation, without the cessation of growth associated with

seed maturation. In many cases, this can occur prematurely (precocious germination) with the same consequences that have been seen when immature zygotic embryos are excised and placed into culture, usually the growth of malformed or aborted plants.

Obviously, embryo development can be remarkably plastic. Its path can be diverted a number of ways. How does this occur? What is the normal sequence of organizational events? How can somatic embryogeny be controlled or modulated? These topics will be explored in this paper drawing mainly on observations with carrot (*Daucus carota* L.), caraway (*Carum carvi* L.) and other members of the Umbelliferae (Ammirato, 1983b, 1986) but with references to other plants when appropriate.

ZYGOTIC EMBRYOGENESIS, SEED DEVELOPMENT AND GERMINATION

The development of a carrot or other umbellifer seed proceeds along a well defined pathway which, for the purposes of this discussion, will be presented in only the most general terms (Figure 1). [For a more detailed presentation of zygotic embryo development, the reader is referred to the excellent discussion by Raghavan (1965, cf. pp. 25-29).] The zygote actively divides and increases in size but retains at first a general spherical shape--the globular stage (Fig. 1A). It is during this stage that the three primary meristems (protoderm, ground meristem and procambium) become visible, delineated by the orientation of cell divisions and, in the ground meristem, by increased vacuolation. The potential shoot and root meristems become recognizable at about this time as partially differentiated meristematic cells.

Eventually, a ring of actively dividing cells appears at the apical end and from it, two meristems emerge. These grow out to form the cotyledons. Their initiation changes the outward appearance of the embryo--it now enters the heart-shaped or cordate stage (Fig. 1B). As the cotyledons develop and the embryo passes through into the torpedo and cotyledonary stages (Fig. 1C), the cells set apart inside the cotyledonary ring divide to form a visible shoot apex. Continued meristematic activity both superficially and internally produces the shoot apical dome. The carrot embryo at this point becomes quiescent with a cessation in mitotic division accompanied by a decrease in water content and the other events associated with the formation of a mature seed. At maturity, the embryo is small (from 1000 to 1500 um, cf. Fig. 2A) and surrounded by a mass of cellular, persistant endosperm (Fig. 1D). It is an albuminous seed.

In carrot and other umbellifers, formation of the mature embryo is characterized by extensive cell division without substantial cell enlargement or cell differentiation (Ammirato, 1985; Halperin, 1966). The cells of the mature embryo are therefore far smaller than the subsequent seedling and have few of the structural features associated with differentiated cells. The various histogens that will generate the tissue layers and associated cells as well as the presumptive apical meristems that will grow into

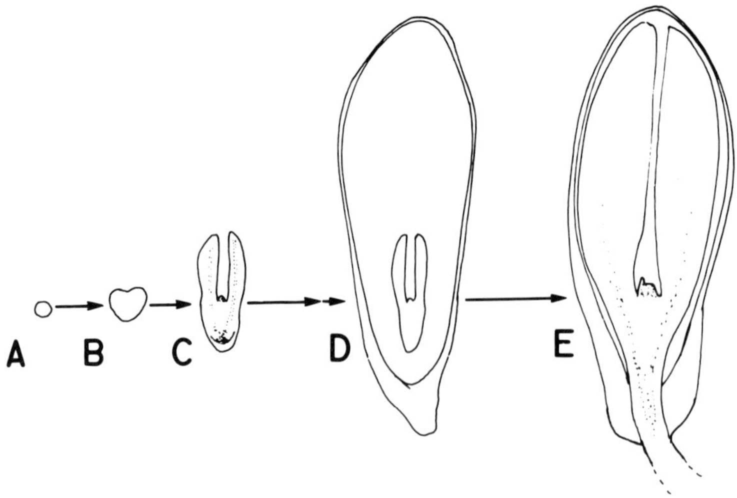

Figure 1. Schematic representation, not drawn to scale, of the key stages in zygotic embryogenesis in carrot.
(A) globular proembryo, (B) heart-shaped stage with the initiation of cotyledonary primordia, (C) torpedo stage with first indications of shoot and root apical meristems and procambium differentiation, (D) mature seed with small embryo surrounded by persistant cellular endosperm, and (E) germinating seed at about five days with enlarged cotyledons, elongated hypocotyl and a large shoot apex and the first leaf primordium.

primary shoot and root are indicated by their shape and position. The cells of the embryonic body are typically arranged in distinct rows. The embryonic shoot apex is a small group of cells. In carrot, the shoot apex at maturity is approximately 6 to 8 cells across.

The imbibition of water initiates the processes of germination which include the resumption of mitotic activity along with cell expansion and differentiation. The cotyledons expand, absorb the endosperm tissue and fill the interior of the seed (Fig. 1E). The shoot apex shows renewed mitotic activity and develops leaf primordia. The hypocotyl expands and the primary root grows out.

There is a definite sequence in the initiation of mitotic activity and a variation in the relative duration of mitoses in the various regions of the germinating embryo. As studied by Feulgen staining (unpublished work of D. Davidson and P. V. Ammirato), the first mitotic spindles are visible at 18 hours after imbibition in the root apical cells. At 24 hours, divisions are visible in the root, hypocotyl and cotyledons. Division figures are first visible in the shoot apex at 48 hours and by 60-62 hours, it is mitotically active. At this point, the cells of the cotyledons and hypocotyl have become greatly enlarged and and the few divisions that are visible are restricted to the vascular tissue (Fig. 2B). The shoot apex, clearly distinguishable as a mass of actively dividing cells surrounded by enlarged, non-dividing cells, is approximately 10-12 cells across and 10-15 cells high. By 120 hours (approximately 5 days), the first leaf primordium is visible (Fig. 2C). This is therefore a temporal difference in the growth of the cotyledons and the reactivation and subsequent activity of the shoot apex. There is also a difference between the relatively small shoot apex that appears during embryo development and seed maturation and the larger shoot apex that forms during germination.

It should be added here that what pertains to carrot, and other umbellifers, need not apply to other plants. It is well known that the development of zygotic embryos can differ substantially from taxon to taxon (Raghavan, 1986). In addition, there will be obvious differences between the sequence that produces a mature seed in a plant such as carrot, which has an albuminous seed with a small embryo and persistant endosperm, and the sequence in albuminous

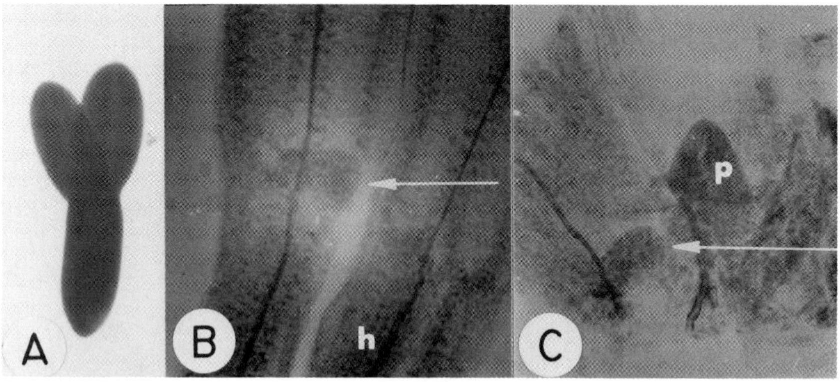

Figure 2. Shoot apex development in carrot zygotic embryos (Feulgen-stained). (A) Excised zygotic embryo 24 hours after beginning of germination. (B) The shoot apical region at 60-62 hours after start of germination. The preparation has been gently squashed to separate cells. The shoot apex (arrow) is a mass of actively dividing cells approximately 10-15 cells across and 10-12 cells high surrounded by the non-dividing cells of the hypocotyl (h) and cotyledons (c). (C) The apical region 120 hours after start of germination with a distinct leaf primordium (p) adjacent to the shoot apex (arrow).

seeds, such as the legumes. In bean, pea and soybean, there is extensive cotyledonary growth, shoot apical development (including the formation of unexpanded leaves) with concomitant changes in biochemistry, such as the synthesis of specific storage proteins (cf. Raghavan, 1986, pp. 72-80).

ORGANIZATIONAL EVENTS IN SOMATIC EMBRYOGENY

The major events in somatic embryogeny may be tentatively divided into several broad categories: induction, early growth, embryo maturation and germination or conversion.

Induction

The induction of embryogenic growth in carrot cultures, as it is in many other genera, appears to occur in one of two ways. Certain cells may only need placement on a simple basal medium to begin embryogenic development while others may require a more complex medium, e.g., an exogenous hormone supply, most often, an auxin. In the former case, the cells may be viewed as already determined for embryogenic development, needing only permissive conditions to allow its expression (Sharp et al., 1980, 1982). Direct embryogenesis proceeds from these pre-embryogenic determined cells or PEDC's. It is most often observed in excised zygotic embryos (Steward et al., 1964) or in young somatic embryos (Kato, 1968) or in the plantlets arising from them, usually from the hypocotyl region (Homes, 1968).

In the second type of response the cells within the explant have differentiated and have become incorporated into specific meristems or mature tissues. An induction treatment is needed to foster redetermination of the differentiated cells and the development of the embryogenically determined cells. Indirect embryogenesis follows the development of induced embryogenically determined cells (IEDC's). This is observed in cultures of carrot petioles or taproots (Halperin, 1966; Steward et al., 1970). [In certain studies, direct embryogenesis has been used to describe the formation of an embryo from a single cell without an intervening callus stage, although the embryo has arisen by means of the dedifferentation of a differentiated cells within the explant (Conger et al., 1983; cf. Raghavan, 1986, pp. 131-132.)]

The earliest cell divisions in the embryogenically determined cells can apparently follow a number of patterns without affecting the final outcome of embryogeny (McWilliam et al., 1974). Similarly, determined cells may operate singly or in groups to form embryos. The reader is directly to a recent article on the coordinated behavior of embryogenic cells (Williams and Maheswaran, 1986).

In mature carrot tissue, such as from the petiole or secondary phloem of the taproot, as in many other genera, the induction treatment both fosters dedifferentiation in the cells and induces embryogenic development. Examination of the developing callus tissue on carrot petioles or the resulting cell suspension shows that young, densely cytoplasmic globular proembryos are visible (Halperin, 1966; McWilliam et al., 1974; Street and Withers, 1974; Ammirato, 1985). When these small proembryos are moved from the auxin-containing to auxin-free (or auxin-reduced) medium, there is usually a one-to-one correspondence between the number of proembryos in the inoculum and the number of somatic embryos that mature. The single cells in the medium, which range in size from round to elongate and from partially to highly vacuolated, do not form embryos when moved to auxin-free medium. They have apparently stopped dividing and differentiated. [However, recent work has shown that they can be re-induced (Nomura and Komamine, 1985. This is discussed in the next section.]

The temporal confluence of the growth-promoting and embryo-promoting steps can be seen if one pulses petioles with auxin and then removes the tissue to auxin-free medium (Ammirato 1985). A one-day pulse of 2,4-D or NAA is sufficient for embryogenic induction. (Interestingly, such short pulses induce embryos on only one end of the petiole, showing the polarity in response seen in many other responses. A longer pulse, for four days, is needed to induce embryogenesis from both ends of the petiole.) In certain cases, a separation of the processes of growth induction and embryogenic determination have been separated. One excellent example is in alfalfa (Walker et al., 1979) where kinetin-NAA medium fosters callus growth but no embryo or shoot formation, even if the callus tissue is moved to media with various combinations of the two hormones. However, if cells are moved to medium containing 2,4-D, a stronger auxin, embryogenic cells are induced and somatic embryos develop when the cells are then transfered to a kinetin-NAA medium.

Accessory or Repetitive Embryogeny

The cells within the small carrot proembryos retain their embryogenic competence and rather than participate in the coordinated growth of one embryo, they can themselves become embryogenic growth centers. This is especially fostered in the auxin-containing medium where proembryos do not grow and mature (Kamada and Harada, 1979). Their constituent cells continue to divide; the proembryos enlarge and either bud-off or separate into additional proembryos (Steward et al., 1970; McWilliam et al., 1974; Street and Withers, 1974). This has been observed in other taxa including celery (Al-Abta and Collin, 1980), *Atropa belladonna* (Konar et al., 1972b), *Ranunculus sceleratus* (Konar et al., 1972a; Thomas et al., 1972) and *Panicum maximum* and *Pennisetum purpureum* (Karlsson and Vasil, 1986).

Thus, the growth within a carrot suspension culture is in effect proembryos or proembryonic masses producing many other proembryos with some cells sloughing off into the medium generating single cells that begin the process of cell differentiation. The subculture of this population provides fresh material for the process to continue and may even induce new embryogenic cells from some of the single cells transferred.

Recently, it has been shown that some of the single cells seen in carrot suspensions, which do not grow into embryos when moved into auxin-free medium, can be induced to grow into embryos (Nomura and Komamine, 1985). The response appears to be restricted to those that are still round and not fully vacuolated. These single cells are selectively transferred to a medium containing 0.01 uM zeatin and 0.5 uM 2,4-D (with 15 mM $CaCl_2$) and then transferred to one with increased cytokinin and reduced auxin concentration, 0.05 uM 2,4-D and 1 uM zeatin (with 0.2 M mannitol). The high auxin treatment appears to reprogram these cells and reinduce embryogenic competence.

Early Development

If the suspension is transferred to an auxin-free medium, each proembryo, whether single and freely floating or part of a clump, will begin to grow and mature as an embryo. Often populations of somatic embryos are a mix of single embryos, twins, double-headed embryos, clusters. The relative proportion of these types in a suspension is determined by a number of factors. One factor is the efficiency by which proembryonic masses release separate proembryos (e.g., Karlsson and Vasil, 1986). If there is substantial isolation of individual proembryos from the proembryonic mass early in their formation, then populations of mainly single structures will follow transfer to auxin-free, maturation medium. This typically occurs in carrot suspensions. If the separation of individual proembryos occurs more slowly, then many more multiple structures will be evident when the population is transferred to embryo maturation medium. This is the case with caraway.

The size of the proembryo is another factor (Figure 3). A small proembryo can readily progress through the embryonic stages. However, if the proembryo has enlarged to form two growth centers before transfer to maturation medium, then twin embryos will develop (Fig. 3B). Even larger proembryos can produce multiple embryos (Fig. 3C). If the enlargement is at the apical end, then embryos with two shoots on one hypocotyl and root will result (Fig. 3D). All of these have been seen in cultures (Ammirato, 1985).

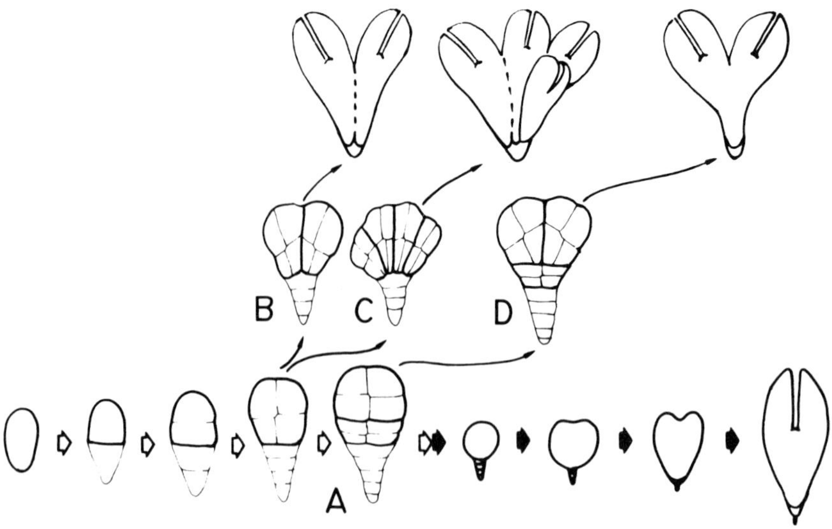

Figure 3. Early stages of somatic proembryo development. (A) The single proembryo passes through all the typical stages. (B) Additional divisions in the proembryo prior to shoot apex and cotyledon initiation create two growth centers that mature into two complete embryos joined at the base. (C) More than two centers generate clusters of multiple embryos. (D) Additional cell divisions and enlargement of the apical region at a latter stage will result in two shoots, each with cotyledons, on one hypocotyl and radicle.

If in suspension cultures, the suspension is sieved so that only the smallest and presumably single proembryos are transferred, then these should mature solely into single somatic embryos. For this reason, and to try to foster some synchrony of maturation in somatic embryo populations, suspension cultures are often highly sieved or otherwise separated into fractions (Ammirato, 1986, cf. pp. 471-472; Giuliano et al., 1983; Fujimura and Komamine, 1984).

Selective sieving or fractionation may not be entirely effective for the inductive process can continue even after the proembryo is transferred to auxin-free medium. In caraway (Ammirato, 1974), highly sieved suspensions of proembryos when transferred to media containing ABA produced almost 90% single embryos. ABA prevents the induction of embryogenic competence but does not inhibit maturation once the embryogenic cell is induced. If the medium did not contain ABA, a large number of multiple embryos developed from the single proembryos. During the early stages of embryonic development, the participating cells can easily be diverted and reinduced into new embryonic growth.

Later Development

Once the proembryo is transferred to maturation medium it is anticipated that the organizational events will proceed as in zygotic embryo maturation: the initiation of cotyledonary primordia, their growth, the early stages of tissue differentation (usually changes in cell shape), and the demarcation of shoot and root apices. Generally, this is the case. Indeed, strikingly normal somatic embryos can be identified in suspension cultures. However, various events may be premature, delayed, absent or extended in time.

Figure 4A shows a mature carrot zygotic embryo excised from a seed after 24 hours imbibition. Figures 4B to 4F show some typical carrot somatic embryos after Feulgen staining (unpublished work of D. Davidson and P. V. Ammirato). One feature of carrot somatic embryos is the potentially large size of some of the proembryos, prior to cotyledon initiation. For example, the mature zygotic embryo (Fig. 4A) is 1200 um long, yet somatic embryos can be 600 um (Fig. 4B) or even 900 um (Fig. 4D) and still show intense staining, indicating uniform cell division. This is particularly apparent if the cells of the Feulgen-stained embryos are tapped apart (Fig. 4G). One feature of somatic embryos is the continuation of cell divisions resulting in embryos larger than their zygotic counterparts.

When the cotyledons begin to grow, they do so from a broad disk-shaped meristem with many cell divisions visible around the periphery and in the center (Figs. 4E and 4F). The cotyledons often appear fused along their margins producing a fasciated, tube-like structure (Figs. 4E and 4F) rather than separate, distinct cotyledons (cf. Fig. 4A).

Finally, there is premature cell enlargement and differentiation. During normal carrot zygotic

---

Figure 4. Comparison of zygotic and somatic embryo development (Feulgen-stained) [Scale in A for A to H = 1 mm; scale in I, for I-L = 20 um, for M = 40 um.] (A) Excised zygotic embryo one day after germination showing extensive cell division, little cell vacuolation and no visible cell differentiation. (B) Proembryo 600 um long, all cells actively dividing and without cotyledonary primordia. (C) Early heart-shaped stage with cotyledonary primordia with premature vacuolation. (D) Elongated heart-shaped stage with initiating cotyledons. (E and F) Cotyledonary stages showing premature vacuolation in hypocotyl and cotyledons, precocious cell differentation in the vascular bundles, multiple, fasciated cotyledons and a broad, darkly stained apical region. The secondary embryo (F) has formed only one cotyledon. (G) A squash of the somatic embryo shown in (D) showing extensive cell division through the embryo and premature differentiation of vascular tissue (arrows). (H) A squash of the somatic embryo shown in (F) with vacuolation and cell enlargement in both the hypocotyl region and cotyledons and differentiated vascular tissue (arrows). (I to M) Mitotic figures within cells shown in (H) showing active cell division in vacuolated cells in the cotyledon (I) and hypocotyl (J). In both, the spindle orientation is with the row of cells, i.e., anticlinal division. (K and L). Mitotic spindle orientation that is periclinal to the row of cells. (M) A row of cells with one showing an periclinal (p) and another an anticlinal (a) orientation.

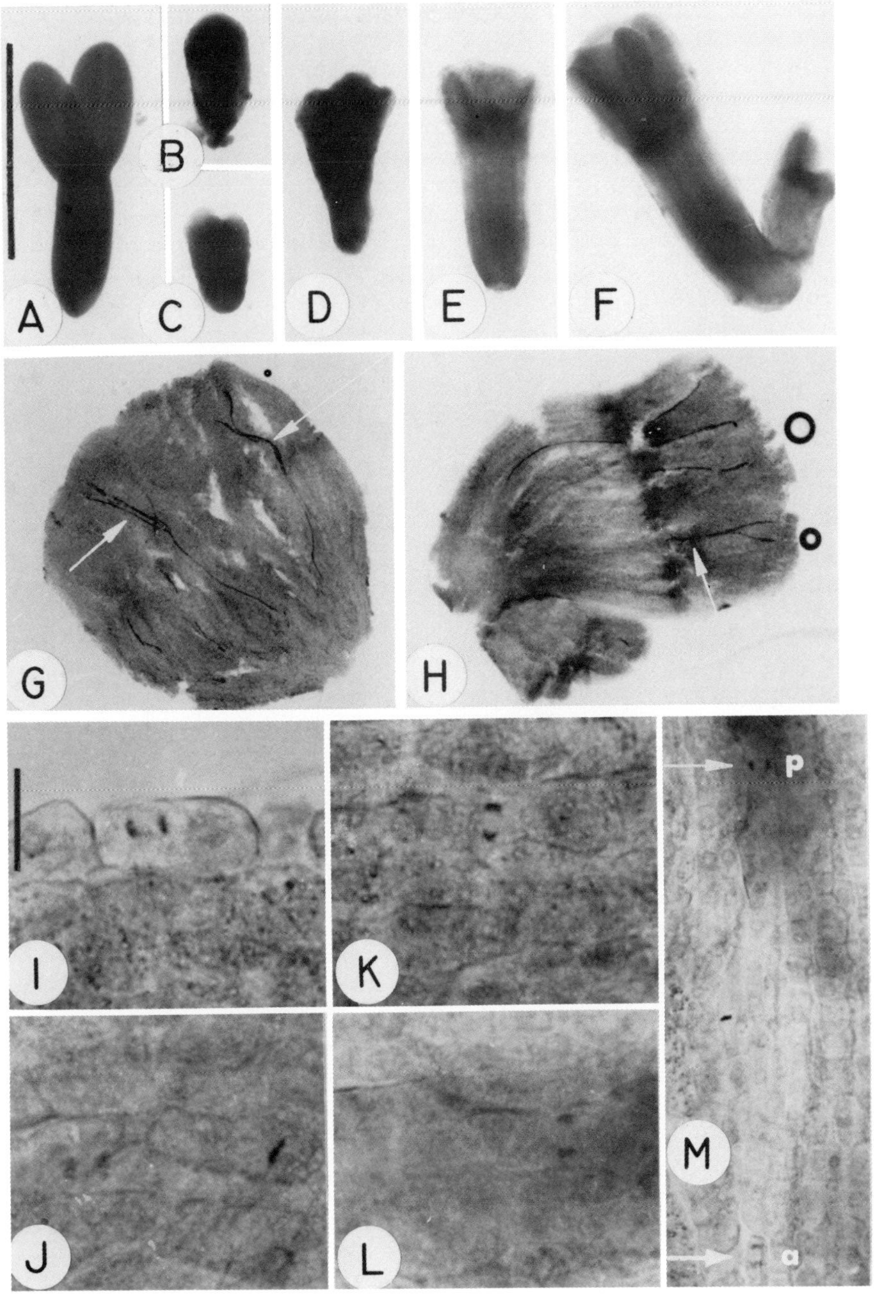
A
B
C
D
E
F
G
H
I
K
M
J
L
p
a

embryogenesis, the cells remain small, densely cytoplasmic and baring some changes in shape, undifferentiated. In somatic embryos, enlarged, highly vacuolated cells are clearly visible by the absence of intense Feulgen-staining, especially in the hypocotyl region, seen in whole embryos (Fig. 4F) and in squash preparations (Fig. 4H).

Premature vacuolation can also be seen in some small embryonic cotyledons (Fig. 4C). This premature vacuolation (which results in highly transparent cotyledons) and inadequate cell division often results in poorly developed cotyledons.

Cell differentiation is most evident in the formation of vascular tissue, especially visible in the hypocotyl region due to cell vacuolation and less intense staining (Figs. 4E and 4F), but present in cotyledons as well, as seen in the squash preparation (Fig. 4H). Vascular tissue is also present in the pre-cotyledonary embryo with numerous cell divisions and little enlargement (Fig. 4D) as shown in the squash preparation (Fig. 4G).

Extensive cell enlargement does not preclude a continuation of cell division. Mitotic figures can be seen in the cotyledonary region (Fig. 4I) and hypocotyl (Fig. 4J). Furthermore, in addition to division being anticlinal to the row, adding new cells to the row (Fig. 4I and 4J), they can also be periclinal, disrupting the linear structure of cells in the organ (Fig. 4K and 4L). Occasionally, both anticlinal and periclinal divisions can be seen simultaneously in one row of cells (Fig. 4M).

Therefore, there can occur early in somatic embryo development (1) extended cell division, (2) precocious vacuolation, (3) continuation of cell division in enlarged cells, (4) altered spindle orientations, (5) early cell differentiation and (6) delayed initiation of cotyledons and shoot apex. What effect do these have on the maturation of the somatic embryo and its subsequent growth into a plantlet?

Consequences to Morphogenesis

The many morphological and developmental abnormalities observed at the end of embryo maturation can be traced to inconsistencies in the normal sequence of organizational events, i.e., in cell division, enlargement and differentiation. The timing of these events in critical in determining the consequences on morphogenesis.

The extended cell divisions in the very young proembryo, with or without intervening cells that enlarge, can result in cells forming new centers of embryonic growth leading to multiple or secondary embryos (cf. Fig. 3).

Changes in cell division, enlargement and differentiation can also affect cotyledon formation. In normal cotyledon formation (Fig. 5A to D), the proembryo undergoes a series of cell divisions and a ring of cells forms a cotyledonary collar. Within that ring, two growth centers

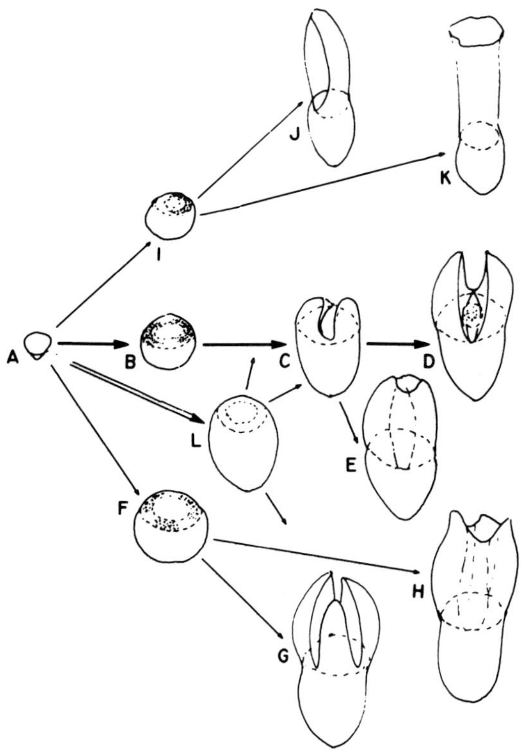

emerge (Fig. 5B). From these, the two cotyledonary primordia grow (Fig. 5C and mature with the presumptive shoot apex formed in the central space (Fig. 5D). If the size of the cotyledonary ring is altered by too many cell divisions and/or premature cell enlargement, then more than two centers can initiate to form multiple cotyledons (Fig. 4G). If there are too few cell divisions (Fig. 5I), then only one cotyledon will form (Fig. 5J). If cell divisions continue in the cotyledonary ring even after the cotyledonary primordia are initiated, then the cotyledons will developed fused or fasciated. This can occur whether there be one (Fig. 5K), two (Fig. 5E), or more than two cotyledons (Fig. 5H). The lack of sufficient cell division or premature vacuolation and differentiation can also abort cotyledon development.

One can credit the variability in cotyledon number and the alterations in cotyledon morphology to changes in the extent of cell division and premature enlargement in the young proembryo beginning at the heart-shaped stage. The relative roles of extended cell divisions vis-a-vis premature cell enlargement (Fig. 5L) in altering the size of the ring and subsequent development remains to be studied.

Figure 5. Cotyledonary abnormalities in somatic embryos traced to altered size of the cotyledonary ring and/or continued cell divisions after primordia initiation. In normal embryogeny, the globular proembryo (A) enlarges via cell divisions and eventually a ring appears at the apical end (B) from which two primordia arise and grow out as separate cotyledons (C). At maturity, the presumptive shoot apex rests in the space between the two cotyledons (D). Continued divisions in the ring between the two primordia will result in connated or fused cotyledons (E). Extended cell divisions in the proembryo before ring initiation can produce more than two centers (F) and multiple cotyledons (G). Continued cell divisions in the entire ring results in fused cotyledons (H). Too few cell divisions in the globular proembryo produces a smaller ring with only one primordium (I) and a single cotyledon (K) which can also be fused (J). Premature cell vacuolation (L) with or without extensive cell division may also enlarge the cotyledonary ring.

The correlation between size and form has been noted in many systems For example, there is an increase in the number of leaves in a whorl in Hippuris vulgaris as the size of the apical meristem and stem increases (cf. Wardlaw, 1965, pp. 153-165). In Pinus ponderosa" embryos, the number of cotyledons is directly proportional to the size of the apical meristem (Buchholz, 1946).

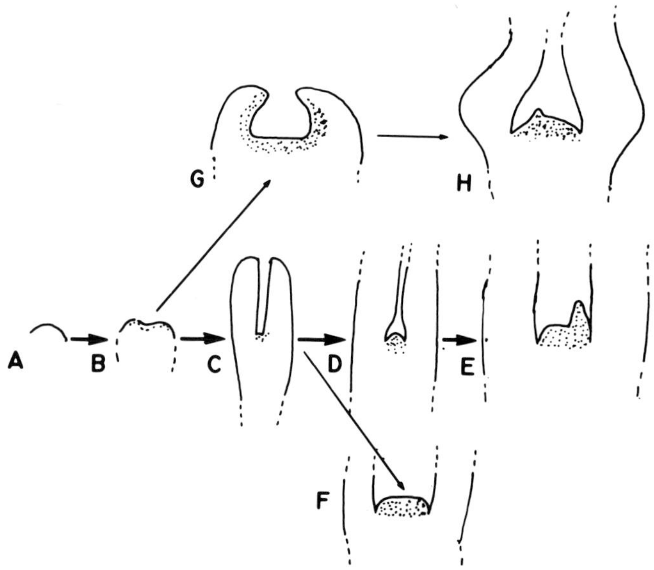

Figure 6. Abnormalities in shoot apical formation in somatic embryos traced to premature conversion of the embryonic shoot apex or extended cell divisions. The normal sequence in zygotic embryos includes the globular stage (A), the formation of the cotyledonary ring (B), the devlopment of the shoot apex during the torpedo stage (C) and the fully-formed, quiescent shoot apex in the mature embryo (D). At germination, the reactivated shoot apex enlarges and develops leaf primordia (E). In somatic embryos, the embryonic shoot apex may continue its development to a seedling shoot apex (F) and precocious germination. This premature differentation may occur when the cotyledons are initiating (G) forming a broad, flat, non-functioning shoot apex (H).

The development of the shoot apex can also be affected by changes in the sequence of cell division, enlargement and maturation (Fig. 6). In the normal sequence, a small number of cells become recognizable as the presumptive shoot apex after the cotyledons have begun their development (Fig. 6C). The shoot apex is small; its cells do not divide and they become quiescent during seed maturation (Fig. 6D). During germination, there is renewed cell division activity in the shoot apex and the formation of a active meristem, with a greater number of cells than the embryonic apex (Fig. 6E). The transition from embryonic apex to seedling apex can, and often does, occur directly, without an intervening quiescent or dormancy phase (Fig. 6F). However, this transition can occur while the cotyledons are being initiated (Fig. 6G). Increased cell division and enlargement at this time can generate an enlarged apical region (Fig. 6H). The cells within the shoot apical meristem can also vacuolate precociously.

Thus, the precocious appearance of a shoot apex more typical of a germinating seedling can arise at the same the cotyledons are developing. Studies of excised zygotic embryos of rapeseed (Brassica napus) showed that immature embryos when germinating precociously showed a mixture of embryogeny and seedling development (Finkelstein and Crouch, 1984).

The resulting apex can also develop so aberrantly that it cannot grow normally. There have been many reports in the literature of apparently well-formed somatic embryos that elongate but fail to produce plantlets (e.g., Kerns et al., 1986). This most likely can be attributed to the malformation of the shoot apical meristem. The disk meristem observed in the Feulgen-stained carrot somatic embryos was also seen in histological sections of caraway somatic embryos (Ammirato, 1985) where mature apices were broad and flat.

The precocious vacuolation, continued cell divisions, and changes in spindle orientation result in the absence of linear rows of cells in hypocotyls and cotyledons when viewed in 1 um plastic sections (Ammirato, 1985).

Precocious vacuolation and cell differentiation have been observed in other cultures. However, there is, as might be expected, some variation in the relative extent and sequence. In celery (Al-Abta and Collin, 1978), there is extended cell divisions and little cell differentiation until the torpedo stage. In opium poppy (Papaver somniferum, Nessler, 1982), mature elements such as laticifers and tracheids were visible in the somatic embryos while they are absent in zygotic embryos and appear only after germination, an example of premature differentiation. Morphologically identificable laticifers, however, are absent in somatic Aslepias embryos but present in somatic embryos--but they can be identified in the somatic embryos by means of fluorescent antibody immunocytochemistry (Dunbar et al., 1986)--an example of incomplete cell differentiation.

MODULATION OF SOMATIC EMBRYOGENY

By manipulating the way somatic embryos are grown, their structure and behavior can be modified. Almost every aspect of the cultural milieu has been shown to affect morphogenesis, i.e., the nutrient medium, exogenous hormones, evolved gases and environmental variables such as light (Ammirato, 1983a, b). The composition of the culture medium is critical and attention has always focused on the sources and levels of nitrogen and especially reduced nitrogen. Recent work has pointed up the beneficial aspects of certain amino acids such as proline in generating somatic embryos in maize (Armstrong and Green, 1985) and in improving the quality (ability to convert into plants) of alfalfa somatic embryos (Stuart et al., 1985). Increasing the osmotic concentration, either by elevating the carbohydrate (sucrose, for example) or by adding hexitols, such as inositol or sorbitol, have been shown to modulate development (Ammirato and Steward, 1971).

Abscisic acid (ABA) has also proven useful in modifying somatic embryogeny in caraway (Ammirato, 1974, 1977), carrot (Ammirato, 1983), pearl millet (Pennisetum americanum, Vasil and Vasil, 1981, 1982) and soybean (Ranch et al., 1985). Interestingly, it modifies the formation of all the developmental abberrant forms described earlier. It can prevent the induction of embryogenic cells, thereby stopping the formation of secondary embryos. Somatic embryos grown with ABA tend to have a high frequency of embryos with two cotyledons. Finally, ABA-treated somatic embryos do not germinate or convert into plants.

When examined in thin sections, ABA-grown caraway somatic embryos had small, domed shoot apices, distinct rows of cells, little vacuolation in cells and the complete absence of mature vascular tissue (Ammirato, 1985). The modulation that ABA could effect on external morphology (number of cotyledons, absence of secondary embryos) could also be manifest at the level of cell division, enlargement, spindle orientation, and cell differentiation.

The effect of treatments that restrict these events should then have different consequences dependant upon the time of their application. In a study of the effect of ABA on caraway somatic embryo maturation (Ammirato, 1974), the effects of adding ABA at different times during the four week maturation period was assessed. The conclusion then was that the sooner the ABA was applied, the more normal the somatic embryo maturation. But the effects are quite specific with time of application. If ABA is added at the time the proembryo suspension is transferred to maturation medium, a vast majority of single embryos with two cotyledons are grown. If ABA is added after one week's growth, the frequency of multiple embryos (twins, clusters) increases, but cotyledon development is normal. If added after two weeks, there is an substantial increase in multiple embryos, and those with accessory embryos, and although two cotyledons have formed, many are abnormal in structure. ABA exerts different effects if applied at different times during development.

Elevated osmolarity can also control somatic embryogenesis in carrot (Ammirato and Steward, 1971) and has a similar effect in _Brassica napus_ (Crouch, personal communication). In carrot, cells within the axis remain small, cotyledons mature but do not elongate, root and shoot apices do not grow out. It may be that the simple restriction of cell expansion can have profound effects in maintaining cell division patterns, preventing premature cell enlargement and differentiation and precocious germination. However, carrot somatic embryos raised in elevated sucrose or hexitols (inositol, sorbitol) did produce large numbers of accessory embryos. ABA can prevent this response. Therefore, in the future, we may be able to identify the roles ABA plays in controlling embryo maturation and separate its effects on somatic embryos in culture and zygotic embryos _in ovulo_. Care must to taken to distinguish between the factors controlling embryogenesis in albuminous seeds like carrot and the clearly more complex factors operating in exalbuminous seeds like soybean.

In all these cases, the developing somatic embryos are responding to different and diverse stimuli, both internal and external. By altering the conditions under which they are grown, the maturation of somatic embryos can be modulated. Vastly different populations of somatic embryos can be grown. In certain cases, the population of somatic embryos appear identical to the mature zygotic embryo removed from the seed. Interestingly, the treatments that cause the most normal development affect almost all of the organizational events that coordinate embryonic maturation, whether at the cell, tissue or organ level. If not diverted from its course, the embryo can pass through all the organizational stages of embryonic development and not reinitiate the process or move onto the juvenile or seedling stage.

CONCLUSIONS

1. The somatic embryo is strikingly plastic in its development. A range of embryonic forms can appear, some which affect its ability to grow into a plant. The somatic embryo may appear normal superficially (external morphology) but may be abnormal in terms of cellular and tissue differentiation.

2. This plasticity is the result of changes in the timing and extent of organizational events. In comparison with the normal sequence, some may be premature, delayed, or missing; their length of expression may be extended or abbreviated. The effects can be expressed at the cellular, tissue or organ levels.

3. Somatic embryos respond to alterations in their cultural milieu. The morphogenetic sequences can be substantially altered with the resulting somatic embryos being strikingly dissimilar or similar to zygotic embryos.

4. Since somatic embryos can be structurally and/or cellularly different from zygotic embryos at apparently comparable stage, these differences will most likely be reflected in their biochemistry and molecular biology.

REFERENCES

Al-Abta S, Collin HA (1978). Cell differentiation in embryoids and plantlets of celery tissue cultures. New Phytol 80:517-521.

Ammirato PV (1974). The effects of abscisic acid on the development of somatic embryos from cells of caraway (Carum carvi L.). Bot Gaz 135:328-337.

Ammirato PV (1977). Hormonal control of somatic embryo development from cultured cells of caraway: Effects of abscisic acid, zeatin and gibberellic acid. Plant Physiol 59:579-586.

Ammirato PV (1983a). The regulation of somatic embryo development in plant cell cultures: Suspension culture techniques and hormone requirements. Bio/Technology 1:68-74.

Ammirato PV (1983b). Embryogenesis. In Evans DA, Sharp WR, Ammirato PV, Yamada Y (eds): "Handbook of Plant Cell Culture, Volume 1," New York: Macmillan, pp. 82-123.

Ammirato PV (1985). Patterns of development in culture. In Henke RR, Hughes KW, Constantin MP, Hollaender A (eds): "Tissue Culture in Forestry and Agriculture," New York: Plenum, pp. 9-29.

Ammirato PV (1986). Carrot. In Evans DA, Sharp WR, Ammirato PV (eds): "Handbook of Plant Cell Culture. Volume 4," New York: Macmillan, pp. 457-499.

Ammirato PV, Steward FC (1971). Some effects of the environment on the development of embryos from cultured free cells. Bot Gaz 132:149-158.

Ammirato PV, Styer DJ (1985). Strategies for large scale manipulation of somatic embryos in suspension culture. In Zaitlin M, Day P, Hollaender A (eds): "Biotechnology in Plant Science: Relevance to Agriculture in the 1980s," New York: Academic Press, pp. 161-178.

Armstrong CL, Green CE (1985). Establishment and maintenance of friable, embryogenic maize callus and the involvement of L-proline. Planta 164:207-214.

Breton AM, Sung ZR (1982). Temperature-sensitive carrot variants impaired in somatic embryogenesis. Dev Biol 90:58-66.

Buchholz, JT (1946). Volumetric studies of seeds, endosperms, and embryos in Pinus ponderosa during embryonic differentiation. Bot Gaz 108:232-244.

Conger BV, Hanning GE, Gray DJ, McDaniel JK (1983). Direct embryogenesis from mesophyll cells of orchardgrass. Science 221:850-851.

Crouch ML (1982). Non-zygotic embryos of Brassica napus L. contain embryo-specific storage proteins. Planta 156:520-524.

Dunbar KB, Wilson KJ, Petersen BH, Diesboer DD (1986). Identification of laticifers in embryoids derived from callus and suspension cultures of Asclepias species (Asclepiadaceae). Am J Bot 73:847-851.

Finkelstein RR, Crouch ML (1984). Precociously germinating rapeseed embryos retain charateristics of embryogeny. Planta 162:125-131.

Fujimura T, Komamine A (1984). Fractionation of cultured cells. In Vasil IK (ed): "Cell Culture and Somatic Cell Genetics of Plants. Vol. 1," New York: Academic Press, pp. 159-166.

Giuliano G, Rosellini D, Terzi M (1983). A new method for the purification of the different stages of carrot embryoids. Plant Cell Rep 2:216-218.

Gupta PK, Durzan DJ (1986). Somatic polyembryogenesis from callus of mature sugar pine embryos. Bio/Technology 4:643-645.

Halperin W (1966). Alternative morphogenetic events in cell suspensions. Am J Bot 53:443-453.

Homes JLA (1968). Influence de la concentration en glucose sur le developpement et la differenciation d'embryons formes dans des tissus de carotte cultives in vitro. In "Les Cultures de Tissus de Plantes". Paris: Centre National de la Recherche Scientifique, pp. 49-60.

Kamada H, Harada H (1979). Studies on the organogenesis in carrot tissue cultures. I. Effect of growth regulators on somatic embryogenesis and root formation. Z Pflanzenphysiol 91:255-266.

Karlsson SB, Vasil IK (1986). Morphology and ultrastructure of embryogenic cell suspension cultures of Panicum maximum (Guinea grass) and Pennisetum purpureum (Napier grass). Am J Bot 73:894-901.

Kato H (1968). The serial observations of the adventive embryogenesis in the microculture of carrot tissue. Sci. Papers Coll Gen Edn Univ Tokyo 18:191-197.

Kerns HR, Barwale UB, Meyer MM Jr, Widholm JM (1986). Correlation of cotyledonary node shoot proliferation and somatic embryoid development in suspension cultures of soybean (Glycine max L. Merr.) Plant Cell Rep 5:140-143.

Konar RN, Thomas E, Street HE (1972a). Origin and structure of embryoids arising from epidermal cells on the stem of Ranunculus sceleratus L. J Cell Sci 11:77-93.

Konar RN, Thomas E, Street HE (1972b). The diversity of morphogenesis in suspension cultures of Atropa belladonna L. Ann Bot 36:123-145.

Lutz JD, Wong JR, Rowe J, Tricoli DM, Lawrence RH Jr (1985). Somatic embryogenesis for mass cloning of crop plants. In Henke RR, Hughes KW, Constantin MP, Hollaender A (eds): "Tissue Culture in Forestry and Agriculture," New York: Plenum, pp. 105-116.

McWilliam AA, Smith SM, Street HE (1974). The origin and development of embryoids in suspension cultures of carrot (Daucus carota). Ann Bot 38:243-250.

Nessler CL (1982). Somatic embryogenesis in the opium poppy, Papaver somniferum. Physiol Plnt 55:453-458.

Nomura K, Komamine A (1985). Identification and isolation of single cells that produce somatic embryos at a high frequency in a carrot suspension culture. Plant Physiol 79:988-991.

Raghavan V (1986). "Embryogenesis in Angiosperms. A Developmental and Experimental Study." Cambridge: Cambridge University Press.

Ranch JP, Oglesby L, Zielinski AC (1985). Plant regeneration from embryo-derived tissue cultures of soybeans. In Vitro Cell Dev Biol 21:653-658.

Sharp WR, Evans DA, Sondahl MR (1982). Application of somatic embryogenesis to crop improvement. In Fujiwara A (ed): "Plant Tissue Culture 1982," Maruzen, Japan, pp. 759-762.

Sharp WR, Sondahl MR, Caldas LS, Maraffa SB (1980). The physiology of in vitro asexual embryogenesis. Hort Rev 2:268-310.

Steward FC, Ammirato PV, Mapes, MO (1970). Growth and development of totipotent cells. Some problems, procedures and perspectives. Ann Bot 34:761-787.

Steward FC, Mapes MO, Kent AE, Holsten RD (1964). Growth and development of cultured plant cells. Science 143:20-27.

Street HE, Withers LA (1974). The anatomy of embryogenesis in culture. In Street HE (ed): "Tissue Culture and Plant Science, 1974," London: Academic Press, pp. 71-100.

Stuart, DA, Nelsen J, McCall CM, Strickland SG, Walker KA (1985). Physiology of the development of somatic embryos in cell cultures of alfalfa and celery. In Zaitlin M, Day P and Hollaender A (eds): "Biotechnology in Plant Science: Relevance to Agriculture in the 1980s," New York: Academic Press, pp. 35-47.

Sung ZR, Okimoto R (1981). Embryonic proteins in somatic embryos of carrots. Proc Nat Acad Sci 78:3683-3687.

Sung ZR, Okimoto R (1983). Coordinate gene expression during somatic embryogenesis in carrot. Proc Nat Acad Sci USA 80:2661-2665.

Terzi M, Giuliano G, LoSchiavo F, Nuti Ronchi V (1982). Studies on plant cell lines showing temperature-sensitive embryogenesis. In: Burger M (ed): "Embryonic Development. Part A," New York: Alan Liss, pp. 521-534.

Thomas E, Konar RN, Street HE (1972). The fine structure of the embryogenic callus of *Ranunculus sceleratus* L.). J Cell Sci 11:95-109.

Vasil V, Vasil IK (1981). Somatic embryogenesis and plant regeneration from suspension cultures of pearl millet (*Pennisetum americanum*). Ann Bot 47:669-678.

Vasil V, Vasil IK (1982). Characterization of an embryogenic cell suspension culture derived from cultured infloresences of *Pennisetum americanum* (Pearl Millet, L. Gramineae). Am J Bot 69:1441-1449.

Walker KA, Wendeln ML, Jaworski EG (1979). Organogenesis in callus cultures of *Medicago sativa*. The temporal separation of induction processes from differentiation processes. Plant Sci Lett 16:23-30.

Wardlaw CW (1965). "Organization and Evolution in Plants." Oxford: Longmans.

Williams EG, Maheswaran G (1986). Somatic embryogenesis: Factors influencing coordinated behavior of cells as an embryogenic group. Ann Bot 57:443-462.

Plant Tissue and Cell Culture, pages 83–93

ANALYSIS OF CARROT SOMATIC EMBRYO GENE EXPRESSION PROGRAMS

Terry L. Thomas and Dayton Wilde

Biology Department, Texas A & M University
College Station, Texas 77843

INTRODUCTION

The molecular events that occur very early during plant differentiation and development are critical to our understanding the processes of determination and differentiation which entrain subsequent morphogenesis of the early plant embryo. Unfortunately, these precise areas have been least productive when molecular biology techniques have been applied. Limited progress in this area is due primarily to the inaccesiblility of early developmental states to molecular analysis because of limiting amounts of staged embryonic material. Plant tissue culture appoaches, in particular those processes that result in somatic embryogenesis, may provide models for these early embryogenic events. Somatic embryogenesis, first demonstrated by Steward (1958) in carrots, has provided invaluable practical information on parameters conducive to plant regeneration in many genera and has also provided significant insights into aspects of plant development (Reviewed in Kohlenbach, 1978). In this paper, we explore some potential applications of somatic embryogenesis as a tool to understand the molecular events of early plant development.

RESULTS

Although not universal, a pervasive model of carrot somatic embryogenesis is that removal of exogenous auxin from carrot suspension cultures initiates new patterns of gene expression which in turn results in the sequential formation of structures morphologically analagous to zygotic globular, heart and torpedo embryo stages. Halperin (1966) identified small cell clusters, termed proembryogenic masses (PEMs), in carrot cultures and described their efficient morphogenesis into plantlets. More recently, Nomura and Komamine (1985) found that subpopulations of single carrot cells about 12 μm in diameter can be isolated from carrot suspension cultures containing 2,4-D (See Fig. 1A).

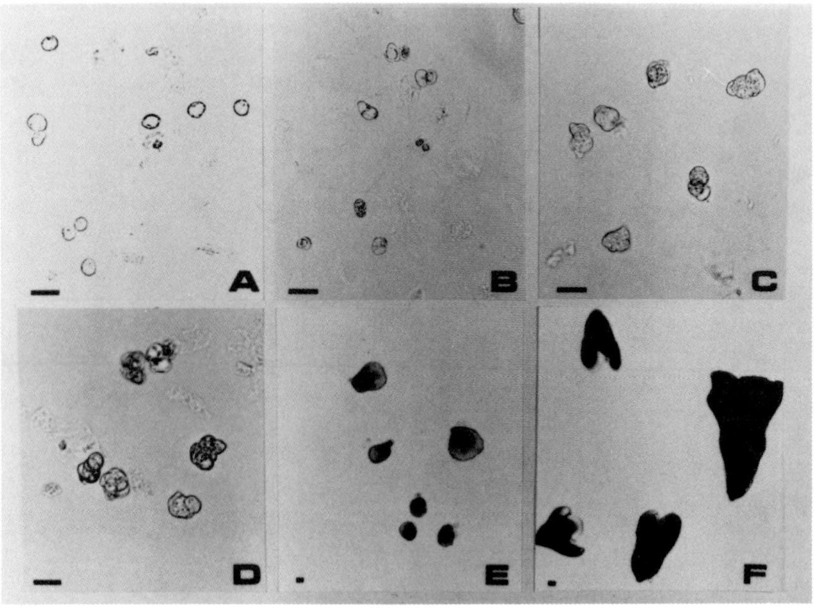

Figure 1. Somatic embryogenesis from single cells. A, 12 μm cells from carrot suspension culture; B, C and D same cells 2 d, 4 d and 7 d after culture in 5 x 10^{-8} M 2,4-D, respectively. E, globular and F, torpedo stage somatic embryos resulting from transfer of cell clusters in D to auxin free medium. Bars indicate 20 μm. Data reproduced with permission from Nomura and Komamine (1985).

Determination and maintenance of embryogenic potential in these cells was shown to require low auxin concentrations ($5x10^{-8}$ M 2,4-D). Subsequent divisions (Fig. 1B, 1C) and differentiation of these small cells in the presence of auxin results in small cell clusters (Fig. 1D), morphologically and developmentally analagous to the PEMs defined by Halperin (1966). On removal of auxin, these clusters or PEMs undergo somatic embryogenesis (See Fig. 1E,1F) at a frequency of 85 90%.

Initial molecular characterization of the transition from unorganized suspension culture through different stages of somatic embryogenesis indicated surprisingly few differences in detectable polypeptides when non-embryonic cultures were compared with cultures of somatic embryos. For example, Figure 2, taken from Choi and Sung (1984), shows a comparison of 2-D polyacrylamide gel electrophoresis (PAGE) patterns of steady state (Panels A and B) and newly synthesized proteins (Panels C and D) from

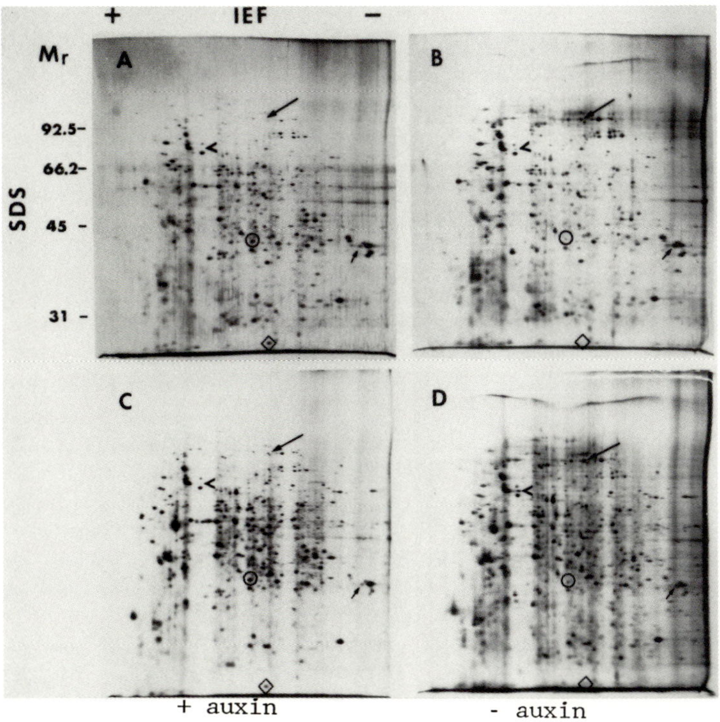

Figure 2. Two dimensional PAGE analysis of proteins extracted from embryogenic (B and D) and non-embryogenic carrot cultures (A and C). Auxin was 2,4-D at $5x10^{-6}$ M. See text for additional details. Data of Choi and Sung (1984) reproduced with permission.

embryonic and non-embryonic carrot cultures. Globular, heart and torpedo stage embryo stages were represented in the cultures grown without auxin for fifteen days. The arrows in Figure 2 indicate a limited number of both steady state (silver stained) and newly synthesized polypeptides labeled with ^{35}S methionine that are represented in embryo cultures but are not represented in non-embryonic cultures. In addition, polypeptides are expressed in non-embryonic cultures that are not found in embryo cultures. However, the most important conclusion derived from the comparison of gel patterns in Figure 2 is the striking similarity of the gene products expressed in embryonic and non-embryonic cultures. Of approxiamtely 800 polypeptides resolved, less than 2%, or fewer than 20, are either embryo specific or embryo modulated.

Comparison of the mRNAs expressed in embryonic and non-embryonic cultures indicate even less diversity in somatic embryo gene expression programs than observed in the preceeding 2-D PAGE analyses (Thomas and Wilde, 1985, Liu et al, 1985). Although technically limited to prevalent and moderately prevalent mRNAs and polypeptides, these experiments, as in other systems, probably reflect the overall pattern of gene expression in carrot somatic embryos and as such represents one end of a spectrum of possible gene expression patterns when compared to other embryonic systems (Reviewed in Thomas and Wilde, 1985).

More detailed investigation of the initiation and maintenance of somatic embryo gene expression programs has involved use of individual, cloned cDNA sequences representing carrot mRNAs in hybridization experiments. In one such experiment RNA expression patterns in embryogenic and nonembryogenic tissues were examined by hybridizing randomly selected cDNA recombinants with cDNA probes representing different mRNA populations. Proembryogenic masses (PEMs) were enriched from carrot suspension cultures grown in the presence of $5x10^{-7}$ M 2,4-D (Giuliano et al, 1983) and poly A RNA was isolated from a portion of the PEM tissue. The remainder of the PEMs were induced to undergo somatic embryogenesis by culturing in auxin-free media; poly A RNA was prepared from resulting cultures containing heart stage embryos. Poly A RNA was also prepared from mature carrot leaves. ^{32}P labeled cDNA probes, representing PEM, heart and leaf poly A RNAs, were prepared and hybridized separately with identical dot blots containing

DNA from randomly selected carrot embryo cDNA recombinants.

The results of these and other hybridization experiments are summarized in Table 1. Of 30 DNA recombinants analyzed, transcripts complementary to 19 were modulated either quantitatively or qualitatively.

TABLE 1. Patterns of gene expression in carrot

	cDNA Recombinant	PEM	Heart	Leaf
	E1	+*	+	-
	E2	+	+	-
EMBRYO	E3	+	+	-
SPECIFIC	E4	+	+	-
	E5	+	ND	-
	E6	+	+	-
	E7	+	+	-
	E8	+	+	-
	E9	++++	++++	++
	E10	+++	+++	+
	E11	++++	++++	+
	E12	++++	++++	+
	E13	+++	+++	+
EMBRYO				
MODULATED	E14	+	+	++++
	E15	+	+	++
	E16	+	+	++
	E17	+	+	+++
	E18	+	+	++++
	E19	+	+	++++

* + indicates clearly distinguishable hybridization at lowest relative prevalence; ++++ indicates highest relative prevalence; - indicates no detectable hybridization.

Surprisingly, the patterns of RNA expression in heart shaped embryos and in PEMs are nearly identical. <u>In contrast</u>, two-thirds of the cDNA recombinants analyzed in this manner exhibit either quantitative or qualitative differences in their representation in embryo or leaf RNA populations. All qualitative differences involve rare transcripts which are represented in both PEM and heart RNA populations but are undetectable in leaf RNA. For example, E3, an embryo specific cDNA recombinant (Thomas and Wilde, 1985), included in these experiments as a control, is not represented in leaf RNA. These results suggest that for the most part patterns of gene expression in somatic embryos are established much earlier than expected and in tissue culture conditions that are typically defined as nonembryogenic.

Initiation and maintenance of RNA expression patterns during establishment of carrot somatic embryo cultures was further investigated by hybridization of E3 and E5 probes (See Table 1) with RNAs representing embryogenic and nonembryogenic carrot tissue. Results of these experiments are shown in Figure 3. Figure 3A shows that E3 transcripts are highly represented in PEM, and surprisingly in callus tissue as well as in globular and torpedo somatic embryo stages. In this experiment, E3 transcripts are represented at detectable but very low levels in cotyledon, hypocotyl and leaf RNA populations. Additional experiments (Data not shown) indicate that E3 transcripts accumulate late in somatic embryogenesis (torpedo stage) and are present in dry seeds.

Hybridization of the E5 cDNA sequence with RNA Dot blots revealed the most striking RNA expression pattern. Carrot suspension cultures, grown in the presence of auxin, were fractionated into three size classes: <48μm, 48-120 μm and >120 μm (Giuliano et al, 1983). The 48-120 μm fraction represents the proembryogenic masses or PEMs described by Halperin (1966). Further culture of PEMs, in the absence of auxin, results in highly synchronous populations of somatic embryos with 80-90 % of the PEMs undergoing embryogenesis. Transcripts complementary to E5 are highly expressed in the PEM RNA population (Fig. 3A) but are nearly undetectable in the >120 μm RNA population. E5 transcripts are also highly abundant in the <48 μm size fraction (Data not shown). This latter result suggests that E5 transcripts may be expressed in the 12 μm cells present in carrot cultures grown at $5x10^{-7}$ M 2,4-D (Nomura and Komamine, 1985). E5 transcripts are represented at practically undetectable levels in hypocotyl, cotyledons, leaves and callus. In the experiment shown in Figure 3A, E5 transcripts are barely detectable in somatic embryos; however, additional data indicates E5 transcripts are present in poly A RNA populations of later embryo stages although at levels lower than observed in PEMs. See, for example, RNA gel blots shown in Figure 3B where 1.7 Kb E5 transcripts are detected in both PEM and torpedo stage RNA.

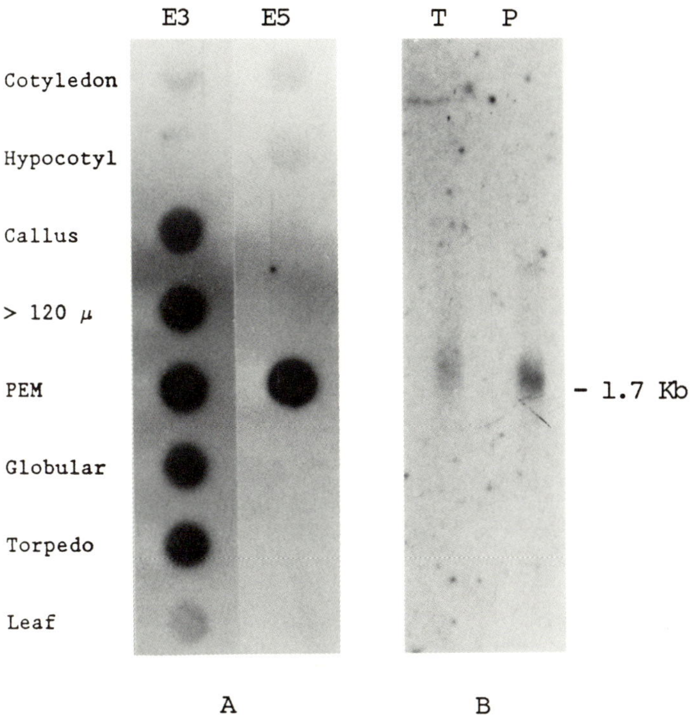

Figure 3. Representation of E3 and E5 transcripts in embryogenic and non-embryogenic carrot tissues. A, Dot blots including RNA from indicated carrot tissues hybridized with nick translated E3 and E5 probes. B, RNA gel blots of poly A^+ RNA from torpedo somatic embryos (T) and proembryogenic mass (P) hybridized with nick translated E5.

These data taken together define E3 and E5 expression patterns in terms of somatic embryogenesis events. E5 expression is indicative of very early events of embryogenic determination and differentiation and may serve as a marker for assessing the embryogenic potential of cell cultures. On the other hand, E3 expression is more characteristic of later somatic embryo stages although it too delineates earlier processes of differentiation.

DISCUSSION

The most striking result of these experiments is the great similarity in RNA expression patterns observed in carrot somatic embryo populations and tissues typically described as nonembryogenic, eg. callus and suspension culture grown in the presence of auxin. Every sequence examined whose expression is qualitatively or quantitatively increased in somatic embryo RNA populations is also represented in proembryogenic mass RNA populations. These results were surprising but not entirely unexpected. An early indication of this phenomenon were the limited number of changes in protein expression patterns during somatic embryogenesis when compared to carrot cultures maintained in the presence of auxin (Choi and Sung, 1984, Thomas and Wilde, 1985). Later, similar observations were made on the changes in mRNA populations during the transitions from "nonembryogenic" suspension culture to different somatic embryo stages (Thomas and Wilde, 1985, Liu et al, 1985). These data suggest that some patterns of gene expression in carrot somatic embryos are established in culture and in callus prior to the initiation of somatic embryogenesis by auxin removal. This observation is contrary to accepted dogma concerning the role of gene expression during somatic embryogenesis, ie. removal of auxin from suspension culture initiates new gene expression programs that are causally coupled to observed morphogenic events.

The diverse observations described above are best interpreted in the context of the results of Nomura and Komamine (1985) and suggest a useful model of pathways and conditions leading to various developmental states in carrot cultures. This model is summarized schematically in Figure 4. Nomura and Komamine (1985) found that small, spherical cells approximately 12 μm in diameter can be isolated from carrot cultures grown in the presence of auxin ($5x10^{-7}$ M 2,4-D) (See,for example,Fig. 1A). The developmental fate of these cells is not determined at this time since culturing conditions affect subsequent developmental events. Embryonic developmental programs appear to be initiated by culturing these 12 μm cells in the presence of low levels of auxin ($5x10^{-8}$ M 2,4-D) where these cells divide to generate cell clusters of 10-12 cells as shown in Figure 1D. Similar, highly embryogenic cell clusters can be isolated directly from carrot cultures maintained in the presence of high auxin levels, but they represent only 2-3% of the culture mass; these cell clusters are termed proembryogenic masses or PEMs (Halperin, 1966). Transfer of PEMs to auxin free media results in somatic embryogenesis at a frequency of 85-90%; while transfer to media containing high levels of auxin ($5x10^{-7}$ M 2,4-D) results in a "developmentally arrested" state. Somatic embryogenesis can be initiated from arrested cultures by transfer to auxin free media; however, the embryogenic potential of PEMs maintained at high auxin concentrations deteriorates substantially with time.

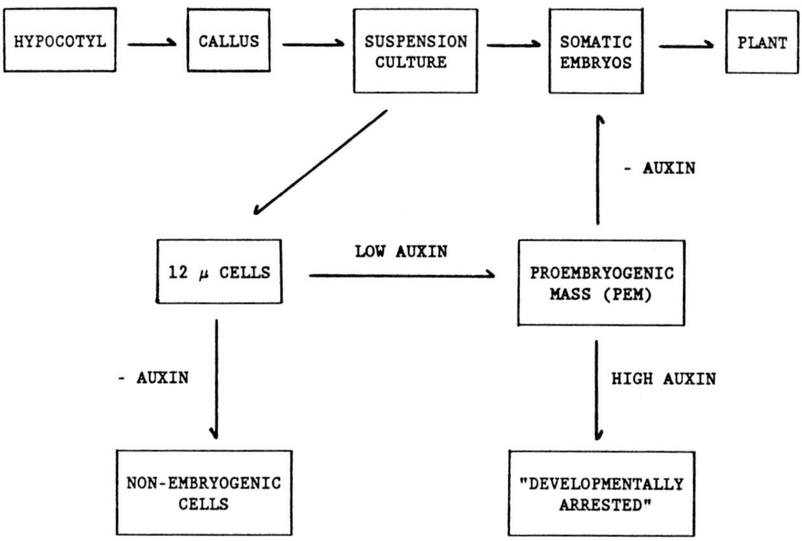

Figure 4. Some pathways and conditions leading to embryogenic and non-embryogenic states in carrot cultures.

Our observations and those of Nomura and Komamine (1985) together suggest that many events of determination and differentiation during somatic embryogenesis occur much earlier than expected and under physiological conditions different from those previously assumed. Determination of developmental potential in single carrot cells requires low levels of exogenous auxin but the expression of this developmental potential in terms of morphogenesis occurs in the absence of auxin and is arrested in the presence of high concentrations of auxin. Since highly embryogenic PEMs are generated in the presence of low auxin levels and since many somatic embryo gene expression programs are established in these tissues, it is possible that these programs are initiated in response to spatial and temporal fluctuations in absolute and relative phytohormone levels. This could explain the high representation of E3 transcripts as well as the low and variable representation of E5 transcripts in callus. For example, phytohormone gradients may be established in callus tissue promoting precocious embryonic domains or regions with the subsequent activation of at least some embryonic gene expression programs. A similar explanation is possible for the occasional observation of precocious somatic embryos on the callus surface. Along these same lines, Halperin (1966) suggested that polarity established in the PEM resulting in polyembryos joined at the radicle end may arise "due to metabolic gradients" established within the PEM. More detailed experiments, including use of hybridization probes such as E3 and E5, are required to determine the onset of embryonic gene expression programs during the establishment of carrot somatic embryo cultures. Investigations of this type may define the role, if any, of auxins or other phytohormones in the initiation and maintenance of these programs.

ACKNOWLEDGEMENTS: The authors gratefully acknowledge the useful discussions of these results with R. D. Allen, C. L. Nessler, T. D. McKnight and W. D. Park. This work was supported by USDA Grant 84CRCR-1-1391 awarded to T.L.T.

REFERENCES

Choi J and Sung ZR (1984). Two-dimensional gel analysis of carrot somatic embryonic proteins. Plant Mol.Biol Reporter 2: 19-23.

Giuliano GD, Rosellini D and Terzi M (1983). A new method for the purification of the different stages of carrot embryoids. Plant Cell Reports 2: 216-218.

Halperin W (1966). Alternative morphogenetic events in cell suspensions. Amer.J.Bot. 53: 443-453.

Kohlenbach HW (1978). Comparative somatic embryogenesis.

In Thorpe TA (ed): "Frontiers of Plant Tissue Culture," Calgary: University of Calgary Press, pp 59-66.

Liu L-S, Borkird C and Sung ZR (1985). Molecular cloning of cDNA complementary to poly A^+ mRNA differentially expressed during somatic embryogenesis. In Terzi M, Pitto L and Sung ZR (eds): "Somatic Embryogenesis," Rome: IPRA, pp 70-76.

Nomura K and Komamine A (1985). Identification and isolation of single cells that produce somatic embryos at a high frequency in carrot suspension cultures. Plant Phys. 79, 988-991.

Steward FC, Mapes MO and Smith J (1958). Growth and organized development of cultured cells I. Growth and division of freely suspended cells. Amer. J. Bot. 45, 693-703.

Thomas T and Wilde D (1985). Analysis of gene expression in carrot somatic embryos. In Terzi M, Pitto L and Sung ZR(eds): Somatic Embryogenesis," Rome: IPRA, pp 77-85.

Section II.

Biochemistry and Physiology

Plant Tissue and Cell Culture, pages 97–117

PRIMARY METABOLISM AND ITS REGULATION

Donald K. Dougall

Department of Botany, The University of Tennessee, Knoxville, TN 37996-1100

INTRODUCTION

"Primary Metabolism and Its Regulation" as a title reflects an area of study which has been actively pursued in plants and plant cell cultures for many years. There is therefore a very large literature detailing these studies which can be referred to. In addition there are others here who can and will describe in detail various aspects of primary metabolism and its regulation. Here I want to take a different approach to the topic. I want to address some topics selected because they are not well described but have the potential for being significant and general. Effects on primary metabolism may increase or decrease the particular aspect of metabolism. The consequences of an increase in the rate of an area of metabolism may be to increase growth rate or it may be to leave growth rate unaltered but to increase the rate of accumulation of a metabolite or phytochemical. A number of alternative possibilities can be laid out, each depending on the details of additional aspects of the metabolism of a specific cell. One of the difficulties in the area of primary metabolism and its regulation is having a sufficient number of well described systems to form the basis of generalization. This difficulty is particularly acute in this presentation which is an attempt to identify aspects of cellular metabolism and regulation which have the potential for improving our control over the performance of cell cultures. I have chosen for discussion today the following:

1) Ethylene

2) Polyamines

3) Ammonia assimilation

4) Aromatic amino acid biosynthesis

5) Control of gene expresion

These choices undoubtably involve some personal bias but hopefully they will stimulate discussion and increase study of metabolism in plant cell cultures.

ETHYLENE

Ethylene represents an uncontrolled variable in plant tissue cell cultures, the significance of which has not been established.

The production of ethylene by plant cell cultures and its accumulation in culture vessels has been documented in a number of cases (e.g. Coleman et al, 1980; Thomas and Murashige, 1979; Tisserat and Murashige, 1977a, b; Lieberman et al, 1979; Chappell et all, 1984; Zobel and Roberts, 1978). The accumulation of ethylene in vessels will be subject to physical factors such as the tightness of the closure which may vary within a batch of cotton plugs and which will vary between types of closures. The levels achieved will also vary with the size of the vessel relative to medium and tissue. In vessels through which gas is flowing such as stirred or airlift fermentors or even callus cultures, the level of ethylene should come into equilibrium with that in the gas being passed through the culture. However, the gasses being used for this purpose may not be free of ethylene. Morison and Gifford (1984) showed that cylinders of CO_2 may contain ethylene as a contaminant. Eastwell et al (1978) show the contamination of air by ethylene and hydrocarbons and used metal catalysts at high temperature to oxidize the contaminants to CO_2.

While ethylene is well known for its effects on various aspects of plant behavior, it is less well known for its effects in culture tissue. Zobel and Roberts (1978) reported that ethylene levels of the order of 0·1ppm prevented both callus initiation and tracheid differentiation in primary explants of lettuce.

Only recently has evidence become available which begins to show sites of action of ethylene. Carrots or potatoes treated with 50ppm ethylene showed a significant increase in fructose-2:6-bisphosphate (Stitt et al, 1986). This compound is of current interest as a regulator of glycolysis and gluconeogenesis in plants (Huber, 1986) in that it appears to stimulate glycolysis and inhibit gluconeogenesis. While its role in various aspects of plant metabolism have been explored, the significance of fructose-2:6-bisphosphate in cell cultures does not appear to have been examined yet.

The recent observations by Apelbaum et al (1985) open a second possibility for effects of ethylene. They showed with pea seedlings that ethylene decreased the activity of arginine decarboxylase. The threshold for this effect was at 0·05 microliters ethylene/litre, was half maximal at 0.12 microlitres/l and was saturated at approximately 5 microlitres/litre. The ethylene effect was reversible. Ethylene increased the apparent Km of the enzyme five-fold and decreased the Vmax ten-fold. The ethylene did not cause changes in the enzyme activity when applied to the crude extracts. This observation appears to rule out a direct effect of ethylene on the enzyme. Arginine decarboxylase is part of one of two routes by which putrescine, the precursor of the polyamines spermidine and spermine, is synthesized in cells. Because of the apparent requirement of polyamines for cell division, the effect of ethylene on arginine decarboxylase activity may result in restriction of cell division via restriction of polyamines. Data on polyamine concentrations were not given in the paper.

Ethylene biosynthesis and polyamine biosynthesis share a common precursor. S-Adenosylmethionine is the precursor of amino cyclopropane carboxylic acid, the immediate precursor of ethylene. S-Adenosylmethionine is also decarboxylate to give the aminopropyl donor for the conversion of putrescine to spermidine and spermine. Further, both biosynthetic routes generate a common product, namely methylthioadenosine. Giovanelli et al (1983) have established the conversion of the methylthioadenosine to adenine and methionine. One would expect some form of regulation to exist between ethylene biosynthesis and polyamine biosynthesis. Slocum et al (1984) have reviewed the evidence from studies with whole plants which show that inhibition of either pathway stimulates the other and that exogenously supplied polyamines inhibited amino cyclopropane carboxylic acid synthase in some cases, and the enzyme converting this intermediate to ethylene in others.

Thus, ethylene is a potential complication in tissue culture systems whose ramifications have not been extensively explored.

POLYAMINES

Polyamines, their biosynthesis and involvement in metabolism have been the subject of increasing interest in plants and plant cell cultures. The current interest in polyamine metabolism is in part the result of recent availability of specific inhibitors of many of the steps of polyamine biosynthesis and of the development of methods to measure polyamines. Much of the available information has been recently reviewed by Slocum et al (1984). In cells grown in culture, involvement of polyamines in three areas has been documented.

The involvement of polyamines in embryogenesis in carrot suspension cultures was examined by Feirer et al (1984) and Fienberg et al (1984). Embryogenesis in carrot suspension cultures was inhibited when arginine decarboxylase was inhibited with difluoromethylarginine and when S-adenosylmethionine decarboxylase and spermidine synthase were inhibited by dicylohexylamine or methyglyoxal bis(guanylhydrazone). Embryogenesis was not decreased when ornithine decarboxylase was inhibited by difluoromethylornithine. The expected effects of these inhibitors on cellular polyamines were found as was reversal of the inhibitions by added polyamines. In both cases embryo formation was more sensitive to the inhibitors used than was growth of the tissue. These data show that polyamines synthesized through the reaction catalysed by arginine decarboxylase are required for embryo differentiation but not growth in carrot cell cultures.

Tiburcio and Galston (1986) have concluded that ornithine decarboxylase is important for growth and that arginine decarboxylase provides putrescine which is incorporated into the pyrrolidine ring of nicotine in tobacco callus tissue. These observations raise questions concerning the metabolism or functions of putrescine derived from arginine decarboxylase and ornithine decarboxylase.

Finally there is the most intersting observations by Muhitch et al (1983) that sucrose plus polyamines extended the viability of stationary phase Paul's Scarlet Rose cells in suspension for 30-45 days. Further Muhitch and Fletcher (1985) have shown with this same system that the spectrum of phenolic compounds separated by HPLC is different in the sucrose and spermidine treatment compared to the controls. The pressing questions related to these observations concern their generality, the mechanism of the apparent protective effect and the events leading to the appearance of the phytochemicals.

AMMONIA ASSIMILATION

The path of assimilation of ammonia by plant cells via glutamine synthetase and glutamate synthase and the role of glutamic dehydrogenase continues to be developed. Recent reviews of this area have been given by Miflen and Lea, 1980; Stewart et al, 1980; and Suzuki and Gadal, 1984. Since the presentation of the kinetic model of ammonia assimilation via glutamine synthase and glutamate synthase in *Lemna minor* by Rhodes et al (1980), Fentem et al (1983a), have showed that in barley seedlings the assimilation of ammonia into glutamine amide could be quantitatively accounted for by a three compartment model. They (Fentem et al 1983b) further concluded from experiments with methionine sulfoximine and azaserine that ammonia is assimilated through glutamine synthetase - glutamate synthase and that neither glutamic dehydrogenase nor asparagine synthetase was involved in ammonia assimilation. These conclusions are consistant with the high affinity of glutamine synthetase for ammonia and the apparently low affinity of glutamic dehydrogenase.

Glutamine Synthetase

The demonstration of glutamine synthetase activity in extracts of tissue along with other enzymes of nitrogen assimilation continues. Recent papers include a comparison of sugar cane cultures under shoot forming and unorganized conditions of growth, i.e. plus and minus 2:4-D (Dwivedi et al, 1984), in suspension cultures of *Silene alba* grown with different nitrogen sources (Brassart et al, 1984), and in root and leaf derived callus from *Bouvardia ternifolia* grown with differnet sources of nitrogen (Sanchez de Jimenez and Fernandez, 1983). In plants, the octameric glutamine synthetase occurs in up to four isoforms. Two of these at least are found in leaves, one cytoplasmic and the other chloroplastic, and show polypeptide heterogeneity in isoelectric focusing (Dwiniak and Dunham, 1983; Lara et al, 1984; McNally et al, 1983; Hirel et al, 1984a and b; Nato et al, 1984). McNally et al (1983) have classified plants on the basis of the relative proportions of the cytoplasmic and chloroplastic forms of glutamine synthetase and shown that tobacco (*Nicotiana tabacum*) contains only the chloroplastic form of glutamine synthetase. Hirel et al (1984a) examined the glutamine synthetase levels in suspension cultures of two mutants of tobacco which had previously been shown to have undetectable levels of ribulose-1:6-bisphosphate carboxylase. Both the cell lines derived from the mutant genotypes and cell line from the dark grown wild-type contained about one sixth of the activity and glutamine synthetase protein as was produced by the green wild-type cells. This behavior of glutamine synthetase paralleled that of ribulose-1:6-bisphosphate carboxylase and suggests that the level of both enzymes is controlled by the same light dependent mechanism.

In alfalfa suspension cultures lines resistant to the glutamine synthetase inhibitor L-phosphinothricin have been selected by Donn et al (1984) and they have shown that the resistance was due to gene amplification. Using a c-DNA clone isolated from the resistant line, Tischer et al (1986) have isolated the amplified gene and determined its nucleotide sequence. DasSarma et al (1986) have reconstructed a complete glutamine synthetase c-DNA, inserted it into a glutamine synthetase deficient mutant of *E. coli* and shown that the gene is expressed and converts the strain to glutamine independence.

Glutamate Synthase

The properties of the several glutamate synthases from many sources have been reviewed by Susuki and Gadal (1984) recently. The ferredoxin dependent glutamate synthases from plants are distinct from the pyridine nucleotide dependent enzymes in several ways. The pyridine nucleotide forms are not precipitated by antibodies to the ferredoxin dependent forms in green tobacco suspension cultures (Susuki et al, 1984). The ferredoxin dependent activities increase in parallel with increases in chlorophyll during greening of tobacco suspension cultures (Susuki et al, 1984) and in *Bouvardiar ternifolia* callus cultures (Murillo and Sanchez de Jimenez, 1985) while the pyridine nucleotide dependant enzyme activities remained constant or declined a little. The activity of the pyridine nucleotide dependent activity was dependent on the nitrogen source supplied to cultures of *Bouvardia ternifolia* (Sanchez de Jimenez and Fernandez, 1983). While the available data on glutamate synthase remains consistant with a role in ammonia assimilation especially in photorespiration, the data is limited and the role of the pyridine nucleotide enzyme is still unproven.

Glutamine Dehydrogenase

Glutamic dehydrogenase, particularly that using NAD, continues to be an enigma in that it is ubiquitous but its function is not clearly understood. Recent demonstration of the effects of nitrogen source on extractable glutamic dehydrogenase (NAD) have been given by Sanchez de Jemenez and Fernandez (1983) and Brassart et al (1984) and for differences in the time course of its activity in the presence (non-shoot forming) and absence (shoot forming) of 2:4-D in sugar cane (Dwivedi et al, 1984). Furuhashi and Takahashi (1982) described the details of the Ca^{++} activation of the aminating reaction catalysed by glutamic dehydrogenase isolated from mitochondria of green tobacco callus cultures. In addition to Ca^{++} activation, the apparent Km for NH_4^+ decreased from 143 to 21 mM as the pH changed from 7.3 to 9.0. The reverse reaction is not affected by Ca^{++}. Yamaka et al (1984) using corn shoots as a source of mitochondrial NAD dependent glutamic dehydrogenase confirmed the Ca^{++} activation and showed that it only occured at high NADH concentrations. The apparent Km for ammonia decreased from 12.5 mM to 4.8 mM as the NADH concentration was lowered from 0.16 mM to 0.01 mM.

Further they measured the intramitochondrial concentrations of substrates and products of the reaction incuding Ca^{++} and simulated the environment of the glutamic dehydrogenase in the mitochondria. Over the pH range of 6.0 to 9.0 this simulation always moved in the direction of amination, the rate of amination reaction was maximal at pH 8.0 and the behavior of this simulation was not altered qualitatively by changing the ammonia concentration in the range 2-100 mM.

Glutamic dehydrogenase appears to be a hexamer made up of two different polypeptide chains giving a maximum of seven isozymes. Laurier and Daussant (1983)using detached wheat leaves have shown that the NH_4^+ induced increase in glutamic dehydrogenase is due to an increase of one of the two polypeptides. Ehmke et al (1984) have shown that heat inactivation of the glutamic dehydrogenase from peas preferentially inactivated the six fast migrating bands and converted these into the slowest migrating form, observations which are compatible with differential thermal stability of the two monomers of the enzyme.

Cammaerts and Jacobs (1985) who have shown in *Arabidopsis thaliana* that glutamic dehydrogenase is controlled by two non-allelic genes and can be resolved into seven bands on electrophoresis, examined the behavior of the isozymes and measured the aminating and deaminating activities in several different environments. They examined these activities during growth and development of the plants, in dark stressed plants and in plants grown on nitrate or ammonia in the presence and absence of sucrose. They also measured the effects of these treatments on the level of glutamine synthetase, in some cases in the presence of methionine sulphoximine to prevent the glutamine synthetase from participating in ammonia assimilation. They conclude that all three enzymatic activities, glutamic synthetase, aminating glutamic dehydrogenase and the deaminating glutamic dehydrogenase were important in ammonia metabolism, that glutamine synthetase assimilates principally physiological levels of ammonia as well as that generated in photorespiration, that the aminating and deaminating activities of glutamic dehydrogenase are associated with different isozymes, that the aminating form deals with high levels of ammonia and that the deaminating form is associated with energy production by the cell, presumably via 2-ketoglutarate. These interpretations are compatible with the conclusions of Fentem et al (1983a, 1983b) on the route of assimilation of ammonia by barley because a 10-15-fold lower concentration of nitrogen sources were used in the barley studies compared to the *Arabidopsis* studies.

AROMATIC AMINO ACID BIOSYNTHESIS

The biosynthesis of aromatic amino acids in plants and plant cell cultures has been recently reviewed by Jensen (1985). This review should be supplimented with some recent papers. (Steinrucken et al, 1986; Mousdale and Coggins, 1985; Nafziger et al, 1984; Smart et al, 1985; Schmauder et al, 1985; Goers and Jensen, 1984; Suzich et al; 1984, Suzich et al, 1985; Bonner and Jensen, 1985). Jensen (1985) discussed the possibilities that 1) the pathway is localized in both the plastid/chloroplast compartment as well as in the cytosol and, 2) that the plastid/chloroplast enzymes are subject to sequential feedback control while the cytosolic enzymes are not. Further Jensen (1985) suggests that L-arogenic acid, the product of transamination of prephenic acid should be considered as the immediate precursor of tyrosine and phenylalanine. The regulation and comparmentation of the biosynthesis of aromatic amino acids is clearly in need of further study.

In addition to the enzymes directly involved in aromatic amino acid biosynthesis a number of plant cells accumulate quinic acid. Quinic acid is derived from 3-dehydroquinic acid which is an intermediate in aromatic amino acid biosynthesis by reversible oxidation/reduction catalyzed by quinate : NAD^+ 3 oxidoreductase. Quinic acid thus appears to be a reservoir compound for intermediates in the pathway but may also be further metabolized, for example to chlorogenic acid. The quinate oxidoreductase from carrot suspension cultures appears to be regulated post transcriptionally. In light grown tissue it is inactivated by dephosphorylated by a F^- sensitive and Mg^{++} activated endogenous neutral phosphatase, 2) activated by phosphorylation with ATP by a Ca^{++} and calmodulin dependent protein kinase and 3) has a molecular weight of 42 kD (Refeno et al 1982 a, 1982b; Ranjeva et al, 1983; Graziana et al, 1983a).

In dark grown tissue the native enzyme has a molecular weight of circa 110 kD, is composed of a 42 and a 60 kD subunit and is directly activated by Ca^{++}. Further, in contrast to the light grown cells, the dark grown native enzyme could not be dephosphorylated and deactivated, apparently because the Ca^{++} binding 60 kD polypeptide in the enzyme protected the bound phosphate from phosphatases (Graziana et al, 1983b, 1984). This case is one of two or three examples of plant enzymes being controlled by protein phosphorylation and shows that regulation of metabolism in plant cells can occur at levels more complex than that available from allosteric effects of small molecules on enzyme activities.

CONTROL OF GENE EXPRESSION

The idea that growth of cells and production of secondary products are competitive processes is wide spread. Evidence for this concept from studies of metabolism have been given by Phillips and Henshaw (1977) and Lindsey (1985). The idea underlies the concept of a two stage production system for phytochemicals, one for growth of biomass and the second for production of the phytochemicals and, at least to a large extent, underlies the use of immobilized cells for the production of phytochemicals.

The study of a number of these systems has shown that the synthesis of at least the first unique enzyme of the biosynthetic pathway for the secondary product is increased as a result of the transition to production of phytochemicals, e.g. phenylalanine ammonia lyase and other enzymes in parsley (Chappell and Hahlbrock, 1984), in soybean (Schmelzer et al, 1984), in bean (Edwards et al, 1985), and tryptophan decarboxylase in _Catharanthus roseus_ cell cultures (Noe and Berlin, 1985). It seems that increases in the transcription of specific structural genes is at least a necessary condition for increased production of phytochemicals in plant cell cultures. As a result of interest in understanding phytochemical production, development, morphogenesis, responses to plant pathogens, and other biological processes involving control of gene activity and of the state of molecular biology, attention has now begun to focus on the behavior and properties of promoters, enhancers and other regions of DNA which are involved in the modulation of gene expression (Simpson et al, 1986; Fluhr et al, 1986; Kaulen et al, 1986; Myers et al, 1986: An, 1986; Fluhr and Chua, 1986). A timely example along these lines is the recent identification of the phenolic compounds produced by wounded plants which activate the virulence locus of _Agrobacterium tumefaciens_. The activation of this locus is an early step in transformation. The identification was achieved using plasmids modified in the virulence locus by insertion of a -galactosidase gene which acted as a readily detected reporter of gene activation and showing that specific compounds, including two isolated and purified from cell cultures, led to increased production of -galactosidase (Bolton et al, 1986; Stachel et al, 1985). Approaches along these lines of study should lead to new opportunities to control metabolism for specific purposes by providing sensitive and specific systems in which to identify the events which initiate the changes in cells from growth to phytochemical production.

REFERENCES

An G (1986). Development of plant promoter expression vectors and their use for analysis of differential activity of nopaline synthase promoter in transformed tobacco cells. Plant Physiol 81: 86-91.

Apelbaum A, Goldlust, A, Icekson, I (1985). Control of ethylene of arginine decarboxylase activity in pea seedlings and its implication for hormonal regulation of plant growth. Plant Physiol 79:635-640.

Bolton GW, Nester EW, Gordon MP (986). Plant phenolic compounds induce expression of the _Agrobacterium tumefaciens_ loci needed for virulence. Science 232:983-985.

Bonner CA, Jenson RA (1985). Novel features of prephenate aminotransferase from cell cultures of _Nicotania silvestris_. Arch Biochem Biophys 238: 237-246.

Brassart A, Rambour S, Bouriquet R (1984). Activites glutamate deshydrogenase et glutamine synthetase d'une suspension cellulaire de _Silene alba_ cultivee en presence de differentes sources d'azote. Physiol Veg 22: 801-809.

Cammaerts D, Jacobs M (1985). A study of the role of glutamate dehydrogenase in the nitrogen metabolism of _Arabidopsis thaliana_. Planta 163: 517-526.

Chappell J, Hahlbrock K, (1984). Transcription of plant defence genes in response to UV light or fungal elicitor. Nature 311: 76-78.

Chappell J, Hahlbrock K, Boller T (1984). Rapid induction of ethylene biosynthesis in cultured parsley cells by fungal elicitor and its relationship to the induction of phenylalanine ammonia-lyase. Planta 161:475-480.

Coleman WK, Huxter TJ, Reid DM, Thorpe TA (1980). Ethylene as an endogenous inhibitor of root regeneration in tomato leaf discs cultured in vitro. Physiol Plant 48:519-525.

DasSarma S, Tischer E, Goodman H M (1986). Plant glutamine synthetase complements a *glnA* mutation in *Escherichia coli*. Science 232: 1242-1244.

Donovan DST, Murashige T (1979). Volatile emissions of plant tissue cultures: II. Effects of the auxin 2,4-D on production of volatiles in callus cultures. In Vitro 15:659-663.

Donn G, Tischer E, Smith JA, Goodman HM (1984). Herbicide-resistant alfalfa cells: An example of gene amplification in plants. J Molecular and App Genet 2: 621-635.

Dwivedi UN, Khan BM, Rawal SK, Mascarenhas AF (1984). Biochemical aspects of shoot differentiation in sugarcane callus: I. Nitrogen assimilating enzymes. J Plant Physiol 117: 7-15.

Dyviniak SD, Dunham VL (1983). Four isoforms of glutamine synthetase in light-grown soybeans. Biochem Biophys Res Comm 114: 604-11.

Eastwell KC, Bassi PK, Spencer ME (1978). Comparison and evaluation of methods for the removal of ethylene and other hydrocarbons from air for biological studies. Plant Physiol 61:723-726.

Edwards K, Cramer CL, Bolwell GP, Dixon RA, Schuch W, Lamb CJ (1985). Rapid transient induction of phenylalanine ammonia-lyase mRNA in elicitor-treated bean cells. Proc Natl Acad Sci USA 82: 6731-6735.

Ehmke A, Scheid H-W, Hartmann T (1984). Glutamate dehydrogenase of *Pisum sativum*: Heat-dependent interconversion of the multiple forms. Z Naturforsch 39: 257-260.

Feirer RP, Mignon G, Litvay JD (1984) Arginine decarboxylase and polyamines required for embryogenesis in the wild carrot. Science 223:1433-1435.

Fentem PA, Lea PJ, Stewart GR (1983a). Ammonia assimilation in the roots of nitrate- and ammonia-grown *Hordeum vulgare* (cv Golden Promise). Plant Physiol 71:496-501.

Fentem PA, Lea PJ, Stewart GR (1983b). Action of inhibitors of ammonia assimilation on amino acid metabolism in *Hordeum vulgare* L. (cv Golden Promise). Plant Physiol 71:502-506.

Fienberg AA, Choi JH, Lubich WP, Sung ZR (1984). Developmental regulation of polyamine metabolism in growth and differentiation of carrot culture. Planta 162:532-539.

Fluhr R, Chua N-H (1986). Developmental regulation of two genes encoding ribulose-bisphosphate carboxylase small subunit in pea and transgenic petunia plants: Phytochrome response and blue-light induction. Proc Natl Acad Sci USA 83: 2358-2362.

Fluhr R, Kuhlemeier C, Nagy F, Chua N-H (1986). Organ-specific and light-induced expression of plant genes. Science 232: 1106-1112.

Furuhashi K, Takahashi Y (1982). Control of glutamate dehydrogenase from green tobacco callus mitochondria by Ca^{2+} and pH. Plant and Cell Physiol 23: 179-184.

Giovanelli J, Datko AH, Mudd SH, Thompson GA (1983). *In vivo* metabolism of 5'-Methylthioadenosine in *Lemma*. Plant Physiol 71:319-326.

Goers SK, Jensen RA (1984). Separation and characterization of two chorismate-mutase isoenzymes from *Nicotiana silvestris*. Planta 162: 109-116.

Graziana A, Ranjeva R, Boudet AM (1983a). Provoked changes in cellular calcium controlled protein phosphorylation and activity of quinate: NAD^+ oxidoreductase in carrot cells. FEBS Letters 156: 325-328.

Graziana A, Ranjeva R, Salimath BP, Boudet AM (1983b). The reversible association of quinate:NAD^+ oxidoreductase from carrot cells with a putative regulatory subunit depends on light conditions. FEBS Lett 163:306-310.

Graziana A, Dillenschneider M, and Ranjeva R (1984). A calcium-binding protein is a regulatory subunit of quinate : NAD^+oxidoreductase from dark-grown carrot cells. Biochem and Biophys Res Comm 125: 774-783.

Hirel B, Nato A, Martin F (1984a). Glutamine synthetase in ribulose 1,5-bisphosphate carboxylase/oxygenase deficient tobacco mutants in cell suspension culture. Plant Cell Reports 3:106-108.

Hirel B, Weatherley C, Cretin C, Bergounioux C, Gadal P (1984b). Multiple subunit composition of chloroplastic glutamine synthetase of *Nicotiana tabacum* L. Plant Physiol 74: 448-450.

Huber SC (1986). Fructose 2,6-bisphosphate as a regulatory metabolite in plants. In Briggs WR (ed), Jones RL, Walbot V (assoc. eds): "Annual Review of Plant Physiology, Volume 37." Palo Alto: Annual Reviews, Inc., pp. 233-246.

Jensen RA (1986). The shikimate/arogenate pathway: Link between carbohydrate metabolism and secondary metabolism. Physiol Plant 66: 164-168.

Kaulen H, Schell J, Kreuzaler F (1986). Light-induced expression of the chimeric chalcone synthase-NPTII gene in tobacco cells. The EMBO J 5: 1-8.

Lara M, Porta H, Padilla J, Folch J, Sanchez F (1984). Heterogeneity of glutamine synthetase polypeptides in *Phaseolus vulgaris* L. Plant Physiol 76: 1019-1023.

Lauriere C, Daussant J (1983). Identification of the ammonium-dependent isoenzyme of glutamate dehydrogenase as the form induced by senescence or darkness stress in the first leaf of wheat. Physiol Plant 58: 89-92.

Lieberman M, Want SY, Owens LD (1979). Ethylene production by callus and suspension cells from cortex tissue of postclimacteric apples. Plant Physiol 63:811-815.

Lindsey K (1985). Manipulation, by nutrient limitation, of the biosynthetic activity of immobilized cells of *Capsicum frutescens* Mill. cv. annuum. Planta 165: 126-133.

McNally SF, Hirel B, Gadal P, Mann AF, Stewart GR (1983). Glutamine synthetases of higher plants. Plant Physiol 72: 22-25.

Miflin BJ, Lea PJ (1980). Ammonia assimilation. In Miflin BJ (ed): "The Biochemistry of Plants: Volume 5, Amino Acids and Derivatives." New York: Academic Press,Inc., pp. 169-202.

Morison JIL, Gifford RM (1984). Ethylene contamination of CO_2 cylinders. Plant Physiol 75:275-277.

Mousdale DM, Coggins JR (1985). Subcellular localization of the common shikimate-pathway enzymes in *Pisum sativum* L. Planta 163: 241-249.

Muhitch MJ, Edwards LA, Fletcher JS (1983). Influence of diamines and polyamines on the senescence of plant suspension cultures. Plant Cell Reports 2:82-84.

Muhitch MJ, Fletcher JS (1985). Influence of culture age and spermidine treatment on the accumulation of phenolic ompounds in suspension cultures. Plant Physiol 78:25-28.

Murillo E, Sanchez de Jimenez E (1985). Glutamate synthase in greening callus of *Bouvardia ternifolia* Schlecht. Planta 163: 448-452.

Myers RM, Tilly K, Maniatis T (1986). Fine structure genetic analysis of a -globin promoter. Science 232:613-618.

Nafziger ED, Widholm JM, Steinrucken HC, Killmer JL. Selection and characterization of a carrot cell line tolerant to glyphosate. Plant Physiol 76: 571-574.

Nato F, Hirel B, Nato A, Gadal P (1984). Chloroplastic glutamine synthetase from tobacco leaves: A glycosylated protein. FEBS Letters 175: 443-446.

Noe W, Berlin J (1985). Induction of de-novo synthesis of tryptophan decarboxylase in cell suspensions of *Catharanthus roseus*. Planta 166: 500-504.

Phillips R, Henshaw GG (1977). The regulation of synthesis of phenolics in stationary phase cell cultures of *Acer pseudoplatanus* L. J Exp Bot 28: 785-794.

Ranjeva R, Refeno G, Boudet AM, Marme D (1983). Activation of plant quinate:NAD^+ 3-oxidoreductase by Ca^{2+} and calmodulin. Proc Natl Acad Sci USA 80: 5222-5224.

Refeno G, Ranjeva R, Fontaine-Delvare S, Boudet AM (1982a). Functional Properties of Protein Kinase(s) and Phosphatase (s) converting quinate:NAD$^+$ oxidoreductase into active and deactivated forms in carrot cell suspension cultures. Plant & Cell Physiol 23; 1127-1144.

Refeno G, Ranjeva R, Boudet AM (1982b) Modulation of quinate: NAD$^+$ oxidoreductase activity through reversible phosphorylation in carrot cell suspensions. Planta 154: 193-198

Rhodes D, Sims AP, Folkes BF (1980). Pathway of ammonia assimilation in illuminated _Lemna minor_. Phytochem 19:357-365.

Sanchez de Jimenez E, Fernandez L (1983). Biochemical parameters to assess cell differentiation of _Bouvardia ternifolia_ Schlecht callus. Planta 158: 377-383.

Schmauder H-O, Groger D, Koblitz H, Koblitz D (1985). Shikimate pathway activity in shake and fermenter cultures of _Cinchona succirubra_. Plant Cell Reports 4: 233-236.

Schmelzer E, Borner H, Grisebach H, Ebel J, Hahlbrock K (1984). Phytoalexin synthesis in soybean (_Glycine max_). FEBS Letters 172: 59-63.

Simpson J, Montagu MV, Herrera-Estrella J (1986). Photosynthesis-associated gene families: differences in response to tissue-specific and environmental factors. Science 233: 34-38.

Slocum RD, Kaur-Sawhney R, Galston AW (1984). The physiology and biochemistry of polyamines in plants. Arch Biochem Biophys 235:283-303.

Smart CC, Johanning D, Muller G, Amrhein N (1985). Selective overproduction of 5-_enol_-Pyruvyl-shikimic acid 3-phosphate synthase in a plant cell culture which tolerates high doses of the herbicide glyphosate. J Biol Chem 260:16338-16346.

Stachel SE, Messens E, van Montagu M, Zambryski P (1985). Identification of the signal molecules produced by wounded plant cells that activate T-DNA transfer in *Agrobacterium tumefaciens*. Nature 318: 1-6.

Steinrucken HC, Schulz A, Amrhein N, Porter CA, Fraley RT (1986). Overproduction of 5-enolpyruvylshikimate-3-phosphate synthase in a glyphosate-tolerant *Petunia hybrida* cell line. Arch Biochem Biophys 244: 169-178.

Stewart GR, Mann AF, Fentem PA (1980). Enzymes of glutamate formation: Glutamate dehydrogenase, glutamine synthetase, and glutamate synthase. In Miflin BJ (ed): "The Biochemistry of Plants: Volume 5, Amino Acids and Derivatives." New York: Academic Press, Inc., pp. 271-327.

Stitt M, Cseke C, Buchanan B (1986). Ethylene-induced increase in fructose-2,6-bisphosphate in plant storage tissues. Plant Physiol 80:246-248.

Suzich JA, Ranjeva R, Hasegawa PM, Herrmann KM (1984). Regulation of the shikimate pathway of carrot cells in suspension culture. Plant Physiol 75: 369-371.

Suzich JA, Dean JFD, Herrmann KM (1985). 3-deoxy-D-*arabino*-heptulosonate 7-Phosphate synthase from carrot root (*Daucus carota*) is a hysteretic enzyme. Plant Physiol 79: 765-770.

Suzuki A, Gadal P (1984). Glutamate synthase : physicochemical and functional properties of different forms in higher plants and in other organisms. Physiol Veg 22: 471-486.

Suzuki A, Nato A, Gadal P (1984). Glutamate synthase isoforms in tobacco culture cells. Plant Sci Let 33: 93-101.

Tiburcio AF, Galston AW (1986). Arginine decarboxylase as the source of putrescine for tobacco alkaloids. Phytochem 25:107-110.

Tischer E, DasSarma S, Goodman HM (1986). Nucleotide sequence of an alfalfa glutamine synthetase gene. Mol Gen Genet 203: 221-229.

Tisserat B, Murashige T (1977a). Effects of ethephon, ethylene, and 2,4-dichlorophenoxyacetic acid on asexual embryogenesis in Vitro. Plant Physiol 60:437-439.

Tisserat B, Murashige T (1977b). Probable identity of substances in citrus that repress asexual embryogenesis. In Vitro 13:785-789.

Yamaya T, Oaks A, Matsumoto H (1984). Characteristics of glutamate dehydrogenase in mitochondria prepared from corn shoots. Plant Physio 76: 1009-1013.

Zobel RW, Roberts LW (1977). Effects of low concentrations of ethylene on cell division and cytodifferentiation in lettuce pith explant. Can J Bot 56:987-990.

Plant Tissue and Cell Culture, pages 119–133

OVERPRODUCTION OF 5-ENOLPYRUVYLSHIKIMATE 3-PHOSPHATE SYNTHASE IN GLYPHOSATE-TOLERANT PLANT CELL CULTURES

Nikolaus Amrhein, Heike Holländer-Czytko, Detlef Johänning, Arno Schulz, Cheryl C. Smart and Hans C. Steinrücken

Lehrstuhl für Pflanzenphysiologie,
Ruhr-Universität Bochum,
D 4630 Bochum, Fed.Rep.Germany

INTRODUCTION

Plant cell cultures offer a number of advantages in studies on the mode of action of herbicides, as well as for the selection of variant cell types in which herbicide resistance is expressed at the cellular level (Chaleff,1981; Meredith and Carlson, 1982). The recent progress in the identification of amino acid biosynthetic enzymes as the molecular targets of herbicide action, as well as the demonstration of target-related herbicide resistance can, at least in part, be attributed to the judicious use of cell cultures (for reviews see La Rossa and Falco, 1984; Chaleff, 1986).

The molecular understanding of herbicide action and herbicide resistance will make the genetic engineering of crop plants resistant to specific herbicides, as well as a more rational design of new herbicides possible (Hardy and Giaquinta, 1984). Herbicides which specifically inhibit enzymes of amino acid biosynthetic pathways comprise glyphosate (Steinrücken and Amrhein, 1980), the sulfonylurea type herbicides such as chlorsulfuron and sulfometuron methyl (Ray, 1984; Chaleff and Mauvais, 1984), the imidazolinones (Shaner et al., 1984), and phosphinothricin (Leason et al., 1982). For each of these herbicides, variant cell strains have been selected in which herbicide-tolerance was clearly correlated with either a qualitative, or a quantitative change in the respective target enzyme (Amrhein et al., 1983; Chaleff and Ray, 1984; Shaner and Anderson, 1985; Donn et al., 1984). These results therefore corroborated the previous identification of the respective herbicide target sites.

In this paper, we discuss the molecular basis of the tolerance of plant cell cultures to the herbicide glyphosate which interferes specifically with the shikimate biosynthetic pathway. While the elucidation of the mode of action of glyphosate in our and other laboratories was greatly aided by studying the action of the herbicide in microorganisms (for review see Amrhein, 1986), the contribution of studies involving plant cell cultures will be emphasized here.

THE SHIKIMATE PATHWAY

The biosynthesis of the aromatic amino acids, phenylalanine, tyrosine and tryptophan, proceeds by the shikimate pathway, the occurrence of which is restricted to plants and microorganisms (Haslam, 1974). While the chemistry of the intermediates of the pathway (Fig.1) had been thoroughly investigated more than 20 years ago, knowledge of the enzymes catalyzing the individual reactions remained fragmentary. This was, and is, particularly true for the shikimate pathway enzymes from higher plants. To date, only the bifunctional shikimate oxidoreductase/dehydroquinate hydrolyase (Fiedler and Schultz, 1985) and the 5-enolpyruvylshikimate 3-phosphate (EPSP) synthase (Mousdale and Coggins, 1984; Smart et al., 1985; Steinrücken et al., 1986; Holländer-Czytko, 1986) have been purified from higher plant sources to apparent homogeneity. Great efforts have recently been directed at the cloning of the genes for the enzymes of the prechorismate pathway of *Escherichia coli*. Most of these genes have been sequenced, and strains which overproduce the corresponding enzymes have been constructed (for review see Coggins, 1986). A wealth of information on the biochemistry and molecular biology of the *E.coli* prechorismate pathway enzymes will therefore be available within a few years.

INTERFERENCE OF GLYPHOSATE WITH THE SHIKIMATE PATHWAY IN PLANT CELL CULTURES

After initial studies with *Lemna gibba* and *Rhizobium japonicum* had suggested that the herbicide glyphosate (N-[phosphonomethyl]glycine; for structure see Fig.2)inhibited the biosynthesis of aromatic amino acids in these organisms (Jaworski, 1972), it was shown that aromatic amino acids alleviate the inhibition of growth by glyphosate in cultured plant cells (Haderlie et al., 1977; Gresshoff, 1979). The massive accumulation of shikimate in glyphosate-treated plant cell cultures (Amrhein et al., 1980, 1983; Berlin and Witte,

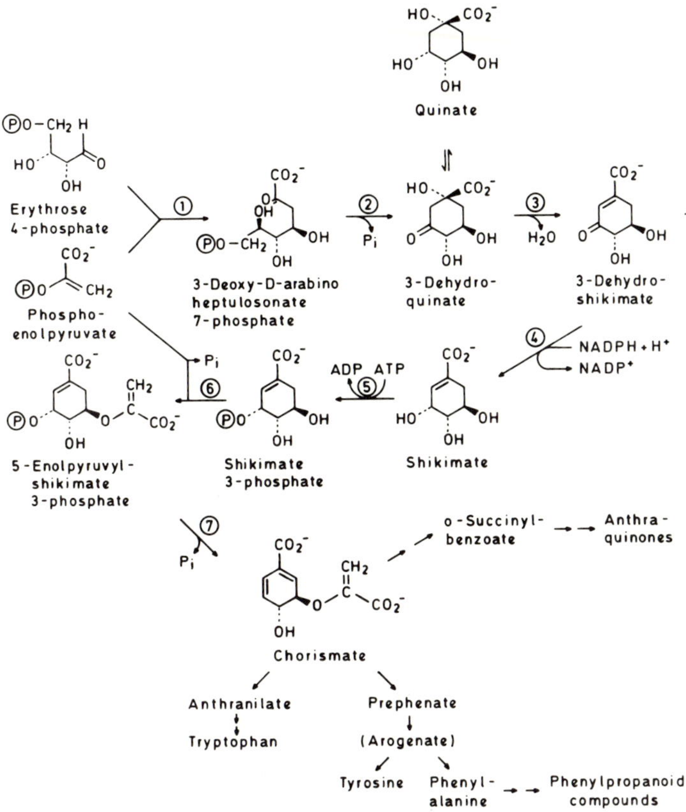

Figure 1. The shikimate pathway. Structures of intermediates are shown only for the prechorismate pathway. Numbers in circles refer to the following enzymes: (1) 3-deoxy-D-*arabino*-heptulosonic acid 7-phosphate (DAHP) synthase (EC 4.1.2.15); (2) 3-dehydroquinate synthase (EC 4.6.1.3); (3) 3-dehydroquinate hydrolyase (EC 4.2.1.10); (4) shikimate: $NADP^+$ oxidoreductase (EC 1.1.1.25); (5) shikimate kinase (EC 2.7.1.71); (6) 5-enolpyruvylshikimate 3-phosphate (EPSP) synthase (EC 2.5.1.19); (7) chorismate synthase (EC 4.6.1.4.)

1981; Ishikura and Takeshima, 1984), and the inhibition of the formation of chorismate-derived anthraquinones (Fig.1) in cultured *Galium mollugo* cells (Amrhein et al., 1980a) were in agreement with the postulate that EPSP synthase, which catalyzes the sixth reaction in the shikimate pathway (Fig.1), is the target of the herbicide (Amrhein et al., 1980b). The fact that shikimic acid, rather than shikimate 3-phosphate, the substrate of EPSP synthase, accumulates in the presence of glyphosate, was shown for cultured buckwheat cells to be due to dephosphorylation of shikimate 3-phosphate and storage of free shikimic acid in the vacuoles (Holländer-Czytko and Amrhein, 1983).

PROPERTIES OF EPSP SYNTHASES FROM CULTURED PLANT CELLS

EPSP synthase catalyzes the transfer of the intact enolpyruvyl (carboxyvinyl) group from phosphoenolpyruvate (PEP) to the 5-hydroxyl group of shikimate 3-phosphate with the concomitant release of inorganic phosphate (Fig.2). Because of its easier accessibility, the enzyme was initially isolated from microorganisms, such as *Klebsiella pneumoniae*, *Escherichia coli*, and *Neurospora crassa* (for reviews see Amrhein, 1986; Coggins, 1986). These studies were greatly aided by modern methods for the cloning and sequencing of genes and for the overexpression of the gene products (Rogers et al., 1983; Duncan et al., 1984a, 1984b; Stalker et al., 1985). In *E.coli* (Duncan et al., 1984b) and *Salmonella typhimurium* (Stalker et al., 1985) the complete polypeptide chain of EPSP synthase consists of 427 amino acids and has an M_r of 46,112. The crystallization of the *E.coli* EPSP synthase has been achieved (Abdel-Meguid et al., 1985).

The first higher plant EPSP synthase to be purified to homogeneity was isolated from light-grown pea seedlings (Mousdale and Coggins, 1984), and a partial purification was described for the enzyme from cultured cells of *Nicotiana silvestris* (Rubin et al., 1984). The availability of EPSP synthase overproducing strains of cultured plant cells (see below), as well as a simple scheme for the rapid and efficient purification of the enzyme (Fig.3), enabled us to purify EPSP synthase from *Corydalis sempervirens* (Smart et al., 1985), *Petunia hybrida* (Steinrücken et al., 1986) and *Fagopyrum esculentum* (Holländer-Czytko, 1986) to homogeneity. Some of the properties of these enzymes, in comparison with those of the enzyme from pea seedlings, are compiled in Table 1.

TABLE 1. Properties of EPSP Synthases from Cultured Cells of *Corydalis sempervirens*[a], *Petunia hybrida*[b], and *Fagopyrum esculentum*[c], as well as from *Pisum sativum* seedlings[d]

Property	*C.sempervirens*	*P.hybrida*	*F.esculentum*	*P.sativum*
Molecular mass (Da)[e]:				
of mature enzyme	45,000 ± 1,500	ca. 49,000	49,000 ± 1,000	ca. 50,000
of enzyme precursor	53,900 ± 900	ca. 55,000	n.d.	n.d.
$K_{m(app)}$ for PEP (μM)	ca. 80	14	21.7	5.2 ± 0.8
K_i for glyphosate (μM) with respect to PEP	0.32	0.17	1.75	0.08
$K_{m(app)}$ for S-3-P (μM)	ca. 21	ca. 8	n.d.	7.7 ± 0.8

[a] Smart et al., 1985; Holländer-Czytko and Amrhein, 1986. [b] Steinrücken et al., 1986; della-Cioppa et al., 1986. [c] Holländer-Czytko, 1986. [d] Mousdale and Coggins, 1984. [e] determined by SDS-PAGE. n.d. = not determined.

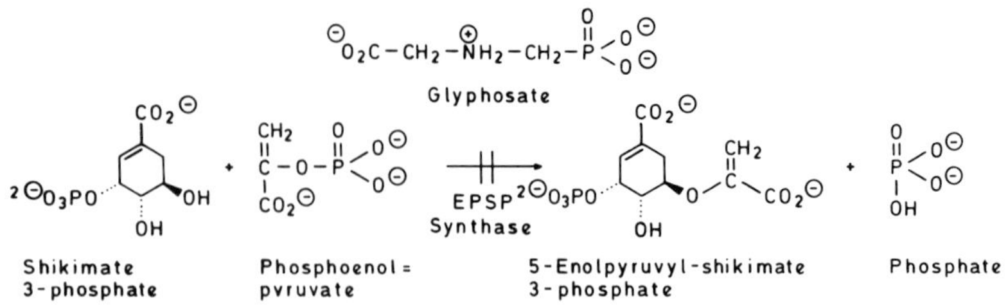

Figure 2. Reaction catalyzed by EPSP synthase and structure of glyphosate (shown in the ionic form prevalent at physiological pH).

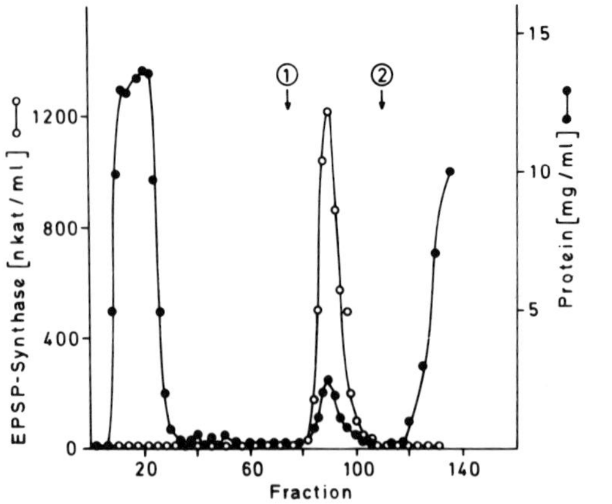

Figure 3. Chromatography of the *Petunia hybrida* EPSP synthase on a phosphocellulose column. After ammonium sulfate fractionation of a crude extract and desalting on a Sephadex G-25 column, the protein fraction was applied to a phosphocellulose column which was then extensively washed with buffer. EPSP synthase was selectively eluted with buffer containing 0.1 mM shikimate 3-phosphate and 0.1 mM glyphosate (arrow 1). Residual proteins were eluted with 1 M phosphate buffer (arrow 2). From Steinrücken et al., 1986.

The higher plant EPSP synthases are monomeric proteins with a molecular mass of 45,000 to 50,000 Da. The *C.sempervirens* and *P.hybrida* EPSP synthases were shown to be synthesized as higher molecular weight precursors (see Fig.4) which indicates that the mature enzyme is organelle-bound (see below). Kinetic characterization of the enzymes revealed a high sensitivity to glyphosate which inhibits EPSP synthase competitively with respect to phosphoenolpyruvate (PEP) and uncompetitively with with respect to shikimate 3-phosphate (S-3-P). This pattern indicates an ordered sequential mechanism, in which either PEP or glyphosate can bind only to an [enzyme·S-3-P] complex. The previously observed binding of glyphosate to the *Escherichia coli* EPSP synthase in the absence of S-3-P (Amrhein, 1986) can now be explained by the observation that S-3-P, which was used in the elution of EPSP synthase from the phosphocellulose column (Fig.3), binds tightly to EPSP synthase and is not easily removed by dialysis.

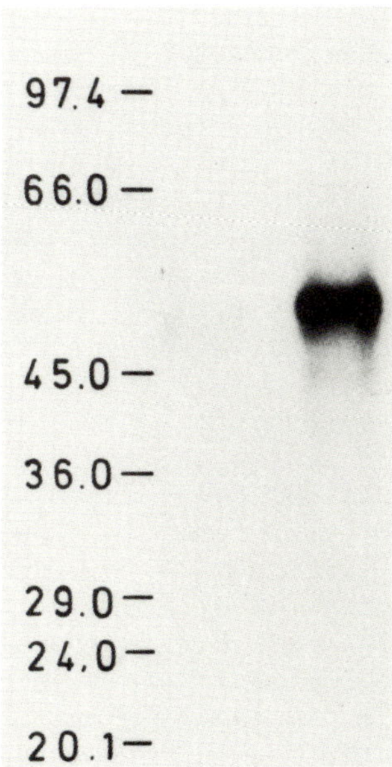

Figure 4. Fluorogram of SDS-PAGE of protein precipitated with anti-(EPSP synthase) antiserum after *in vitro* translation of total RNA from glyphosate-sensitive (center lane) and glyphosate-tolerant (right lane) cell cultures of *Corydalis sempervirens*; left lane: molecular weight markers (kD). The mature enzyme has an M_r of 45,500 $\pm$ 1,500 (Table 1).

MOLECULAR BASIS OF GLYPHOSATE TOLERANCE IN PLANT CELL CULTURES

While tolerance to glyphosate could be acquired at the cellular level by its reduced uptake, its degradation or conjugation, the two possible target-related mechanisms are particularly informative, because they allow additional and firm conclusions with reference to the mode of action of the herbicide. These two mechanisms are:
(1) Insensitivity of the target enzyme (EPSP synthase) to glyphosate due to a mutation, or
(2) Overproduction of the target enzyme.
The first mechanism, acquisition of a glyphosate-insensitive form of EPSP synthase through mutation, has so far only been found in bacteria (Comai et al., 1983; Schultz et al., 1984). The gene coding for the glyphosate-resistant *Salmonella typhimurium* EPSP synthase has been cloned and sequenced (Stalker et al., 1985). It was shown that a single amino acid substitution in the enzyme conferred resistance to the herbicide. This gene has been used to transform tobacco plants which exhibited increased glyphosate tolerance (Comai et al., 1985).

Cultured cells of *Daucus carota* (Nafziger et al., 1984), *Corydalis sempervirens* (Amrhein et al., 1983; Smart et al., 1985), *Petunia hybrida* (Steinrücken et al., 1986), and *Fagopyrum esculentum* (Holländer-Czytko, 1986) have been adapted to growing in high concentrations of glyphosate by transfer into progressively higher concentrations of the herbicide. The herbicide tolerance thus acquired was related to up to 40-fold (*C.sempervirens)* increased levels of the extractable activity of EPSP synthase, while the activities of other shikimate pathway enzymes were not affected. Immunoblotting, as well as a comparison of physical, kinetic and immunological properties of the EPSP synthases from the glyphosate-sensitive and -tolerant cell strains established that the increased EPSP synthase activity resulted from the overproduction of a herbicide-sensitive form of the enzyme. In *C.sempervirens* cultures, the proportion of EPSP synthase protein of the total soluble protein was 2.6 % in the glyphosate tolerant cells, as compared to 0.06 % in the glyphosate sensitive cells (Smart et al., 1985). In the same cultures, the rate of translatable mRNA activity for EPSP synthase is tenfold higher in the glyphosate tolerant, as compared to the glyphosate sensitive cells (Fig.4, Table 2) (Holländer-Czytko and Amrhein, 1986). Likewise, transcript levels for EPSP synthase, as determined by hybridization of poly(A) RNA with an oligonucleotide probe that had been deduced from the

TABLE 2. Molecular Basis of Overproduction of EPSP Synthase in a Glyphosate Tolerant Cell Culture of *Corydalis sempervirens*

Parameter	Relative abundance: Glyphosate tolerant / Glyphosate sensitive
EPSP synthase activity[a]	30 - 40
EPSP synthase protein[b]	30 - 40
Translatable activity of mRNA for EPSP synthase[c]	ca. 10
EPSP synthase transcript[d]	8 - 12

[a] determined by enzymatic assay; [b] determined by immunoblot and ELISA; [c] determined by cell-free translation; [d] determined by hybridization of mRNA with oligonucleotide probe (from Smart et al., 1985; Holländer-Czytko and Amrhein, 1986; Holländer-Czytko, unpublished).

amino acid sequence of the N-terminus of the *C.sempervirens* EPSP synthase, had a ca. tenfold increased level in the glyphosate tolerant cells (Table 2) (Holländer-Czytko and Johänning, unpublished). The discrepancy in the relative abundance of the EPSP synthase protein and the EPSP synthase transcript in the glyphosate tolerant cells (Table 2) strongly suggests that EPSP synthase overaccumulation in the glyphosate tolerant *Corydalis sempervirens* cells is regulated at the transcriptional as well as the posttranslational level. While gene amplification (Stark and Wahl, 1984) has been found to be the molecular basis for glyphosate tolerance in the *P.hybrida* cells (Shah et al., 1985), we cannot yet give a definitive answer to this question for the *C.sempervirens* cells.

THE SUBCELLULAR LOCALIZATION OF EPSP SYNTHASE IN CULTURED *Corydalis sempervirens* CELLS

While there is convincing evidence for the localization of the enzymes of the shikimate pathway in plastids, the possible co-occurrence of the entire, or possibly a truncated shikimate pathway in the cytosol remains controversial (for

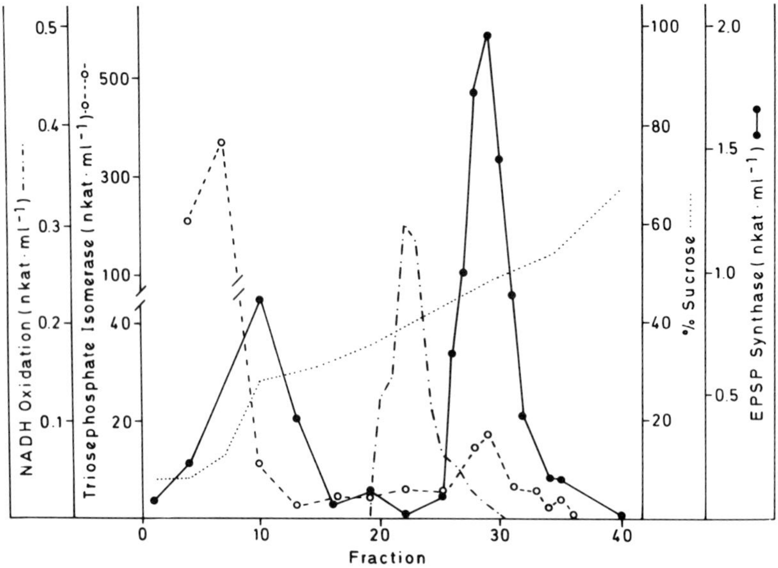

Figure 5. Sucrose density gradient centrifugation of organelles from cultured *Corydalis sempervirens* cells. Mitochondria were identified by NADH oxidation, plastids by their triosephosphate isomerase activity.

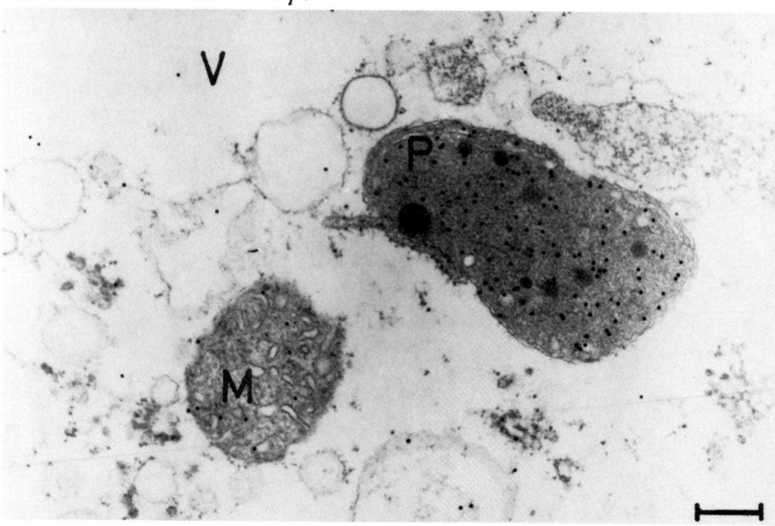

Figure 6. Protein A-gold labelling of EPSP synthase in glyphosate-tolerant cells of *Corydalis sempervirens*. (M = mitochondrion; P = plastid; V = vacuole). Bar represents 0.5 μm.

discussion see Mousdale and Coggins, 1985; Fiedler and Schultz, 1985). The synthesis of EPSP synthase in cultured *C.sempervirens* and *Petunia hybrida* cells as a higher molecular weight precursor (Holländer-Czytko and Amrhein, 1986; della-Cioppa et al., 1986) is typical for nuclear-encoded plastidic proteins which require a "transit peptide" to allow the protein to be transported through the plastid envelope (Ellis, 1981). The plastidic localization of EPSP synthase in *C.sempervirens cells* is evident from cell fractionation studies in which EPSP synthase activity cosedimented with the organelle-bound activity of triosephosphate isomerase (Fig.5) as well as from protein A-gold immunocytochemistry with an affinity-purified antiserum against *C.sempervirens* EPSP synthase which allowed the direct visualization of the enzyme's subcellular localization (Fig.6; Smart and Amrhein, 1986). Efforts to transfer the genetic information for a glyphosate resistant EPSP synthase into higher plants (Comai et al., 1985) must therefore make provision for the delivery of the resistant enzyme to the plastidic compartment.

ACKNOWLEDGEMENTS

Support of our work by Deutsche Forschungsgemeinschaft, Bonn, Fonds der Chemischen Industrie, Frankfurt,(literature provision), and Monsanto Company, St.Louis, is gratefully acknowledged.

REFERENCES

Abdel-Meguid SS, Smith WW, Bild GS (1985). Crystallization of 5-enolpyruvylshikimate 3-phosphate synthase from *Escherichia coli*. J Mol Biol 186:673.

Amrhein N (1986). Specific inhibitors as probes into the biosynthesis and metabolism of aromatic amino acids. In Conn EE (ed):"The Shikimic Acid Pathway", New York: Plenum Press, pp.83-117.

Amrhein N, Deus B, Gehrke P, Steinrücken HC (1980a). The site of the inhibition of the shikimate pathway by glyphosate. II. Interference by glyphosate with chorismic acid formation *in vivo* and *in vitro*. Plant Physiol 66:830-834.

Amrhein N, Schab J, Steinrücken HC (1980b). The mode of action of the herbicide glyphosate. Naturwissenschaften 67:356-357.

Amrhein N, Johänning D, Schab J, Schulz A (1983). Biochemical basis for glyphosate-tolerance in a bacterium and a plant tissue culture. FEBS-Lett 157:191-196.
Berlin J, Witte L (1981). Effects of glyphosate on shikimic acid accumulation in tobacco cell cultures with low and high yields of cinnamoyl putrescines. Z Naturforsch 36c: 210-214.
Chaleff RS (1981). "Genetics of Higher Plants. Applications of Cell Culture." Cambridge: Cambridge University Press.
Chaleff RS (1986). Herbicide resistance. In Day PR (ed): "Biotechnology and Crop Improvement and Protection", Thornton Heath: British Crop Protection Council, pp.111-121.
Chaleff RS, Mauvais CJ (1984). Acetolactate synthase is the site of action of two sulfonylurea herbicides in higher plants. Science 224:1443-1445.
Chaleff RS, Ray TB (1984). Herbicide-resistant mutants from tobacco cell cultures. Science 223:1148-1151.
Coggins JR (1986). Enzymology and molecular biology as aids for the invention and improvement of herbicides. In Day PR (ed): "Biotechnology and Crop Improvement and Protection", Thornton Heath: British Crop Protection Council, pp. 101-110.
Comai L, Sen LC, Stalker DM (1983). An altered *aro*A gene product confers resistance to the herbicide glyphosate. Science 221:370-371.
Comai L, Facciotti D, Hiatt WR, Thompson G, Rose RE, Stalker DM (1985). Expression in plants of a mutant *aro*A gene from *Salmonella typhimurium* confers tolerance to glyphosate. Nature 317:741-744.
della-Cioppa G, Bauer Sc, Klein BK, Gasser CS, Tumer NE, Shah DM, Fraley RT, Kishore GM (1986). Translocation of the precursor of 5-enolpyruvylshikimate-3-phosphate (EPSP)synthase into chloroplasts of higher plants *in vitro*. Plant Physiol 80: S 127.
Donn G, Tischer E, Smith JA, Goodman HM (1984). Herbicide resistant alfalfa cells: An example of gene amplification in plants. J Mol Appl Genet 2:621-635.
Duncan K, Lewendon L, Coggins JR (1984a). The purification of 5-enolpyruvylshikimate 3-phosphate synthase from an overproducing strain of *Escherichia coli*. FEBS-Lett 165:121-127.
Duncan K, Lewendon A, Coggins JR (1984b). The complete amino acid sequence of *Escherichia coli* 5-enolpyruvylshikimate 3-phosphate synthase. FEBS-Lett 170:59-63.
Ellis JR (1981). Chloroplast proteins: Synthesis, transport, and assembly. Ann Rev Plant Physiol 32:111-137.

Fiedler E, Schultz G (1985). Localization, purification, and characterization of shikimate oxidoreductase-dehydroquinate hydrolyase from stroma of spinach chloroplasts. Plant Physiol 79:212-218.

Gresshoff PM (1979). Growth inhibition by glyphosate and reversal of its action by phenylalanine and tyrosine. Aust J Plant Physiol 6:177-185.

Haderlie LC, Widholm JM, Slife FW (1979). Effect of glyphosate on carrot and tobacco cells. Plant Physiol 60:40-43.

Hardy RWF, Giaquinta RT (1984). Molecular biology of herbicides. BioEssays 1:152-156.

Haslam E (1974). "The Shikimate Pathway". London: Butterworths.

Holländer-Czytko H(1986).Glyphosate tolerance in a buckwheat cell suspension culture. Abstract No.2460. 17th FEBS-Meeting, Berlin.

Holländer-Czytko H, Amrhein N (1983). Subcellular compartmentation of shikimic acid and phenylalanine in buckwheat cell suspension cultures grown in the presence of shikimate pathway inhibitors. Plant Sci Lett 29:89-96.

Holländer-Czytko H, Amrhein N (1986). 5-Enolpyruvylshikimate 3-phosphate synthase, the target enzyme of the herbicide glyphosate, is synthesized as a precursor in a higher plant. Plant Physiol: in press.

Ishikura N, Takeshima Y (1984). Effects of glyphosate on caffeic acid metabolism in *Perilla* cell suspension cultures. Plant Cell Physiol 25:185-189.

Jaworski EG (1972). Mode of action of N-phosphonomethylglycine: inhibition of aromatic amino acid biosynthesis. J Agric Food Chem 20:1195-1198.

LaRossa RA, Falco SC (1984). Amino acid biosynthetic enzymes as targets of herbicide action. Trends Biotechnol 2:158-161.

Leason M, Cunliffe D, Parkin D, Lea PJ, Miflin BJ (1982). Inhibition of pea leaf glutamine synthetase by methionine sulphoximine, phosphinothricin and other glutamate analogues. Phytochemistry 21:855-857.

Meredith CP, Carlson PS (1982). Herbicide resistance in plant cell cultures. In LeBaron HM, Gressel J(eds):"Herbicide Resistance in Plants", New York: Wiley pp.275-291.

Mousdale DM, Coggins JR (1984). Purification and properties of 5-enolpyruvylshikimate 3-phosphate synthase from seedlings of *Pisum sativum* L.. Planta 160:78-83.

Mousdale DM, Coggins JR (1985). Subcellular localization of the common shikimate-pathway enzymes in *Pisum sativum* L. Planta 163:241-249.

Nafziger ED, Widholm JM, Steinrücken HC, Killmer JL (1984). Selection and characterization of a carrot cell line tolerant to glyphosate. Plant Physiol 76:571-574.

Ray TB (1984). Site of action of chlorsulfuron. Inhibition of valine and isoleucine biosynthesis in plants. Plant Physiol 75:827-831.

Rogers SG, Brand LA, Holder SB, Sharps ES, Brackin MJ (1983). Amplification of the *aro*A gene from *Escherichia coli* results in tolerance to the herbicide glyphosate. Appl Environ Microbiol 46:37-43.

Rubin JL, Gaines CG, Jensen RA (1984). Glyphosate inhibition of 5-enolpyruvylshikimate 3-phosphate synthase from suspension-cultured cells of *Nicotiana silvestris*. Plant Physiol 75:839-845.

Schulz A, Sost D, Amrhein N (1984). Insensitivity of 5-enolpyruvylshikimate 3-phosphate synthase to glyphosate confers resistance to this herbicide in a strain of *Aerobacter aerogenes*. Arch Microbiol 137:121-123.

Shaner DL, Anderson PC (1985). Mechanism of action of the imidazolinones and cell culture selection of tolerant maize. In Zaitlin M, Day P. Hollaender A, Wilson CM (eds):"Biotechnology in Plant Science. Relevance to Agriculture in the Eighties", Orlando: Academic Press, pp.287-299.

Shaner DL, Anderson PC, Stidham MA (1984). Imidazolinones. Potent inhibitors of acetohydroxyacid synthase. Plant Physiol 76:545-546.

Shah DM, Winter JW, Sanders PS, Hironaka CM, Steinrücken HC, Amrhein N, Fraley RT (1985). Gene amplification is the molecular basis for glyphosate resistance in *Petunia* cells. Abstract No. OR-09-01. 1st Int Congr Plant Mol Biol,Savannah.

Smart CC, Amrhein N (1986). Ultrastructural localization by protein A-gold immunocytochemistry of 5-enolpyruvylshikimic acid 3-phosphate synthase in a plant cell culture which overproduces the enzymes. Planta: in press.

Smart CC, Johänning D, Müller G, Amrhein N (1985). Selective overproduction of 5-*enol*-pyruvylshikimic acid 3-phosphate synthase in a plant cell culture which tolerates high doses of the herbicide glyphosate. J Biol Chem 260:16338-16346.

Stalker DM, Hiatt WR, Comai L (1985). A single amino acid substitution in the enzyme 5-enolpyruvylshikimate 3-phosphate synthase confers resistance to the herbicide glyphosate. J Biol Chem 260:4724-4728.

Stark GR, Wahl GM (1984). Gene amplification. Ann Rev Biochem 53:447-491.

Steinrücken HC, Amrhein N (1980). The herbicide glyphosate is a potent inhibitor of 5-enolpyruvylshikimic acid-3-phosphate synthase. Biochem Biophys Res Commun 94:1207-1212.

Steinrücken HC, Schulz A, Amrhein N, Porter CA, Fraley RT (1986). Overproduction of 5-enolpyruvylshikimate-3-phosphate synthase in a glyphosate-tolerant *Petunia hybrida* cell line. Arch Biochem Biophys 244:169-178.

Plant Tissue and Cell Culture, pages 135–149

PLANT GROWTH REGULATORS AND THE CONTROL OF GROWTH AND DIFFERENTIATION IN PLANT TISSUE CULTURES

Roger Horgan

Department of Botany and Microbiology, The University College of Wales, Aberystwyth SY23 3DA, U.K.

INTRODUCTION

Within the limited space allotted to this article it is not possible to give a comprehensive account of the role of plant hormones in the growth and development of plant tissue cultures. From the point of view of virtually all plant tissue and cell cultures the cytokinins and auxins are the most important classes of plant hormones. The importance of these hormones in plant tissue culture is so great that even a review of their functions in this context is impossible. This article will therefore take a highly selective and personal view of some of the aspects of cytokinin and auxin biochemistry and physiology in tissue and cell cultures. Because of the author's own research interests it will concentrate mainly on certain aspects of cytokinin biosynthesis and metabolism which appear relevant to the action of these compounds in maintaining the growth of tissue cultures.

It is particularly satisfying to be able to discuss cytokinins within the context of plant tissue cultures. Firstly because tissue cultures have proved to be such valuable systems for the study of cytokinin biochemistry and secondly because in the authors biased view the cytokinins are very much tissue culture hormones. Indeed their discovery is inexorably linked with the early studies of plant tissue culture. With regard to cytokinins tissue cultures fall very clearly into two classes, those that require an external supply of cytokinins and those that do not. In addition cytokinins often exhibit spectacular effects on the differentiation of tissue cultures. Thus there is little room for argument about the relevance of studying some of the basic aspects of cytokinin biochemistry and physiology in relation to the growth and development of tissue cultures.

CYTOKININ METABOLISM

The response of tissue cultures to externally supplied cytokinin is clearly a complex function of several biochemical processes. These include uptake and transport, metabolic deactivation and possibly metabolism to an active form, interaction of the active form of the cytokinin with a putative receptor and finally the transduction of this interaction into some biochemical process or processes involved with cell division.

Principally as a result of work by Laloue and his coworkers there is considerable detailed information on the uptake and metabolism of cytokinins by tobacco cell suspension cultures. In tobacco cells cytokinins are accumulated as the 5'nucleotide and 7-glucoside. The 7-glucoside appears to be an inactive endproduct but the nucleotide regulates the level of free base, which is considered to be the active form of the cytokinin. The level of free base is always low as it can diffuse freely through the cell membrane whereas the membrane is impermeable to nucleotides (Laloue and Pethe, 1982).

From the physiological view point cytokinin metabolism may be classified under three headings.
(1) Irreversible loss of the N-6 side chain leading to loss of biological activity.
(2) Irreversible conjugation with sugars or amino acids leading to loss of or reduced biological activity.
(3) Reversible conjugation leading to compounds which may in themselves possess biological activity or function as storage forms ie. are potential sources of active cytokinins.

With regard to the biological activity of externally applied cytokinins in tissue culture systems, side chain cleavage is a key reaction since it leads directly to loss of cytokinin activity. The enzyme responsible for this step, 'cytokinin oxidase' has been isolated and partially purified from several plant sources. Cytokinin oxidase acts only on cytokinins possessing a Δ^2-double bond in the side chain as shown in Fig. 1. The substrate specificity of cytokinin oxidase for a wide variety of cytokinins has been studied (McGaw and Horgan, 1982). The synthetic cytokinins most frequently used in tissue culture applications, benzyladenine and kinetin, are resistant to cytokinin oxidase although long term metabolic studies reveal that these compounds do eventually lose the N-6 side chain by as yet uncharacterised enzymatic reactions.

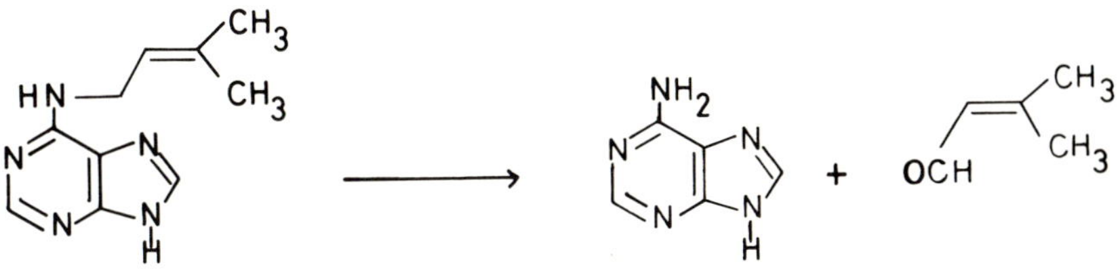

Figure 1. Side chain cleavage of Δ^2-isopentenyl adenine by cytokinin oxidase.

Naturally occurring cytokinins such as zeatin and isopentenyladenine are excellent substrates for cytokinin oxidase and yet are often more active in promoting cell division in tissue cultures than synthetic cytokinins. Since side chain cleavage certainly attenuates the activity of externally applied natural cytokinins one may speculate that the real activity, ie. the response elicited at the putative receptor, of these compounds may be relatively much greater than that of their synthetic counterparts as judged by simple dose response relationships.

Fig. 2 shows the growth of two cell suspension cultures derived from the same batch of soybean cotyledon callus tissue. A mid log phase culture produced as a result of 12 subcultures from the original culture was divided into 18 subcultures using benzyladenine at 0.45 μM as the cytokinin source. These cultures all showed the growth characteristics typified by Z005 in Fig. 2. However the subculture of mid log phase aliquots of the 18 cultures into a medium containing zeatin at 0.45 μM resulted in somewhat reduced growth of all subcultures eg. Z005 (Fig.2). However one subculture designated B023 (Fig.2) exhibited greatly reduced growth on zeatin. When mid log phase zeatin grown cells were re-cultured on benzyladenine normal growth was restored. A comparative study of zeatin metabolism in B023 and Z005 revealed a greater rate of zeatin inactivation (side chain cleavage) in the poor zeatin responder B023 than in the good zeatin responder Z005. Fig.3 shows the qualitative and quantitative pattern of ^{14}C zeatin metabolism in these cultures.

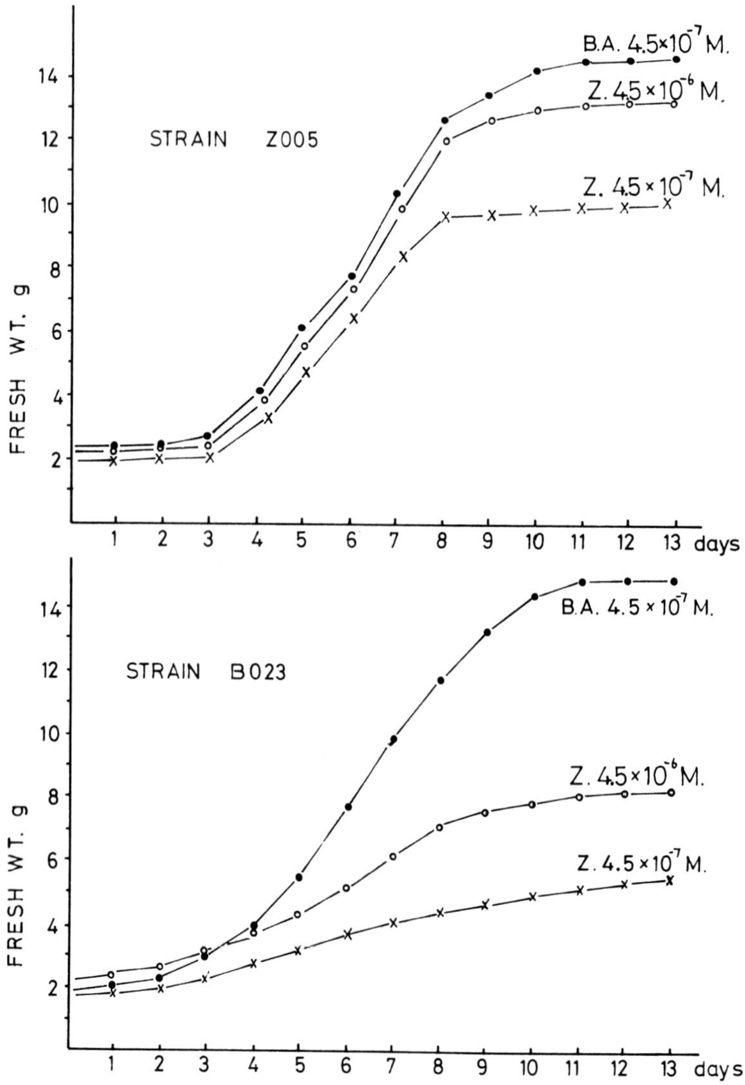

Figure 2. Growth of two strains of soybean cells (Z005 and B023) in liquid culture on benzyladenine and zeatin.

These differences in growth were not apparent in callus cultures re-isolated from the cell suspension cultures. In addition in the callus cultures from all sources zeatin was more active than benzyladenine in promoting growth.

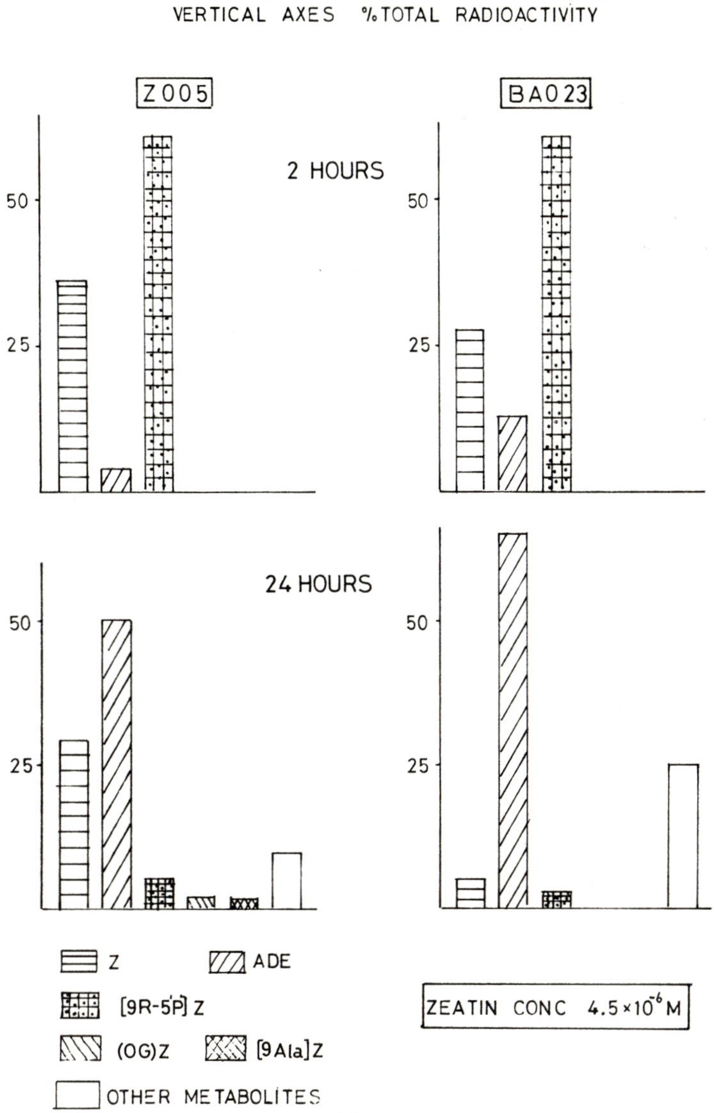

Figure 3. Metabolism of 8-^{14}C zeatin by a zeatin responsive (Z005) and a zeatin non-responsive (BA023) strain of soybean cells grown in liquid culture.

The physiological and biochemical differences between Z005 and B023 were not stable. After several subcultures B023 began to grow on zeatin to the same extent as the other cultures. Since none of the material used in these experiments was cloned this may have occurred simply as a result of a change in balance from cells with a high capacity to degrade zeatin in favour of cells with a lower degradative capacity. However an alternative explanation of this phenomenon will be considered later.

The results described above demonstrate clearly that degradative metabolism can be a very important factor in determining the response of cell cultures to applied cytokinin. In this case the differences in response were not observable in callus tissue grown on solid medium. This is almost certainly due to differences in metabolism caused by different rates of supply of cytokinin to the growing cells. In cell suspension cultures all cells receive a uniform dose of cytokinin and uptake is rapid. In solid grown cultures the cells in intimate contact with the medium receive a high initial dose of cytokinin but this is considerably attenuated across the tissue. We have shown that in contrast to cell suspension cultures, where zeatin is predominantly degraded to adenine and related compounds, soybean callus cultures show less degradative metabolism and more conversion of zeatin to the O-glucoside and lupinic acid, when zeatin is supplied through the agar. This provides an explanation for the different growth patterns of cell suspension cultures and callus tissues described above.

Similar correlations between cytokinin degradation and biological activity have been observed in callus tissues derived from two different genotypes of *Phaseolus*. Mok and Mok, and their co-workers (Mok *et al*., 1978) have shown that callus cultures derived from *P. vulgaris* grow better on dihydrozeatin, which lacks a Δ^2-double bond in the side chain, and its riboside than on zeatin and zeatin riboside. The reverse was found to be true for callus cultures derived from *P. lunatus*. They were able to shown an increased rate of breakdown of Δ^2-isopentenyl cytokinins by *P. vulgaris* callus (Mok *et al*., 1982). This presumably involved cytokinin oxidase to which dihydrozeatin and its riboside are resistant. Callus tissue from a *P. vulgaris* x *P. lunatus* hybrid exhibited intermediate responses to the two classes of cytokinin and metabolism characteristic of both parents.

CYTOKININ BIOSYNTHESIS AND AUTONOMOUS TISSUE CULTURES

It is generally assumed that tissue cultures which require an external supply of cytokinin for growth do not possess the capacity to biosynthesise these compounds. Likewise cytokinin autonomous cultures are assumed to produce sufficient of these hormones to maintain cell division and cell expansion.

Autonomous tissue cultures may be divided broadly into three types:
(1) Crown gall tissues.
(2) Tissues derived from 'genetic tumors'.
(3) Habituated tissues which either never have, or eventually lose the requirement for cytokinin and/or auxin.

Crown Gall Tissues

One characteristic of crown gall tissues is their ability to grown on defined media in the absence of externally supplied cytokinins and auxin. The demonstration that this is due to the presence on the T-DNA of genes directing the synthesis of (Δ^2-isopentenyl)adenosine-5'-monophosphate ([9R-5'P]iP) and IAA has been an outstanding achievement for plant molecular biology. Considerable work by several groups using transposon mutagenesis revealed the presence of two loci on the T-DNA which markedly affected crown gall tumor morphology. Insertional inactivation of a locus designated *tms* gave rise to tumors consisting mainly of leafy shoots. Insertional inactivation of a locus designated *tmr* produced tumors consisting mainly of roots. The marked similarity between the morphology of these tumor tissues and the differentiation produced in normal, hormone requiring, tobacco tissue cultures by varying the cytokinin auxin ratio led to the suggestion that the *tms* and *tmr* loci controlled the ratio of endogenous auxin to cytokinin in crown gall tumors (Hooykass *et al*., 1982). It was suggested that the *tmr* locus controlled the level of cytokinin and the *tms* locus the level of auxin. Measurement, by radioimmunoassay, of gross 'zeatin like' cytokinin levels in *tms* and *tmr* mutants of tobacco crown gall tissue revealed a very large reduction in the level of these cytokinins in *tmr* tissue when compared with a wild type tumor. *Tms* mutant tissue on the other hand had considerably elevated level of 'zeatin like' cyotkinins (Akiyoshi *et al*., 1983).

The results of these studies suggested that the *tmr* gene was directly involved in cytokinin biosynthesis. Cloning of the *tmr* region in an *E. coli* plasmid, designated pMON230, resulted in the direct demonstration that the *tmr* gene coded for a protein which catalysed the key reaction of cytokinin biosynthesis, shown in Fig.4, between 5'-AMP and Δ^2-isopentenyl pyrophosphate to give [9R-5'P]iP (Barry *et al.*, 1984). This result, which was independently confirmed by two other groups (Akiyoshi *et al.*, 1984; Buchman *et al.*, 1985), unequivocally established the function of the *tmr* gene.

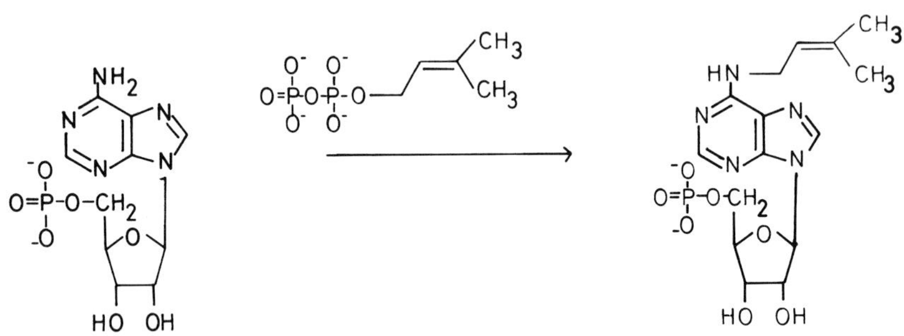

Figure 4. Biosynthesis of [9R-5'P]iP catalysed by the *tmr* gene product.

Although insertional inactivation of the *tms* gene appeared to affect cytokinin levels more than auxin levels it was suggested that its primary function was to control auxin biosynthesis. It was shown that the *tms* region contained two genes, designated gene 1 and gene 2, which from sequence analysis probably produced protein products of 83.7 kD and 49.6 kD respectively (Klee *et al.*, 1984). It has now been demonstrated (Schroder *et al.*, 1984; Van Onckelen *et al.*, 1985; Van Onckelen *et al.*, 1986) that these genes function in an analagous manner to the auxin biosynthetic genes of *Pseudomonas savastonoi*. This organism converts tryptophan to indole-3-acetamide (IAM) via tryptophan-2-monooxygenase, and IAM to IAA via an amidohydrolyase.

It can be seen from the above brief account that hormone autonomy in crown gall tumors depends on the presence of ocogenes coding for cytokinin and auxin biosynthesis. However the exact role of these genes in controlling growth and development of crown gall tissues is still not entirely clear. Although concerned with auxin biosynthesis the *tms* genes clearly influence cytokinin levels which are greatly elevated in tissue in which these genes have been inactivated. Table 1 illustrates the effect of *tmr* and *tms* mutations on the full cytokinin complement of crown gall tissue.

TABLE 1. Cytokinin levels in Tobacco crown gall tissues carrying mutations in the *tms* (auxin) and/or *tmr* (cytokinin) genes (McGaw, Horgan, Wullems and Schilperoort, unpublished data). Measurements were made by the method of Scott and Horgan (1984).

	Cytokinin levels (ng.g^{-1} fresh wt.)			
Tissue	[9R-5'P]Z	[9R]Z	Z	[7G]Z
LBA 4001 (wt)	25	17	22	10
LBA 4210 (tmr)	10	19	33	2
LBA 4060 (tms)	183	62	74	779
LBA 1501-1 (tms)	526	77	34	316
TSO 119 (tmr, tms)	10	26	35	8
SRI 4013-3	91	51	31	70
SRI (control)	10	16	7	4

The effect of *tms* mutations on cytokinin levels is clearly seen, as is the fact that the qualitative and quantitative spectrum of cyotkinins follows closely the pattern of metabolism of externally applied compounds. These data also suggest that the morphological changes observed in *tms* tissues are predominantly controlled by cytokinin levels.

Whilst studies of the crown gall system have provided invaluable information on the molecular biology of cytokinin and auxin production they also raise the important question as to what extent the crown gall system provides a model for cytokinin and auxin production in normal plant tissues?

In the case of auxin the crown gall genes introduce a novel pathway which is probably not relevant to the plant's own production of auxin.

In the case of the cytokinins studies of other autonomous tissue cultures ie. genetic tumors and habituated cultures may help to shed light on this issue.

Genetic Tumors.

Genetic tumors eg. in tobacco, are produced by the combination of genetically distant genomes and characterised by growth in tissue culture without the addition of cytokinin and auxin. Naf (1958) has placed the parental species of tobacco crosses leading to genetic tumors into two groups called 'plus' and 'minus'. It has been suggested that members of the 'plus' group carry genes for a factor (factor 1) which regulates tumor induction whereas members of the 'minus' group carry genes for a factor (factor ee) which controls the degree of tumor expression (Ahuja, 1968). Hybrids of members in the same group are not tumorous whereas inter-group hybrids produce tumors. Unfortunately the molecular nature of these factors in unknown although it has been suggested that in combination they produce some form of biochemical and/or physiological imbalance which leads to tumor formation. One component of this 'imbalance' is presumably an increase in auxin and cytokinin production to a level necessary to sustain autonomous growth. Unfortunately there is no reliable information on cytokinin levels in genetic tumor tissues.

Habituated Tissues.

The term 'habituation' refers to the loss of the exogenous requirement for auxin and/or cytokinin by plant tissues and cells in culture. Meins and his co-workers have made extensive studies of cytokinin habituation in tissue cultures of the tobacco variety 'Havana 425' (Meins, 1982; Meins and Foster, 1985). Cortical cells of this variety are cytokinin autonomous when initially isolated whereas pith tissues appear to contain two cell types. One cell type which becomes cytokinin autonomous when exposed to kinetin or incubation at 35° is classified as 'inducible'. The other cell type which does not respond to these treatments is said to be 'non inducible'. 'Non inducible' cells can change into 'inducible' cells at rates greater than 5×10^{-3} per cell generation which is 100-1000 times faster than the reported rates of somatic mutations in tobacco. Since this process is also reversible Meins has argued that it involves 'epigenetic' changes. The term 'epigenetic' finds little usage in modern molecular biology and can be taken as simply meaning a fairly long term change in gene expression brought about by some uncharacterised mechanism. A similar process may be invoked to explain the changes in cytokinin metabolism in soybean cells reported in the first section of this paper.

Meins has also described a second form of habituation in Havana 425 which involves stable and heritable genetic change (Meins *et al*., 1983). Plants regenerated from cloned leaf and cortex cells of one particular plant differed in their cytokinin habituation. Leaf tissues from leaf cell derived plants were non habituated whereas leaf tissues from cortical cell derived plants were habituated. A preliminary genetic analysis indicated that in this system cytokinin autonomy is being controlled by a single Mendelian gene exhibiting incomplete dominance.

Mok *et al*. (1980) have studied the cytokinin requirement of hypocotyl tissue cultures from plants derived from crosses between a genotype of *P. vulgaris* (P.I. 200960), which exhibits cytokinin dependence in tissue culture, and two cytokinin independent genotypes (G 50 and P.I. 286303). Hypocotyl tissue cultures from F1 plants exhibited an intermediate degree of cytokinin autonomy. The F2 generation contained both parental and intermediate phenotypes. The frequency distribution of cytokinin requiring progeny in F2 and backcross populations suggested, as for the tobacco system described above, the presence of a single set of alleles controlling cytokinin autonomy.

The above brief description of cytokinin autonomy in tissue cultures is included in this paper because in the authors opinion such systems have much to offer to future studies on cytokinin biochemistry and physiology. Although cytokinin and auxin autonomy is not yet explicable in molecular terms it is possible to make some limited speculations based on our current knowledge of the biosynthesis and metabolism of these compounds.

CONCLUSIONS

There are basically two likely explanations for the acquisition of cytokinin autonomy by plant tissue cultures. Firstly an increase in active pool size of these compounds relative to that in non-autonomous tissues. This in turn may be brought about by a 'switch on' or increased rate of biosynthesis and/or a reduced rate of metabolism. Secondly the tissue may acquire an increased ability to respond to the growth substance. This latter explanation seems unlikely in the case of tissues which already respond to externally applied growth substances and become autonomous by losing this requirement.

It is most likely that growth substance autonomy is caused by biochemical changes involving the biosynthesis and/or the metabolism of the compounds concerned. That changes can occur in tissue cultures with regard to their metabolism of externally applied cytokinins has been demonstrated in the first section of this paper. Some evidence that the rate of breakdown of endogenous cytokinins may in part determine a tissue's requirement for externally applied cytokinin is provided by the recent finding by Laloue (M. Laloue personal communication) that diphenyl urea derivatives which are active as cytokinins in tissue culture systems, but bear no structural similarity to N-6 substituted purines, are potent inhibitors of cytokinin oxidase.

An important question pertinent to any understanding the role cytokinins in the growth of tissue cultures is what precise endogenous amounts of these substances are required for autonomous growth? It is clear from the few studies that have been carried out on other autonomous tissue cultures that quantitatively the crown gall phenomenon is a case of 'hormonal overkill', certainly with regard to cytokinins. Using bioassays Nakajima *et al.* (1979) estimated the [9R]Z content of cytokinin autonomous tobacco tissue to be at least 400 fold less than that of crown gall tissue. Using mass spectrometry and stable isotope dilution methods. Scott and Horgan (unpublished results) found the levels of several zeatin related cytokinins in a cytokinin automonous, non crown gall, tobacco tissue to be at least 100 fold lower than in crown gall tissue. Thus it would appear that the endogenous level of cytokinin required to sustain growth, at least in tobacco pith cultures is vastly lower than that which might be estimated from the supply of exogenous material.

In the opinion of the author the value of tissue culture systems in the study of cytokinin biochemistry resides in the possibility that they may shed light on the mechanisms that control cytokinin levels in plants. If cytokinins do in fact function as one level of control of cell division in plants it is reasonable to assume that their biosynthesis and possibly their metabolism will be carefully regulated. The T-DNA cytokinin biosynthesis gene does not appear to have any inbuilt regulatory mechanism and so its introduction into a plant cell results in overproduction of cytokinin. The other forms of cytokinin autonomy described in this article probably result from relatively subtle changes in the plant's own mechanism(s) for regulating cytokinin levels. Cytokinin autonomy in genetic tumors and habituated tissues may be explicable in terms of changes in, as yet unidentified genes, which regulate the activity of the cytokinin biosynthetic genes. Impaired function of these 'regulator' genes may switch on cytokinin production. This would seem a likely explanation for cytokinin autonomy in the *Phaseolus* hypocotyl and tobacco leaf tissue cultures described above. In these cases the impared function may be due to a classical mutation. In the case of genetic tumors it may arise via recombination.

With regard to the possible activation of plant cytokinin biosynthesis genes, it is of considerable interest to note that Stachel *et al.* (1986) have shown that expression of the *vir* region of the *Agrobacterium tumefaciens* Ti plasmid which also carries a cytokinin biosynthetic gene, is activated by acetosyringone and α-hydroxyacetosyringone. These molecules are released from plant cells on wounding. It may be hypothesised that the formation of wound callus results from activation of the plant's own cytokinin biosynthetic genes by wound derived compounds. This could explain the cytokinin activity of dihydroconiferyl alcohol (Lee *et al.*, 1981) which is structurally similar to the syringones but has no structural similarity to N-6 substituted purines.

REFERENCES

Ahuja MR (1968). An hypothesis and evidence concerning the genetic components controlling tumor formation in Nicotiana. Mol Gen Genet 103:176-184.

Akiyoshi DE, Morris RO, Hinz R, Mischke BS, Kosuge T, Garfinkle DJ, Gordon MP, Nester EW (1983). Cytokinin auxin balance in crown gall tumors is regulated by specific loci in the T-DNA. Proc Nat Acad Sci USA 80:407-411.

Akiyoshi DE, Klee H, Amasino RM, Nester EW, Gordon MP (1984). T-DNA of *Agrobacterium tumefaciens* encodes an enzyme of cytokinin biosynthesis. Proc Nat Acad Sci USA 81:5994-5998.

Barry GF, Rogers SG, Fraley RT, Brand L (1984). Identification of a cloned cytokinin biosynthetic gene. Proc Nat Acad Sci USA 81:4776-4780.

Buchmann I, Marner FJ, Schroder G, Waffenschmidt S, Schroder J (1985). Tumor genes in plants: T-DNA encoded cytokinin biosynthesis. EMBO J 4:853-859.

Hooykaas PJJ, Ooms G, Schilperoort RA (1982). Tumors induced by different strains of *Agrobacterium tumefaciens*. In Kahl G, Schell JS (eds): "Molecular Biology of Plant Tumors", New York: Academic Press, pp 373-390.

Klee H, Montoya A, Horodyski F, Lichtenstein C, Garfinkle D, Fuller S, Flores C, Peschon J, Nester EW, Gordon M (1984). Nucleotide sequence of the tms genes of the pTiA6NC octopine Ti plasmid: Two gene products involved in plant tumorigenisis. Proc Nat Acad Sci USA 81:1728-1732.

Laloue M, Pethe C (1982). Dynamics of cytokinin metabolism in tobacco cells. In Wareing PF (ed): "Plant Growth Substances 1982", London: Academic Press, pp 185-195.

McGaw BA, Horgan R (1983). Cytokinin oxidase from *Zea mays* kernels and *Vinca rosea* crown gall tissue. Planta 159: 30-37.

Meins FJr (1982). Habituation of cultured plant cells. In Kahl G, Schell JS (eds) "Molecular Biology of Plant Tumors" New York: Academic Press, pp 3-31.

Meins FJr, Foster R (1985). Reversible, cell-heritable changes during the development of tobacco pith tissues. Devel Biol 108:1-5.

Meins FJr, Foster R, Lutz JD (1983). Evidence for a Mendelian factor controlling the cytokinin requirement of cultured tobacco cells. Devel Genet 4:129-141.

Mok MC, Mok DWS, Armstrong DJ (1978). Differential cytokinin structure activity relationships in *Phaseolus*. Plant Physiol 61:72-75.

Mok MC, Mok DWS, Armstrong DJ, Rabakoarihanata A, Kim S-G, (1980). Cytokinin autonomy in tissue cultures of Phaseolus: A genotype-specific and heritable trait. Genetics 94:675-686.

Mok MC, Mok DWS, Dixon Sc, Armstrong DJ, Shaw G (1982). Cytokinin structure activity relationships and the metabolism of N-(Δ^2-isopentenyl)adenosine-8-^{14}C in *Phaseolus* callus tissues. Plant Physio 70:173-178.

Naf U (1958). Studies in tumor formation in *Nicotiana* hybrids. I. The classification of parents into two etiologically significant groups. Growth 22:167-180.

Nakajima H, Yokota T, Matsumoto T, Noguchi M, Takahashi N (1979). Relationships between hormone cells cultured in suspension. Plant Cell Physiol 29:1489-1499.

Lee TS, Purse JG, Pryce RJ, Horgan R, Wareing PF (1981). Dihydroconiferyl alcohol-A cell division factor from Acer species. Planta 152:571-577.

Schroder G, Waffenschmidt S, Weiler EW, Schroder J (1984). The T-region of Ti plasmids codes for an enzyme synthesizing indole-3-acetic acid. Eur J Biochem 138: 387-391.

Scott IM, Horgan R (1984). Mass spectrometric quantification of cytokinin nucleotides and glycosides in tobacco crown gall tissues. Planta 161:345-354.

Stachel SE, Messens E, Van Montagu M, Zambryski P (1986). Identification of the signal molecules produced by wounded plant cells that activate T-DNA transfer in *Agrobacterium tumefaciens* Nature 318:624-629.

Van Onckelen H, Rudelsheim P, Inze D, Follin A, Messens E, Horemans S, Schell J, Van Montagu M, De Greef J (1985). Tobacco plants transformed with the Agrobacterium T-DNA gene 1 contain high amounts of indole-3-acetamide. FEBS Let 181:373-376.

Van Onckelen H, Prisen E, Inze D, Rudelsheim P, Van Lijsebettens M, Follin A, Schell J, Van Montagu M, De Greef J (1986). Agrobacterium T-DNA gene 1 codes for tryptophan 2-monooxygenase activity in tobacco crown gall cells. FEBS Let 198:357-360.

Plant Tissue and Cell Culture, pages 151–165

PHOTOSYNTHESIS IN CELL AND TISSUE CULTURE SYSTEMS

Karl-Hermann Neumann and Ludwig Bender

Institut fuer Pflanzenernaehrung, Abt. Gewebekultur der Justus-Liebig-Universitaet Giessen, Suedanlage 6, 6300 Giessen, F.R.-Germany.

INTRODUCTION

Although generally carbohydrates are supplied to the nutrient media, usually in the form of sucrose, as sources of carbon and energy, many plant cell and tissue culture systems turn green and some, if not most, are at least potentially able to develop a functional photosynthetic system. Such cultures grow with a mixotrophic nutritional system with a considerable contribution of photosynthetic CO_2-fixation to growth (Neumann et al., 1982). However, out of originally mixotrophic cultures of some plant species, cell strains could be selected which were able to grow photoautotrophically in sugar free nutrient media for long periods of up to several years. For substantial growth of such autotrophic cultures usually an increase in the atmospheric CO_2-concentration of up to 2% is required, although for some species like _Gossypium hirsutum_ (L.C. Blair and J.M. Widholm, personal communication) _Arachis hypogea_ and _Daucus carota_ (Bender et al., 1981) photoautotrophic growth can be observed also at ambient CO_2-concentrations. Still, also in these latter two cultures the daily growth rate can be increased six- to seven-fold using the higher atmospheric CO_2-concentration (Gross, unpublished data of our laboratory). In spite of considerable efforts, however, until now autotrophic cell cultures could only be obtained from so-called C3-plants. However, recently a photoautotrophic suspension culture of

Amaranthus powellii has been initiated (L.C. Blair and J.M. Widholm, personal communication).

The most outstanding common denominator of mixotrophic and autotrophic cell culture systems screened in detail so far seems to be a considerable contribution of C4-carboxylation mediated by PEPCase with PEP as the carbon acceptor. However, as was shown by Yamada et al. (1982) malate produced by this C4-fixation mechanism seems not to be used as a carbon source for the Calvin cycle as was shown for C4 or CAM plants. This C4-fixation mechanism is most pronounced during the exponential phase of cell division (Huesemann et al., 1984) when demand on carbon skeletons for biosynthesis is high. Consequently, the most obvious interpretation of this phenomenon seems to be that the carboxylation of PEP to yield oxaloacetic acid and subsequently malate could serve as an anaplerotic reaction to keep the tricarboxylic acid cycle going in spite of a heavy drainage, e.g., of oxoglutarate as a precursor for the synthesis of glutamic acid.

Some PEPCase activity can be found in most, if not all, C3-plants and as was shown in a recent paper by Aoyagi and Bassham (1986), e.g., also in the meristematic regions at the base of developing wheat leaves (C3-plant) PEPCase has a higher share in the total carboxylation capacity of the cells than in the more mature areas at the leaf tip. So possibly a high PEPCase activity could be a more general characteristic of actively dividing cells to meet the high demand of the synthetic machinery to produce protein etc.

Though the number of photoautotrophic cell culture systems increases, still most illuminated green cell cultures grow faster in a sugar supplemented medium. An evaluation of the significance and contribution of photosynthesis to growth and the chemical composition of such mixotrophic cells requires the consideration of interactions of carbon coming from exogenous carbohydrates with inorganic carbon incorporated by photosynthesis or also by dark fixation. However, as will be shown, mixotrophic and autotrophic cultures are basically quite similar and deviations are primarily of a quantitative nature, mostly based on a higher chlorophyll concentration of the latter. However, some authors cautioned against selecting autotrophic cell strains based only on chlorophyll concentration (Sato et al., 1979).

DEVELOPMENT OF THE PHOTOSYNTHETIC APPARATUS

In most investigations on photosynthesis of cultured cells, subcultured strains often isolated many years ago are used. Therefore, only limited information is available on de novo development of the photosynthetic machinery in freshly isolated material using the inital stages in culture. Though a characterization of chloroplasts in most systems was made which indicated a well developed thylacoidal system, aside from the earlier investigations in the late sixties, e.g., by Israel and Steward (1967) and Laetsch and Stetler (1965) recently only two more elaborate studies on this problem were published. For carrot root explants initially containing chromoplasts a sequence of serial developmental stages with amylo-chromoplasts and amyloplasts as intermediates leading to mature chloroplasts was described (Kumar et al., 1983; 1984). Here ten days in a sucrose containing medium in light suffices to establish a photosynthetic machinery able to support the growth of these cultures in a sugar free medium in a 5 liter laboratory fermentor for at least another six weeks. The transformation of chromoplasts into chloroplasts requires a supplement of IAA, inositol and/or kinetin, otherwise the sequence of plastid development will be arrested at the amyloplast stage also in the light. The transformation of chromoplasts into chloroplasts occurs also in a sugar free medium. However, the number of chloroplasts, i.e., the quantitative extent of the photosynthetic apparatus per cell is strongly increased in a nutrient medium originally supplied with sucrose. Apparently, a sucrose supplement stimulates proliferation of plastids per cell. However, the transformation of amyloplasts, abundant also in a hormone free medium, into chloroplasts seems to be under hormonal control.

In the other system cultured cells (established tobacco cultures originally isolated from the pith) containing well developed chloroplasts were transferred from light to darkness for several weeks. During the dark period the chloroplasts turned into starch containing plastids defined by the authors as proplastids which, however, more closely resemble amyloplasts (NATO et al., 1985). The further development of these initial structures follows closely the developmental path as observed in the carrot system. In both systems no etioplasts characteristic of dark-grown seedlings often used in studies on chloroplast development were observed.

A problem not yet satisfactorily understood is the propagation of plastids. As published many years ago (Neumann et al., 1969) and as also reported from others (e.g., Nato et al., 1981) fission of chloroplasts can be observed in cultured cells. The number of fissions observed, however, is rather low and by inspecting several thousands of electron microscopic samples not more than 35 to 40 examples of dividing plastids could be observed, a number hardly enough to explain the tremendous increase in plastid numbers in actively dividing cultured cells. Unless a narrow circadian rhythm of plastid division exists which we missed, the high rate of plastid proliferation observed in actively dividing cultures cannot be satisfactorily explained as being due to division only. Though admittedly no other explanation seems to be at hand presently to account for plastid proliferation, nevertheless this problem should be subjected to more systematic investigations (see also Possingham, 1980).

LIGHT REACTION SYSTEM

Considerable chlorophyll concentrations, in particular of autotrophic cell cultures, were reported ranging from about 0.5-1.5 mg/g dm for mixotrophic to 4-7 mg/g dm for autotrophic cultures - the latter coming close to that of leaves. Only a few reports are available on the light reaction system in cultured cells. In these the photosynthetic capacity of the light reaction system as measured by means of ferricyanide reduction or oxygen evolution is comparable to leaves on a chlorophyll basis. However, on a weight or per cell basis only in autotrophic cultures with highest chlorophyll concentrations does photosynthetic capacity come close to that of leaves (Neumann and Raafat, 1973; Sato et al., 1979, Seeni and Gnanam, 1982; Nato et al., 1985).

Measurements of variable fluorescence as an indicator of electron flow through both photosystems as well as chlorophyll a/b ratios revealed qualitative similarities of the light reaction system in leaves and cultured cells (Bender et al., 1981; Seeni and Gnanam, 1982). However, already in earlier and preliminary investigations on the distribution of ^{14}C and from $H^{14}CO_3$ into various organic fractions of explants and leaves it became clear that profound qualitative variations in carbon metabolism between these two systems should exist (s.a. Neumann et al., 1978).

MIXOTROPHIC CULTURES

Before details on the fixation and metabolism of inorganic carbon are discussed, first some data on the contribution of light to growth of cultured mixotrophic cells will be discussed. In an investigation reported many years ago it was shown that light contributed considerably to growth (Newmann and Raafat, 1973). Here, fresh root explants after some preculture in a sucrose containing medium during which the photosynthetic apparatus develops including pastid proliferation per cell (s. above) and without selection procedure for autotrophy, carrot tissue cultures were able to grow on a sugar free medium for at least 3 weeks, mostly by cell division. In terms of cell number increment per explant the treatment without carbohydrates was almost identical with the sample cultured in a 2% sucrose medium. However, in terms of fresh and dry weight the sugar free treatment was much lower than that of the treatment supplied with sucrose, approaching the data for the cultures grown in the dark on 2% sucrose (s.a. Nishida et al., 1980). Apparently cellular growth and not cell division activity is promoted by sucrose supply. Still, if the high starch content of plastids as usually observed on sucrose containing media is considered, one wonders how much of the dry weight of these cells really accounts for actual growth in terms of biosynthesis of new material of the cytoplasm in such experiments and not just for starch accumulation, a question raised already several years ago by M. Fowler. Still, the data indicated a heavy contribution of light to growth of mixotrophic cells, though some light influences on growth may be also attributed to interactions with the hormonal system, notably with IAA as a supplement to the nutrient medium (Bender and Newmann, 1978) or the phytochrome system.

PHOTOSYNTHETIC EFFICIENCY OF AUTOTROPHIC CULTURES

The data on photosynthetic efficiency for various species calculated as uM CO_2 fixed per mg chlorophyll per hour are of the same order of magnitude from 50-80 uM for <u>Chenopodium rubrum</u> (Huesemann et al., 1984) and tobacco (Nato et al., 1985; Yamada et al., 1982) to about 200 uM CO_2 for <u>Arachis</u> (Seeni and Gnanam, 1982). Mostly differences are due to measurements at various stages of the growth cycle, a problem to be dealt with later.

Interestingly, values on CO_2-fixation capacity on a chlorophyll basis (Table 1) are practically the same for mixotrophic and autotrophic cultures as shown clearly for tobacco (Nishida et al., 1980) and for _Chenopodium rubrum_ (Huesemann et al., 1979). The major difference between the two systems seems to be again a higher concentration of chlorophyll in autotrophic systems. Negative influences of sucrose on chlorophyll concentrations were reported for all systems investigated to this end, a problem which still lacks a physiologically and biochemically plausible explanation.

Table 1: Comparison of CO_2-fixation capacity of photomixotrophic and photoautotrophic cell cultures on a chlorophyll basis.

	$uM\ CO_2 \cdot mg\ Chl^{-1} \cdot h^{-1}$ photomixotrophic	photoautotrophic
Nicotiana tabaccum*	94.5	132
Chenopodium rubrum**	190	185

*Nishida et al., 1980; **Huesemann et al., 1979.

Not many data are available to compare the photosynthetic CO_2-fixation of cultured cells and leaves of the same species employing the same experimental conditions. In Table 2, some data on carrots are summarized for two experimental sets; one to compare leaves and cultured cells with $H^{14}CO_3$ as the substrate and in the other, gaseous $^{14}CO_2$ was used (Bender, 1985). Unfortunately, the data are somewhat hampered because in these two experiments different ^{14}C-activities were employed. Still, it is quite obvious that leaves and cultured cells are able to use both substrates. However, for leaves CO_2 seems to be the preferred inorganic carbon source while in tissue cultures the fixation of HCO_3^- (e.g., Cooper and Wood, 1971). The main reason for the observed differences between leaves and cultured cells should be an extensive participation of PEPCase in the incorporation of inorganic carbon in cultured cells, whether mixotrophic or autotrophic (Yamada et al.,

Table 2: Fixation and metabolism of $H^{14}CO_3^-$ and $^{14}CO_2$ by carrot tissue cultures and carrot leaves (dpm · 10^4 · mg Chl^{-1}, exp. 4 min.) (Bender, 1985).

	$H^{14}CO_3^-$ (10 mCi · 1 medium-1)		$^{14}CO_2$ (1.7 mCi · 1 air-1)	
	Tiss.Cult.	Leaves	Tiss.Cult.	Leaves
Total	2374	169	892	3689
p-Ester	430	9	64	n.d.
Sugars	174	47	126	1743
Neutr. Amino Acids	47	35	26	6
Malate+ Aspartate	513	7	305	72

1982; Huesemann et al., 1979; Kumar et al., 1983; Bender et al., 1981; Nato et al., 1981; Seeni and Gnanam, 1982). The involvement of PEP-carboxykinase as another candidate to catalyze carbon fixation could be ruled out (Yamada et al., 1978).

PEPCase AND CARBON INCORPORATION

Particularly in short time experiments, aside from Calvin cycle derivatives, a high ^{14}C labelling can be observed in C4-fixation products like aspartate and in particular in malate (s. Table 3) in cultured cells.

If the labelling of glycine and serine is taken as an indicator for the occurrence of photorespiration, then in this system RuBPOase activity should exist. However, results as to the significance of photorespiration are controversial. Berlyn et al. (1978) found considerable rates of photorespiration in tobacco callus cells with autotrophic material exceeding that of mixtotrophic cultures three times. Also, Nato et al. (1985) reported about 17% of total ^{14}C in serine/glycine during exponential and 31% during stationary growth of tobacco cells, while Yamada et al. (1982) reported only 1-2% of total radioactivity in glycolate and glycine/serine. The latter results are in agreement with data of Bender et al. (1985), who reported 1.7% and 3.6% of total ^{14}C in glycine/serine for carrot and 4.5% for peanut cultures.

Table 3: Carbon incorporation from CO_2 into various metabolites in 21-day-old photosynthetically active carrot tissue cultures after different incubation periods in $H^{14}CO_3^-$ containing nutrient solution (pH 7). Data given as uMol carbon atoms · 10^{-4} · mg Chl^{-1}.

	15 sec.(light)	4 min.(light)	4 min.(dark)
PGA	10.4	88	0
Gluc-P	0	214	0
Fruc-P	0	72	0
Glucose	0	108	0
Fructose	0	74	0
Sucrose	0	0	0
PEP	0	22	0
Glycerate	trace	47	0
Glycine+Serine	0	20	0
Alanine	0	19	0
Malate	7.5*	188**	17***
Aspartate	3.5	198	21
Citrate	2.4	81	13
Glutamate	0	17	2.3
^{14}C in C4-Pos.:	*87%	**72%	***86%

Again in all photosynthetic cell culture systems, mixotrophic or autotrophic, changes in the ratio of RuBPCase/PEPCase activities were observed during the growth cycle with highest participation of PEPCase in carbon fixation during the exponential phase of cell division activity. Here, PEPCase can even be considered as the dominating carboxylase (Huesemann, 1981; Nato et al., 1981, 1985). Employing an immunoassay, Nato et al. (1985) could show that not only the activity but also the concentration of RuBPCase decreases during exponential growth while the activity and the concentration of PEPCase is increased. At the transition from log phase to stationary phase, during which PEPCase activity and concentration decreases, the increase in RuBPCase activity is due to de novo synthesis of RuBPCase molecules. Until now, nothing is really known about the regulatory mechanisms of this inverse relation of the two carboxylating enzymes. Data published by Huesemann et al. (1979) clearly indicate, however, that in mixotrophic cultures PEPCase is more prominent than in autotrophic (x 2.1), but still its activity in the latter is considerably higher than in leaves (x 3.3).

At present the elaboration of details of the photosynthetic system in cultured cells seems to be most advanced for five species, i.e., *Glycine max* (Horn et al., 1983; S.M.D. Rogers and J.M. Widholm, personal communication), *Nicotiana tabacum*, *Daucus carota*, *Chenopodium rubrum* and *Arachis hypogaea*. In summary, from all we know, a high activity of PEPCase seems to be clearly a characteristic feature of photosynthetically active cultured cells of these C3-plants (mixo- or autotrophic) and the most promotive conditions to increase its activity are mixotrophic nutrition and a high cell division activity during exponential growth of the cultures. In most recent studies by Sato et al. (1986) to characterize PEPCase it became clear that the enzyme of Nicotiana cultures is a C3-type PEPCase with a K_m (PEP) of 60 uM as compared to 1000 uM for PEPCase of C4-plants (O'Leary, 1982) and consists of four subunits with identical molecular weight (110 kd). Pyruvate Pi-d-kinase was also detected in these cultures. A rather low K_m (PEP) of PEPcase from cultured *Pinus radiate* cotelydons was also found by Thorpe's group (Kumar et al., in preparation).

To evaluate the role of PEPcase in illuminated cultured cells the possibility of a direct influence of light should also be considered. As was shown by *in vitro* PEPCase activity measurements, light stimulates PEPCase activity severalfold and ^{14}C incorporation into the C4 position of malate in light and darkness *in vivo* were consistent with the influence of light on the *in vitro* PEPCase activity (Bender, 1985).

PEPCase activity requires divalent cations as cofactors and Mn^{2+} shows a much lower K_m than Mg^{2+} (O'Leary, 1982). The influence of Mn^{2+} on CO_2-fixation by PEPCase was recognized rather accidentally in an early experiment with photosynthetic cell cultures (Neumann 1962, s. Table 4). In these experiments with carrot root explants the supplement of Mn^{2+} to a micronutrient limited medium resulted in a strong increase in $^{14}CO_2$-fixation in the light. However, its stimulation of dark fixation was, in relative terms, even stronger. Since carboxylation in the dark should only be catalyzed by PEPCase, this stimulation by manganese could be seen in the context of PEP carboxylation, aside from the well known influences of this micro nutrient on the light reaction system.

Table 4: The influence of a manganese supplement to a micronutrient limited medium (supplied with 10% coconut milk) on $^{14}CO_2$-fixation of cultured carrot explants in light and darkness (three weeks of culture, cpm·10^3/g fr.wt., Neumann 1962).

	Light	Darkness
-Mn	17.7	2.5
+4.5 ppm $MnSO_4$	21.3	8.0

INFLUENCES OF EXOGENOUS SUCROSE

On the basis of the data discussed so far, it appears that photosynthetic cultured cells of C3-plants are able to fix inorganic carbon by two carboxylation mechanisms, i.e., one mediated by RuBPCase to yield PGA and triose-P and a second by PEPCase to yield OAA, malate and aspartate. The metabolites of the two pathways will certainly mingle in due course of metabolism, and in autotrophic cultures of course, the acceptor molecule for beta-carboxylation, PEP, has to be derived from breakdown of Calvin cycle products. However, as was shown by Bender (1985) and Bender et al., (1985) in mixotrophic cultures, almost 90% of labelled malate extracted from carrot cultures after $H^{13}CO_3^-$ fixation (15 sec.) was synthesized from PEP derived from the glycolytic breakdown of exogenous sugars taken up from the nutrient medium (s.a. Table 3). The absence of labelled PEP indicated that the two carboxylation mechanisms are, to some degree, independent of each other, possibly also in respect to the regulation of their individual activities. However, a metabolic link between the two pathways seems to exist, since after longer fixation (4 min. and more) labelled PEP, which should be derived from Calvin cycle products, was detected (Bender, 1985; Bender et al., 1985). To follow this up, some data on the metabolism of ^{14}C from exogenous sugar and from inorganic carbon will finally be discussed.

In most publications dealing with this problem, it is assumed that the sucrose molecules supplied to the medium are split into glucose and fructose by a cell wall or plasmalemma located invertase (in our system free invertase could not be found in the medium), and the two hexoses are taken up by the cells (see e.g., Fowler and

Stepan-Sarkissian, 1985). As was shown recently (Bender et al., 1986), after uptake glucose and fructose apparently are phosphorylated and converted into each other. These sugar phosphates are then either channeled into metabolism via glycolysis, or, if an excess of supply over demand exists, are coupled again to yield sucrose first. Later the sucrose is apparently hydrolyzed agian by an endogenous invertase into glucose and fructose, which now accumulate in the cells either as such or as starch. Some of these free hexoses, as well as sucrose at a lower rate, are even excreted into the nutrient medium again. This seemingly useless cycle of sugar molecules from the nutrient solution may represent an important pathway for cells still integrated in the body of an intact plant. Somehow, this may still go on in cultured cells without an obvious benefit for their metabolism. Perhaps more such inherent metabolic processes may continue in cultured cells.

Following a sucrose supplement both fixation mechanisms are decreased. However, the extent of this decrase is more pronounced on the Calvin cycle (-43%) than on PEP carboxylation (-23%, Bender, 1985). These results may indicate that the negative influences of sucrose on the total fixation of inorganic carbon is predominantly on the operation of the Calvin cycle.

The major products of photosynthesis exported from the chloroplasts to the cytoplasm by the phosphate translocator are triosephosphates and PGA. This requires inorganic phosphate in the cytoplasm to be imported into the chloroplasts on an equimolar basis to keep a delicate balance of Pi/organic P in the cytoplasm and in the plastids (Heber and Heldt, 1981). A low ratio of Pi/organic P in the cytoplasm will slow down assimilate export from the chloroplasts, starch will accumulate and, after the capacity of starch storage in the plastids is exhausted, product feedback inhibition eventually should reduce the activity of the Calvin cycle.

Inorganic phosphorus can be liberated from P-esters used to synthesis sucrose in the cytosol. The enzymes to mediate sucrose synthesis could use hexose phosphates either derived from exogenous sugar or those synthesized from triose-P exported from the chloroplasts. Some competition for the enzyme molecules may be created and phosphorylated Calvin cycle derivatives may pile up. The phosphorus in this pool is not liberated as Pi and again available for the phosphate translocator. This situation may then lead to some endogenous deficiency of Pi for an optimal operation of the phosphate translocator, and after starch accumulation capacity is reached within the plastids, the $^{14}CO_2$-fixation via the Calvin cycle will be reduced by product feedback. This is just an hypothesis and many data are still lacking to check its validity. However, data on the influence of phosphorus deficiency induced by a limitation of phosphorus supplied to the medium clearly indicate an impairment of the "phosphate translocator" with consequences on the metabolism of chloroplasts as described above. Here, P-deficiency strongly reduces CO_2-fixation by the Calvin cycle, with a concomitant increase of starch synthesis. At the same time, carbon assimilation via PEPCase is strongly increased, and total ^{14}C-fixation remains then more or less the same at low and high phosphate levels (Bender, 1985; Bender et al., 1985). It remains to be seen whether influences of phosphorus on the two carboxylating systems are independent of each other or are directly related.

Besides the phosphate translocator, also a dicarboxylate shuttle to translocate reducing equivalents from the chloroplasts to the cytoplasm is described (Heber and Heldt, 1981). In this model OAA derived from PEP carboxylation is transferred to the chloroplasts, then reduced to malate with photosynthetically generated NADPH, which then is exported to the cytosol. Here malate is either taken up by mitochrondira to fuel the TCA cycle in an anaplerotic reaction or it can be oxidized by the cytoplasmic MDH to yield OAA and NADH. Finally, mediated by the cytosolic NADP-dependent malic enzyme, pyruvate can also be produced via oxidative decarboxylation to yield NADPH. This mechanism could perhaps serve to export reducing equivalents from the chloroplasts if the export of triose-P is impaired, e.g., under phosphorus deficiency conditions or following a slow-down of Calvin cycle activity at a high supply of exogenous sucrose as described above. Interestingly, the activity of this shuttle seems to be stimulated by cytokinins like kinetin (Bender et al., 1985) indicating also some control of its activity by the hormonal system.

REFERENCES

Aoyagi K, Bassham JA (1986). Appearance and accumulation of C4 carbon pathway enzymes in developing wheat leaves. Plant Physiol. 80:334-340

Bender L, Neumann KH (1978). Investigations on the indole-3-acetic acid metabolism of carrot tissue cultures (Daucus carota L.). Z. Pflanzenphysiol. 88:209-217

Bender L, Kumar A, Neumann KH (1981). Photoautotrophe pflanzliche Gewebekulturen in Laborfermentern. In: Fermentation (RM Lafferty, Ed.) Springer-Verlag Wien, New York pp 193-203

Bender L, Kumar A, Neumann KH (1985). On the photosynthetic systems and assimilate metabolism of Daucus and Arachis cell cultures. In: Primary and Secondary Metabolism of Plant Cell Cultures. (Neumann et al., Ed.) Springer-Verlag Berlin, Heidelberg, New York, Tokyo pp 24-42

Bender L (1985). Photosynthese und Assimilatverwertung bei Gewebekulturen und intakten Pflanzen von Daucus carota L. Habilitation thesis, Justus-Liebig-Universitaet Giessen

Bender L, Pauler B, Neumann KH (1986). On carbohydrate metabolism of cultured carrot root explants. Plant Cell Tissue Organ Culture (submitted)

Berlyn MB, Zelitch I, Beaudette PD (1978). Photosynthetic characteristics of photoautotrophically grown tobacco callus cells. Plant Physiol. 61:606-610

Cooper TC, Wood HG (1971). The carboxylation of phosphoenol pyruvate and pyruvate. II. The active species of "CO_2" utilized by phosphoenol pyruvate carboxylase. J. Biol. Chem. 246:5488-5490

Fowler MW, Stepan-Sarkissian G (1985). Carbohydrate source, biomass productivity and natural product yield in cell suspension cultures. In: Primary and secondary metabolism of plant cell cultures (Neumann et al., Ed.) Springer-Verlag Berlin, Heidelberg, New York, Tokyo pp 66-73

Heber U, Heldt HW (1981). The chloroplast envelope: Structure, function, and role in leaf metabolism. Ann. Rev. Plant Physiol. 32:139-168

Horn ME, Sherrard JH, Widholm JM (1983). Photoautotrophic growth of soybean cells in suspension culture. I. Establishment of photoautotrophic cultures. Plant Physiol. 72:426-429

Huesemann W, Plohr A, Barz W (1979). Photosynthetic characteristics of photomixotrophic and photoautotrophic cell suspension cultures of Chenopodium rubrum. Protoplasma 100:101-112

Huesemann W (1981). Growth characteristics of hormone and vitamine independent photoautotrophic cell suspension cultures from Chenopodium rubrum. Protoplasma 109:415-431

Huesemann W, Herzbeck H, Robenck H (1984). Photosynthesis and carbon metabolism in photoautotrophic cell suspensions of Chenopodium rubrum from different phases of batch growth. Physiol. Plant. 62:349-355

Israel HW, Steward FC (1967). The fine structure and development of plastics in cultured cells of Daucus carota. Ann. Bot. 31:1-18

Kumar A, Bender L, Pauler B, Neumann KH, Senger H, Jeske C (1983). Ultrastructural and biochemical development of the photosynthetic apparatus during callus induction in carrot root explants. Plant Cell Tissue Organ Culture 2: 161-177

Kumar A, Bender L, Neumann KH (1984). Growth regulation, plastid differentiation and the development of a photosynthetic system in cultured carrot root explants as influenced by exogenous sucrose and various phytohormones. Plant Cell Tissue Organ Culture 3:11-28

Laetsch WM, Stetler DA (1965). Chloroplast structure and function in cultured tobacco tissue. Am. J. Bot. 52:798-804

Nato A, Mathieu Y, Brangeon J (1981). Heterotrophic tobacco cell cultures during greening. II. Physiological and biochemical aspects. Physiol. Plant. 53:335-341

Nato A, Hoarau J, Brangeon J, Hirel B, Suzuki A (1985). Regulation of carbon and nitrogen assimilation pathways in tobacco cell suspension cultures in relation with ultrastructural and biochemical development of the photosynthetic apparatus. In: Primary and Secondary Metabolism of Plant Cell Cultures (KH Neumann et al., Ed.) Springer-Verlag Berling, Heidelberg, New York, Tokyo pp 43-57

Neumann KH (1962). Untersuchungen ueber den Einfluss essentieller Schwermetalle auf das Wachstum und den Proteinstoffwechsel von Karottengewebekulturen. Ph.D. thesis, Justus-Leibig-Universitaet Giessen

Neumann KH, Cireli E, Cireli B (1969). Untersuchungen ueber Beziehungen zwischen Zellteilung und Morphogenese bei Gewebekulturen von Daucus carota. III. Der Einfluss des Kinetins auf die Ultrastruktur der Zellen. Physiol. Plant. 22:787-800

Neumann KH, Raafat A (1973). Further studies on photosynthesis of carrot tissue cultures. Plant Physiol. 51-685-690

Neuman KH, Pertzsch C, Moos M, Kroemmelbein G (1978). Untersuchungen zur Charakterisierung des Ernaehrungssystems von Karottengewebekulturen: Photosynthese, Kohlenstoff- und Stickstoffernaehrung, Mineralstoffverbrauch. Z. Pflanzeneranehr. Bodenk. 141:299-311

Neumann KH, Bender L, Kumar A, Szegoe M (1982). Photosynthesis and pathways of carbon in tissue cultures of *Daucus* and *Arachis*. Proc. 5th Intl. Cong. Plant Tissue and Cell Culture, Tokyo, 251-252

Nishida K, Sato F, Yamada Y (1980). Photosynthetic carbon metabolism in photoautotrophically and photomixotrophically cultured tobacco cells. Plant and Cell Physiol. 21: 47-55

Possingham JV (1980). Plastic replication and development in the life cycle of higher plants. Ann. Rev. Plant Physiol. 31:113-129

O'Leary M (1982). Phosphoenolpyruvate carboxylase: An enzymologist's view. Ann. Rev. Plant Physiol. 33:297-315

Sato F, Asada K, Yamada Y (1979). Photoautotrophy and the photosynthetic potential of chlorophyllous cells in mixotrophic cultures. Plant and Cell Physiol. 20:193-200

Sato F, Koizumi N, Takeda S, Yamada Y (1986). Photoautotrophic culture of tobacco cells and their character: Carboxylation enzyme and herbizide resistance. Abstract, VI. Int. Cong. Plant Tissue and Cell Culture, Minneapolis

Seeni S, Gnanam A (1982). Growth of photoheterotrophic cells of peanut (*Arachis hypogaea L.*) in still nutrient medium. Plant. Physiol. 70:815-822

Yamada Y, Sato F, Hagimori M (1978). Photoautotrophism in green cultured cells. In: Frontiers of Plant Tissue Culture, (TA Thorpe, Ed.). Proc. 4th Int. Congress Plant Tissue and Cell Culture, Calgary, pp 453-462

Yamada Y, Sato F, Watanabe K (1982). Photosynthetic carbon metabolism in cultured photoautotrophic cells. Proc. 5th Intl. Cong. Plant Tissue a. Cell Culture, Tokyo, pp 249-250

Section III.
Secondary Metabolites

Plant Tissue and Cell Culture, pages 169–185

SECONDARY METABOLITES FROM PLANT CELLS -- PHARMACEUTICAL APPLICIATIONS AND PROGRESS IN COMMERCIAL PRODUCTION

Yasuhiro Fujita and Mamoru Tabata*

Bioscience Research Center, Mitsui Petrochemical Ind.,Ltd. Yamaguchi 740, Japan
* Faculty of Pharmaceutical Science, Kyoto University, Kyoto 606, Japan

INTRODUCTION

There have been many attempts to establish tissue culture methods for the commercial production of useful plant metabolites. We at Mitsui Petrochemical Industries are pleased that our first commercial production of shikonin by cell cultures of *Lithospermum erythrorhizon* in 1983 has encouraged other researchers in this field.

Dr. Zenk reported on target metabolites for production by plant cell culture at the 1975 congress and Dr. Staba at the last Congress in 1979. Recently, we checked the number of Japanese patents taken out on useful secondary metabolites. The number for the five years from 1980 to 1984 was 39; 32 of which (82%) were for pharmaceutical compounds such as indole alkaloids, morphine alkaloids, and ubiquinone. In addition, three were for food additives and two for pigments. In comparison, the patents for pharmaceutical compounds made up only 50% of the patents applied for in the five years from 1975 to 1979 (total number, 20). This large increase in the number of patents applied for shows that commercial interest in the production of pharmaceutical compounds has increased remarkably year by year, and it remains high.

But, in spite of this high interest, there has been little industrial-scale production of pharmaceutical compounds, the most valuable target metabolites. One reason is that much of the technology needed for full-scale industrial production is still lacking. There is also the economic factor. How can the production costs of target metabolites produced by plant cell cultures be reduced? The basic problem here is how to increase productivity, commonly expressed as the product yield per unit volume of culture medium (for example, mg/liter). An increase in this value means not only a reduction in the cost of the media used, but major savings in the cost of the plant and its equipment which can be miniaturized.

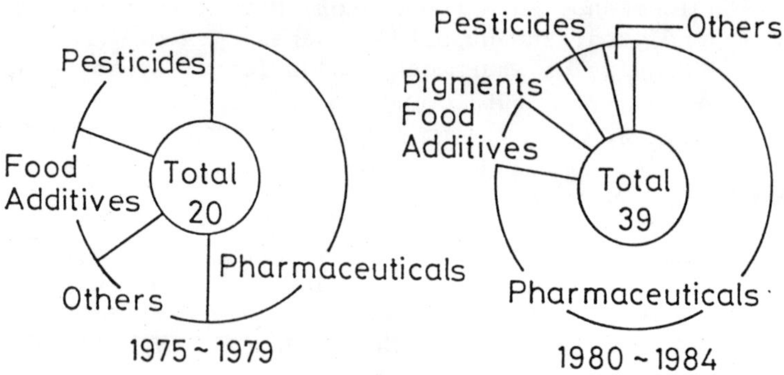

Figure 1. Japanese Patents Related to Secondary Metabolites Produced by Plant Tissue Culture

The basic ways to increase productivity are 1) the acquisition of high-producing strains by selection or some other method, 2) establishment of culture procedures and conditions that ensure the maximum productivity of the cell strain, and 3) development of a culture tank that is suitable for large-scale production. I will use our own experiences in developing an industrial process for the production of shikonin and in our research on the commercial production of berberine to show how to reduce costs and raise productivity.

MEDIA

The composition of the components of the medium and the phytohormones used generally are controlled in order to produce secondary metabolites by cell culture. To date, there have been various modifications of such established media as that of Linsmaier and Skoog (LS) (1) and that of Gamborg (B-5) (2), which originally were developed for cell growth. We, however, have aimed at developing a medium that can induce maximum productivity of secondary metabolites by a cell line because we believe that the medium used for cell growth and for the production of secondary metabolites must differ because the two are completely different phenomena.

An example is the shikonin production by our *L. erythrorhizon* cells. When these *Lithospermum* cells were cultured in liquid LS medium suitable for growth, they produced no shikonin; but, shikonin was produced in White's medium (3) which has a lower nutrient concentration than LS medium and which is not suitable for cell growth (4). The reason why shikonin was produced with White's medium is that it contains no NH_4^+. Also, shikonin could be produced in LS medium when NH_4^+ was omitted; but, the yield in this modified LS medium was less than that in White's medium. This shows that the concentrations of the other LS medium nutrients also are unsuitable for shikonin production.

In addition to shikonin production, there are many examples that show secondary metabolites can be produced by reducing the concentration of the main medium components; particularly, nitrate, ammonium, and phosphate: For instance, production of the spice compound, capsaicin, when the nitrogen concentration is reduced (5), production of ajmalicine and serpentine when the phosphate concentration is lowered (6), and production rosmarinic acid when both the nitrogen and phosphate concentrations are reduced. (7). As shown here, the direct use of one of the established medium used to promote cell growth usually is unfavorable for secondary metabolites production; therefore, a suitable production medium for each metabolite or plant species must be developed.

Zenk was one of the first to use the idea of a two-stage culture, in which a growth medium and a production medium are combined, for the culture of higher plants (8). He has succeeded in producing serpentine by using an established medium for cell growth and another, greatly modified, medium for production. In our research on shikonin, we developed the MG-5 growth medium from LS medium (9) and the M-9 production medium from White's medium (10) by optimizing the concentrations of the nutrients in each medium. We also optimized a growth and a production medium for berberine production by *C. japonica* cells, but their nutrient compositions were very similar, and there was little increase in berberine production even when the two media were combined. We believe that two-stage culture is the best way to produce secondary metabolites, but the common one-stage culture can be used when growth and production media have very similar effects on the productivity of a cell line, as shown with berberine production.

In addition to the medium's nutrients, phytohormones are important factors in the production of secondary metabolites. Up to now, auxins and cytokinins have been widely studied, but we have found only negative effects of gibberellin on the production of secondary metabolites reported in the literature. Our studies on the effect of gibberellin on *L. erythrorhizon* and *C. japonica* cell cultures, have shown that berberine production by *C. japonica* cells increases remarkably when gibberellin is added to the medium (11); whereas, shikonin production is inhibited (12). Further research on phytohormones is necessary because their effects differ with the plant species as shown in our experiences with gibberellin.

CELL DENSITY

The increase in the production of a secondary metabolite per unit volume of culture tank is a very important factor for commercial production. An increase in the cell density in the tank is the most direct method by which to increase production. The maximum cell density is reached when the tank is filled with cells; 30 to 120g dry weight (DW) of cells being packed in one liter of tank space. This is a rough value because the water content of the cells varies with the plant species and the culture conditions used.

In the usual culture of cells the maximum cell yield is about 15 g/liter when one of the established media with 3% sucrose is used. Therefore, the yield when a tank is filled with cells is 2 to 8 times that for the usual culture methods. This 'cell-filling' culture method may eventually be of commercial use. Nitto Electric Co. in Japan has applied for a patent for a procedure in which cells are inoculated in a culture vessel without medium then cultured by spraying them with a mist of medium (13). But if the cell-filling culture method is to have commercial value, many engineering problems must be solved; for example, how to prevent the destruction of the lower layers by the weight of the upper layers, how to provide an equal supply of medium and air to every cell, and how to harvest the cells. The cell-filling culture method, however, is well worth the challenge of making it commercially useful because with it cell density is at a maximum.

The maximum cell density of a culture in the commonly used aerated, agitation type jar is determined by the physical conditions under which the cell can be agitated without their destruction. Consider our *C. japonica* cells. When these cells were caused to sediment in the medium. about 100 gDW/liter of cells became packed because the ratio of open space to aggregated cells was about 50 and the water content of the cells was 80%. Thus, culture was impossible because no agitation could take place. But, when the cell density was 70 gDW/liter, successful culture was possible although it was not easy at this cell density. Therefore, we established the following criteria:

1. adequate agitation without cell destruction,
2. a sufficient supply of oxigen,
3. a supply of nutrients suitable for culture at a high cell density.

Agitation will be discussed later. When the cell density is raised, the oxygen supply must be increased. But increased aeration causes the suspended cells to bubble up, and many of them adhere to the upper part of the jar wall. To prevent this, the aeration rate must be kept low by mixing oxygen with the aeration gas. The mixing rate of oxigen affects the dissolved oxygen concentration in the medium; therefore, the growth rate of the cells and the cells' production of secondary metabolites can be controlled. For this reason a suitable mixing rate of oxygen for the cell species being cultured must be determined.

The nutrient supply is the most critical element for culturing cells at a high density. When the nutrient concentration increases in the inoculum to produce a high cell density, cell growth is controlled. In particular, the production of secondary metabolites is prevented when the nutrient concentrations become very high. An example is the *C. japonica* cells shown in figure 2. When the concentration of each component in the medium was increased in proportion to the increase in the size of the inoculum, the cell yield increased proportionally as well in the range between 2 and 4 gDW/l of the inoculum -- but, the berberine productivity, as expected, in spite of the increased cell yield.

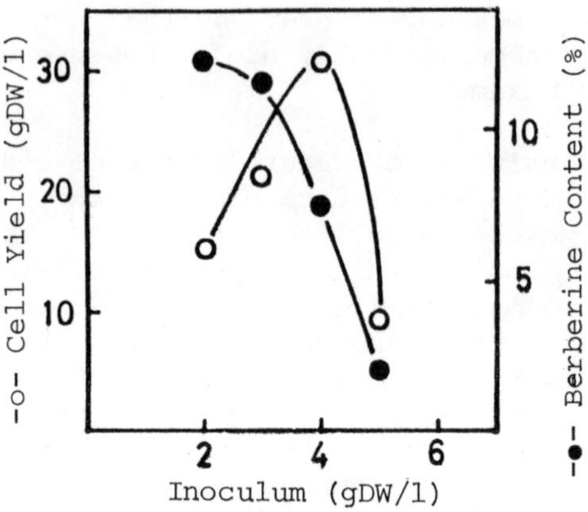

Additions of nutrients present in the basic medium were made in proportion to the amount of the inoculum.
(LS medium for 2g inoculum)

Figure 2. Culture at an Increased Cell Density——Berberine Production by *C. japonica* Cells

One solution to this problem is the 'fed-batch culture', in which culture of the increased inoculum is begun in a medium that has the usual nutrient concentration, after which these same nutrients are supplied to the culture in proportion to cell growth. Because this method does not raise the concentration of the individual nutrients, inhibition of the secondary metabolites is largely prevented. If the most suitable concentration for each nutrient in the medium can be established for a given time, tight control of the nutrient supply is possible. Using the fed-batch culture method, we successfully cultured *C. japonica* cells, getting a 70 gDW/liter cell yield that had about a 10% content of berberine.

PRODUCTION PROCEDURES

The culture procedure used with secondary metabolite-producing cell lines should be determined from the characteristics of the individual lines; for example, does the time course of cell growth sychronize with that of secondary metabolite production, and is the product released from the cell into the medium. In many cell cultures, cell growth first takes place, then secondary metabolites are produced. In some cases, the medium suitable for cell growth and that for production are completely different, as in the production of shikonin derivatives by *L. erythrorhizon* cells described earlier. We therefore used a two-stage culture method and succeeded in increasing the production of shikonin.

When secondary metabolites are released into the medium by the producing cells, a procedure that utilizes immobilized cells is possible. For example, when a solution of medium is circulated over immobilized cells of *Thalictrum minus*, the berberine produced accumulates in the medium because this species of cell releases most of the berberine it produces (figure 3) (14).

This type of procedure also is suitable for the production of secondary metabolites by biotransformation; but it cannot be used when metabolites accumulate only inside the producing cells. Therefore, the induction of a mutant cell line which releases its metabolites, or the use of some substance which stimulates the secretion of metabolites, would be necessary for the widespread use of immobilized cells.

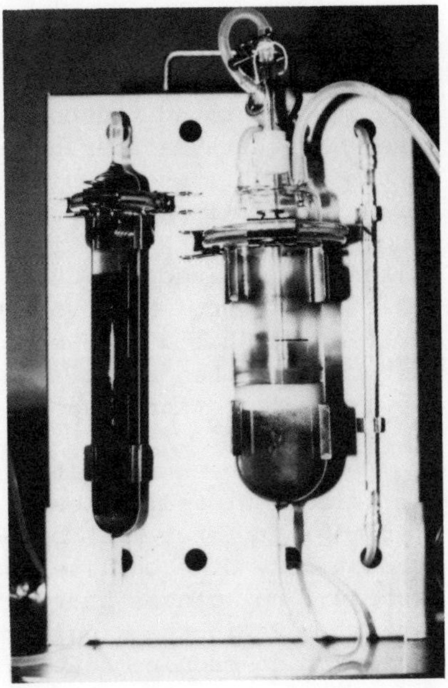

Figure 3. Apparatus for the Production with Immobilized Cells

SELECTION AND INDUCTION OF HIGH-PRODUCING CELL STRAINS

Use of a high-producing cell lines, which yield large amounts of the target compounds, usually have been obtained by selection methods.

The most widely used selection procedure is one in

which the parent cell line first is separated into small cell aggregates that are recultured. A high-producing cell line then is selected from the second generation cell lines derived from the parental aggregates. The efficiency of this selection method, however, is not very great, and the productivity of a selected cell line often is unsuitable because cells with different productivities exist in the original aggregate (15)(16).

The best method therefore would be to select from cell lines derived from a single cell. Two such methods actually are in operation; one that uses the protoplast (17), which is essentially a single cell, and one that uses cells obtained by filtration through a mesh (18), a procedure which mainly separates single cells.

We recently reported selection of high-producing cell lines for shikonin derivates from protoplast cultures of *Lithospermum erythrorhizon* cells (17). When we examined the shikonin production of our protoplast-derived cell lines, we found that productivities were distributed widely and equally around the corresponding value for the parent line, and that the average productivity was similar to that of the parent. When a high-producing cell line (4.2 g shikonin/g of inoculum/23 days) was the parent, a superior daughter line with 1.5 times higher productivity (6.5 g shikonin/g of inoculum/23 days) could be obtained by protoplast selection. Moreover, we found that this protoplast-derived line has a more stable productivity during subculture than the parent line which had been obtained by the usual selection method.

One drawback is that much labor is involved in individual selection; cell lines derived from an aggregate or from a protoplast first must be obtained by culture, then the productivity of each line obtained must be examined. This procedure now can be shortened by the selection of only superior cells in the single cell stage. It is done with a cell-sorting machine that was developed for the analysis and separation of small particles.

Figure 4 shows an analysis done with the cell sorter on *Coptis japonica* protoplast which produce berberine, an important medicinal compound. Each berberine-producing cell is recognizable by its fluorescence intensity, an indication of the content of berberine present in individual cells. As the machine originally was developed to separate small particles such as chromosomes and small cell organelles, improvements are necessary because the plant protoplast is a relatively large particle, and it is easily destroyed. Research groups at the University of Nebraska (19) and at Nottingham University (20) are leading the way in studies of the application of the cell sorter to plant cells. It is exciting that the selection of high-producing cell lines by the cell sorter method should be possible in the near future.

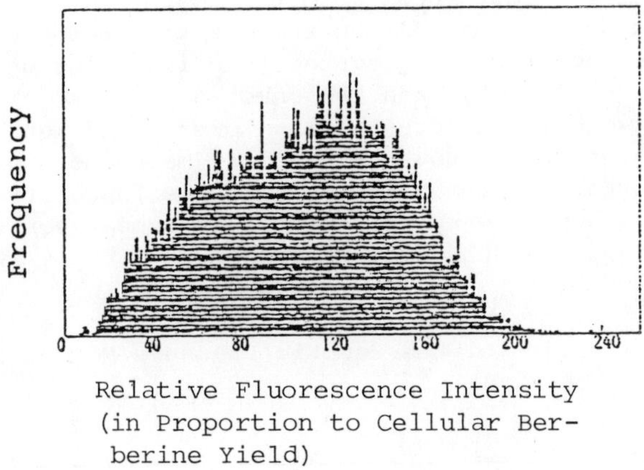

Figure 4. Cell Sorter Analysis of Cultured *C. japonica* Cells

As started earlier, selection is an effective method by which to obtain high-producing cell lines for secondary metabolite production; but, it will never be possible to obtain a cell line that is superior to the line derived from the best-producing cell in the parental cell aggregate. Therefore, new means must be introduced if we are to obtain truely superior cell lines. To this end, plant cell fusion, DNA recombination and mutagenesis are being investigated. I would like to discuss the first two techniques.

Plant cell fusion is a new aspect of biotechnology that has generated intensive interest, especially in the field of plant breeding. It is one of the most promising methods for the establishment of a truely high-producing cell line. Fusion takes place when a fusion-promoting agent, such as polyethylene glycol (PEG) is added (21), or when an electric pulse (as in the electrofusion method) is given (22). It now is possible to obtain fused mesophyll protoplasts, but their fusion is sometimes difficult when the source of the protoplasts are cultured cells. This is particularly true of using the established methods, we have not been able to obtain fused cells of *L. erythrorhizon* that produce shikonin derivatives or of *C.japonica* that produce berberine.

In our search for a fusion method for protoplasts of cultured cells, we have been able to greatly increase the fusion frequency by controlling the osmotic pressure. Protoplasts are induced to aggregate by adding a fusion-promoting agent such as PEG, then the osmotic pressure of the surrounding solution is drastically lowered and kept very low long enough to destroy the boundary membranes between the protoplasts. As shown in table 1, the fusion frequency was markedly increased by this method in comparison to the values obtained with the usual methods.

Table 1. Application of the Osmotic Pressure Method to Many Plant Species

Fusion Combination	Fusion Frequency (%)	
	PEG-method$^{1)}$	Osmotic Pressure-method$^{2)}$
Lithospermum X *Lithospermum*	1	52
Lithospermum X *Coptis*	1	68
Lithospermum X *Duboisia*	1	46
Coptis X *Duboisia*	5	62
Coptis X *Dianthus*	7	61
Coptis X *Brassica*$^{4)}$	13	35$^{3)}$
Brassica$^{4)}$ X *Brassica*$^{4)}$	23	36$^{3)}$
Lilium$^{4)}$ X *Lilium*$^{4)}$	4	68

Osmotic Pressure of Ca^{2+} Solution: 1) 16 atm, 2) 6 atm, 3) 9 atm, 4) Mesophyll Protoplasts

Because our method can be used with many plant species, and many fused cells can be obtained in one operation, the osmotic pressure method should have wide applications. We already have obtained *L. erythrorhizon* calli from fused protoplasts derived from cells that have high shikonin contents and those that show high growth. We now are examining the shikonin productivities of these calli.

In many instances, when establishing high-producing cell lines by fusion, two types of lines from the same plant species may be fused, as with *L. erythrorhizon*. The selection of fused protoplast-derived cell lines is very important, but further selection, after the regeneration of these lines or the production of mutants, is different with fused cell lines from the same parent plant species. Therefore, a selective method for isolating fused cells immediately after fusion must be established. The cell sorter promises an effective means of isolating such fused cells (19) (20); for this purpose, our research group also is investigating the use of this machine.

The creation of high-producing cell lines by DNA recombination technology is very difficult at present because secondary metabolites of plants are produced in biosynthetic pathways in which many enzyme systems are operating. Moreover, few of these enzymes have been identified and characterized. But, what we are able to do today with selection and fusion is far beyond the techniques available 10 to 20 years ago; therefore, we believe that as more basic research on enzymes and genes, as well as new innovations and advances in techniques are reported, the technology that today is difficult, or impossible, will become commonplace in both the basic and applied plant sciences.

CULTURE TANKS

In the mass culture of plant cells, which are sensitive to shear stress, the most important factor is how to supply sufficient oxygen with minimum impact on the metabolite-producing cells. Several studies have been done in which culture tanks suitable for growing plant cells were used. Tanaka (23) has cultured *Maclura tricupidata* cells in several types of culture tanks and studied which is the most suitable for plant cell growth at a high cell density. He found that an aeration-agitation type jar with a modified paddle impellar (fig. 5) was best when the impact to the cells was equal in each type of tank studied. This type of tank gives the highest volumetric oxygen transfer coefficient. In contrast, Wagner (24), who made a similar study on the production of anthraquinones by *Morinda citriforia* cells, reported that an airlift type tank gave the best production of these plant metabolites. This type tank has been studied by Fowler as well (25).

We cultured *L. erythrorhizon* cells in both an aeration-agitation type tank with a modified paddle impellar and an airlift type. In the former, cells grew well but the yield of shikonin was lower than in the usual flask culture. The greater the rotation speed of the impellar, the more the shikonin yield decreased because of injuries to the cells. In the latter type tank, the cells bubbled up, even at the usual cell density of 15 gDW/liter. Consequently, we could not culture the cells at a high cell density. Furthermore, as the cells adhered to the tank wall because of the bubbling up, this lowered the yield of shikonin.

On the basis of these results, we designed and made a rotary drum type tank which is suitable for the culture of *L. erythrorhizon* cells (fig. 6). In our rotary drum type tank, there is little cell injury because the culture mix-

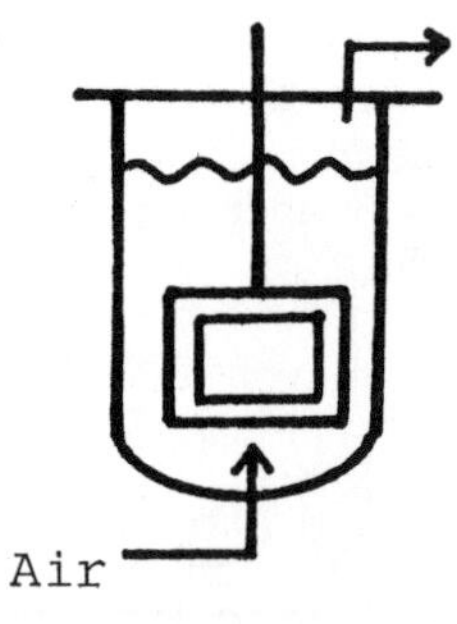

Figure 5. Modified Paddle Impellar Tank

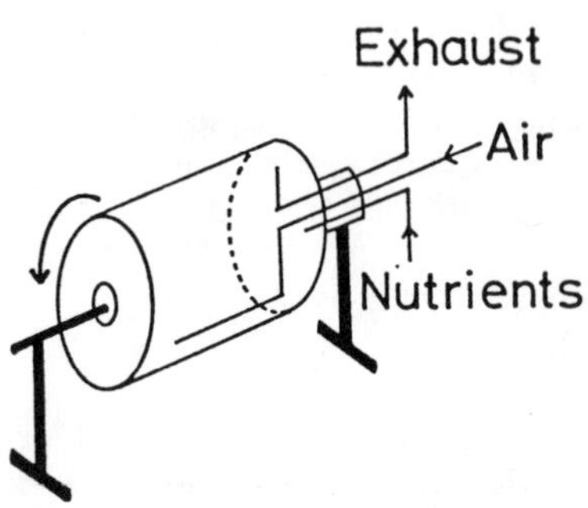

Figure 6. Rotary Drum Culture Tank

ture is moved gently by the revolution of the tank, itself, Moreover, the inside wall of the tank is always washed by the culture mixture; therefore, there is no adhesion of the cells to the wall. Culture at a high cell density also is possible with our rotary drum tank, and we have been able to scale up the volume to 1 m^3 with no decrease in the yield of shikonin (fig. 7).

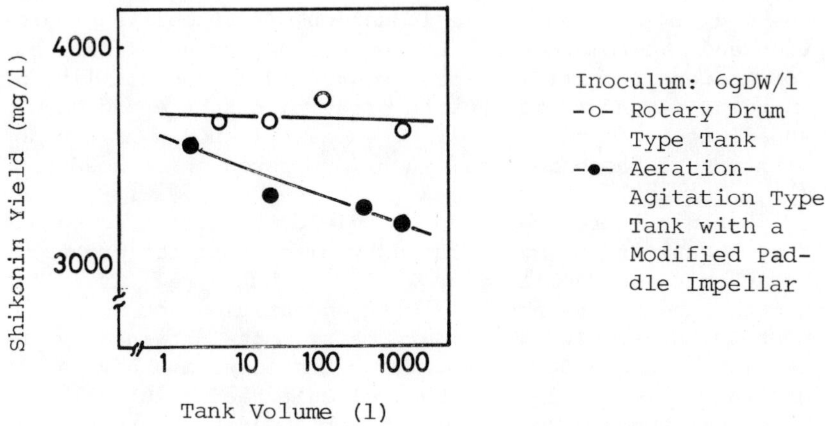

Figure 7. Scale up of Suspension Cultures of *Lithospermum erythrorhizon* Cells

CONCLUSION

I have shown how to improve the production of secondary metabolites by plant cell cultures using the results of our research with *L. erythrorhizon* and *C. japonica* as examples. When producing cell lines for the target compounds have been obtained, several effective procedures can be used to increase their productivities; optimization of the medium, culture at a high cell density and development of a culture procedure and tank that allow the lines maximum productivity of the desired metabolites. In addition, such procedures as the selection and induction of high-producing cell lines improve the productivity of the cell line, itself. Our experiences in establishing industrial methods for the production of secondary plant metabolites, we hope will be of value to other researchers for their studies.

These procedures, however, are applicable only when producing cell lines already have been obtained. Like the naturally produced vincristine and vinblastine that accumulate in *Catharanthus roseus* cells, there are other secondary metabolites that can not be produced by cultured cells, in spite cf much research and repeated trials. The acquisition of a producing cell line is, of course, the first necessity in all cases, but as yet we have no clear understanding of how to obtain metabolites that cannot be produced by cultured cells. Control of the key component in the medium may lead to the induction of metabolite production and, in some cases, mutagenesis may prove effective. Some recent reports (26) have suggested the possibility of inducing secondary metabolite production by the addition of an elicitor compound. So far, success is not guaranteeable with any of these methods; much more research is needed.

When cultured cells do not yield the target compounds, they sometimes can be produced by differentiated organs. Organ cultures generally grow slowly, but a relatively fast growth rate can be obtained with adventitious or somatic root cultures. In particular, hairy roots obtained by infection with *Agrobacterium rhizogenes* sometimes have a growth rate equivalent to that of cultured cells (27). When the target compounds can be produced by hairy root culture, this method should be considered.

DNA recombination technology also has the potential to provide a method for obtaining 'difficult' metabolites. To what extent the genetic information related to secondary metabolite production will have a place in future methods has yet to be determined -- it may, indeed, be the key to this problem. because of its newness, there has been little so far reported in the literature, but this situation should soon change as DNA recombination studies are being done in many parts of the world. I hope that at the next IAPTC Congress, four years from now, many results of research in this promising area will be reported.

Even when it is possible to obtain high-producing cell lines, high productivity often cannot be maintained for long periods. One solution to this problem is the selection of stable cell lines. In fact, we know already that some protoplast-derived cell lines have stable productivities over a long period. Cryopreservation also is being used to maintain cell stability, but cryopreservation procedures for high-producing cell lines have yet to be fully established. Therefore, subculturing should be kept at a minimum, and the culture conditions (e.g. the nutrient composition) and the culture temperature should be modified to maintain stable productivity.

Procedures for the separation and purification of desired metabolic compounds produced by cultured cells are basically the same as those used with field-grown plants. Separation-purification processes, however, are expensive, and a reduction in their costs is very important. To make these processes cheaper, the content of the target compound should be as high as possible. Also, the production of side products that are difficult to remove must be prevented. When too much of a side product(s) is produced, a cell line that gives only the desired compound must be the object of selection. When the target compound is released from the cultured cells into the medium, the separation-purification process is greatly simplified if the compound can be recovered easily; for example, by extraction.

Several major problems remain unsolved: Can a cell line that will release all of its desired secondary metabolite be selected or induced, and can additives that release metabolic end products without damage to the producing cells be found? These problems must be solved because the number and species of natural plants that produce useful metabolites are diminishing rapidly; therefor, plant tissue culture daily becomes more and more important. Many of us here are engaged in the search for speedier and more wide-range methods for plant cell and tissue culture that can be used to ensure the efficient production of useful plant metabolites. It is my belief that the end of that search will come in the foreseeable future. Thank you.

ACKNOWLEDGEMENTS

We thank Professor Arasuke Nishi, Toyama Medical and Pharmaceutical University and Professor Yasuyuki Yamada, Kyoto University for their most helpful advice and through discussions of our research. Thanks also are due to the members of our laboratory staff whose dedication and hard work made our successful research possible.

REFERENCES

(1) Linsmaier EF, Skoog F (1965). Physiol Plant 18: 100-127.
(2) Gamborg OL, Miller RA, Ojima K (1968). Exp Cell Res 50: 151-158.
(3) White PR (1954). "The Cultivation of Animal and Plant Cells." New York: The Ronald Press Co., p 403
(4) Fujita Y, Hara Y, Ogino T, Suga C (1981). Plant Cell Reports 1: 59-60.
(5) Lindsey K (1985). Planta 165: 126-133.
(6) Knobloch KH, Berlin J (1983). Plant Cell Tissue Organ Culture 2: 333-340.
(7) De-Eknamkul W, Ellis BE (1985). Plant Cell Reports 4: 46-49.
(8) Zenk MH, El-Shagi H, Arens H, Stöckigt J, Weiler EW, Deus B (1977). "Plant Tissue Culture and Its Biotechnological Application." Springer-Verlag, pp 27-43.
(9) Yamada Y, Fujita Y (1984). "Handbook of Plant Cell Culture." Vol.1. New York: Macmillan Publishing Co., pp 717-728.
(10) Fujita Y, Hara Y, Suga C, Morimoto T (1981). Plant Cell Reports 1: 61-63.
(11) Morimoto T, Hara Y, Yoshioka T, Fujita Y, Yamada Y (1986). "Proc. 6th Intl. Cong. Plant Tissue Culture." Vol.1.
(12) Yoshikawa N, Fukui H, Tabata M (1986). Phytochemistry 25: 621-622.
(13) Nitto Electric Co. (1984). Japanese Patent Application 1984-45873
(14) Nakagawa K, Konagi A, Fukui H, Tabata M (1984). Plant Cell Reports 3: 254.
(15) Zenk MH (1978). "Frontiers of Plant Tissue Culture 1978," ed. by Thorpe T.A., IAPTC, Cargary, 1978, p 1.
(16) Constabel F *et al.* (1981). Plant Cell Report 1: 3-5
(17) Fujita Y, Takahashi S, Yamada Y (1985). Agric. Biol. Chem. 49: 1755-1759.
(18) Ogino T, Tabata M (1978). Phytochemistry 17: 1907-1910.
(19) Afonso CL, Harkins KR, Thomas-Compton MA, Krejci AE, Galbraith DW (1985). BIO/TECHNOLOGY 3: 811-816
(20) Alexander RG, Cocking EC, Jackson PJ, Jett JH (1985). Protoplasma 128: 52-58.
(21) Kao KN, Michayluk MR (1974). Planta 115: 355-367.
(22) Zimmermann U, Scheurich P (1981). Planta 151: 26-32.
(23) Tanaka H (1981). Biotechnol. Bioeng 23: 1203.
(24) Wagner F, Vogelmann H (1978). "Plant Tissue Culture and

its Biotechnological Application," Heidelberg; Springer Verlag, p 254.
(25) Fowler NW (1981). Progr Ind Microbiol 17: 207.
(26) Eilert U, Kurz WGW, Constabel F (1985). J. Plant Physiol. 119: 65.
(27) Hamill JD, Parr AJ, Robins RJ, Rhodes MJC (1986). Plant Cell Reports 5: 111.

Plant Tissue and Cell Culture, pages 187–197

BIOSYNTHESIS OF CARDENOLIDES IN CELL CULTURES OF *DIGITALIS LANATA* – THE RESULT OF A NEW STRATEGY

Martin Luckner and Beate Diettrich

Section of Pharmacy, Martin-Luther-University Halle-Wittenberg, DDR-4050 Halle, Weinbergweg 15, GDR

1. ELICITING SECONDARY PRODUCT BIOSYNTHESIS IN RECALCITRANT PLANT CELL CULTURES

Biosynthesis and accumulation of most secondary products are limited to certain developmental stages of the producer organisms and to special types of cells built during the formation of certain tissues and organs (Luckner, 1984). This restriction is brought about by the phase-dependent biosynthesis of the enzymes and other proteins involved in secondary metabolism, i.e., is the result of differential gene expression (Luckner et al., 1977). Differential gene expression leading to the formation of specialized cells, tissues and organs proceeds in a programmed manner (Nover et al., 1982). The expression of secondary metabolism is integrated in these programs which

- coordinate secondary metabolism with other biochemical and morphological activities of the producer organism

- coordinate the different aspects of secondary product formation, e.g., synthesis of secondary metabolic enzymes, formation of the facilities necessary for channeling precursors and intermediates to the sites of these enzymes, formation of the means necessary for release and storage of secondary products, etc.

- account for the phases of susceptibility to signals which influence secondary metabolism, and

- originate the determination of secondary product formation by signals which do not act directly on secondary metabolism (Luckner, 1984).

As yet the principles integrating the expression of secondary metabolism in the differentiation programs of plant cells are nearly unknown. It exists therefore no rationale for the "artificial" triggering of the biosynthesis of secondary products in plant cell cultures. Some types of cell strains form secondary products already in unspecialized parenchyma-like cells. In refractory cases, however, secondary metabolism must be elicited "naturally" as part of the differentiation programs leading to the formation of those specialized cells, tissues and organs which in the mother plant comprise the formation of the desired secondary product. A useful strategy for the development of cell strains in recalcitrant systems thus comprehends:

(a) the elucidation of the program(s) which in the mother plant include the formation and accumulation of the secondary product

(b) selection of cell strains able to elicit this program in vitro to make sure that they really contain the genetic material necessary for the expression of the secondary pathways in question, and

(c) selection of daughter strains with simpler pattern of gene expression after treatment with mutagens or more easily by using the genetic heterogeneity characteristic for most cell strains.

In the following this strategy will be exemplified with the expression of cardenolide biosynthesis in cell cultures of D. lanata. Though this plant forms considerable amounts of cardenolide glycosides in vivo all attempts failed to establish cell strains in the classical way able to form cardenolides in vitro (for a summary s. Luckner and Diettrich, 1985).

2. THE MESOPHYLL CELLS: MAIN SITE OF CARDENOLIDE BIOSYNTHESIS AND ACCUMULATION IN *D. LANATA* PLANTS

The analysis of different tissues and organs of *D. lanata* plants showed that the leaves contained most of the cardenolides present (Vogel and Luckner, 1981; Weiler and Zenk, 1976). This agreed with the usage of leaves as raw material for the commercial production of cardenolides. Maximum cardenolide contents were found in the mesophyll (Vogel and Luckner, 1981). Excised leaves were shown to form cardenolides de novo (Tschesche and Lilienweiss, 1964). The same do mesophyll protoplasts obtained from leaves of *D. lanata* (unpublished results) and *D. purpurea* (Hagimori et al., 1984). These results indicated that mesophyll cells are most important for cardenolide biosynthesis and accumulation in *D. lanata* and that the formation of this cell type should be triggered in the cell cultures.

3. FORMATION OF MESOPHYLL-LIKE CELLS AND OF CARDENOLIDES AS PART OF THE DEVELOPMENTAL PROGRAM "SOMATIC EMBRYOGENESIS"

D. lanata cell strains, obtained from different explants, which formed morphologically unorganized cultures with parenchyma-like isodiametric cells, did not synthesize cardenolides de novo (for summary s. Luckner and Diettrich, 1985). The absence of cardenolide biosynthesis in these cultures was caused by the lack of the enzymes catalyzing cardenolide formation rather than by the deficiency of the respective precursors, since feeding of cholesterol, pregnenolone and progesterone, known to be intermediates in cardenolide biosynthesis, did not trigger the formation of cardenolides (Elze et al., 1974; Graves and Smith, 1967). In addition with morphologically unorganized *D. purpurea* cell cultures (in contrast to leaves of adult plants) the enzyme system forming pregnenolone from cholesterol was not measurable (Pilgrim, 1972).

Stable cardenolide biosynthesis was achieved, however, in cultures regenerating shoots or forming somatic embryos, whereas cultures building non-embryoid brown or green tissues or regenerating roots contained either no or only small amounts of cardenolides (Diettrich et al., 1986). In both, the shoots as well as in the somatic embryos, mesophyll-like cells were shown to exist. For studies on the regulation of cardenolide formation the early globular embryo structures, due to their simple morphological structure, were most suitable. Cardenolide formation in cultures of somatic embryos will therefore be discussed in the following in more detail.

A. The Development of Somatic Embryos

The formation of somatic embryos was investigated most thoroughly in cultures of the embryogenic D. lanata strain VII (Diettrich et al., 1986; Kuberski et al., 1984). In cultures of this strain large numbers of somatic embryos were formed from the parenchyma-like cells grown in media with a high auxin-cytokinin ratio. Embryo development was elicited by lowering this ratio. It included the following stages:

- formation of small clusters of meristematic cells within the wall of individual parenchyma-like cells present either in cell colonies or laying apart

- development of the clusters of meristematic cells to separate entities containing a core of small, polyedric, plasma-rich cells surrounded by a layer of larger vacuolated cells disjoining the cluster from nearby parenchyma-like cells

- development of more or less round, smooth embryoid globules (Ø about 0.5 - 1 mm) which at the beginning were white, UV-sensitive and unable to build chloroplasts (stage-I-globules), but later turned yellow and formed chloroplasts upon illumination (stage-II-globules), and

- formation of bipolar heart-shaped and torpedo-shaped embryo structures which developed to plantlets with roots and leaves.

Embryo formation and development were influenced by the carbon source of the medium (most suitable was maltose which was slowly degraded in <u>D. lanata</u> cell cultures) and by the nitrogen source (optimum was a ratio of NO_3^-/NH_4^+ of 5-10 : 1, i.e., a reduced level of NH_4^+) (Kuberski et al., 1984).

B. Expression of Cardenolide Biosynthesis during Embryo Development

Analysis of somatic embryos at different stages demonstrated an increase in the cardenolide level during development (Luckner and Diettrich, 1985, Scheibner et al., in press b):

- rapidly growing colonies of parenchyma-like cells and clusters of meristematic cells contained only very small cardenolide amounts (<0.001 μmol g^{-1} dry wt.)

- non-illuminated embryoid globules had cardenolide contents at least 10-times higher (ca. 0.01 μmol g^{-1} dry wt.)

- embryoid stage-II-globules illuminated occasionally with low doses of white light (which did not yet cause chlorophyll accumulation) contained cardenolide levels of ca. 0.1 μmol g^{-1} dry wt., and

- embryoid stage-II-globules and bipolar embryo structures illuminated with optimum light intensities and light periods accumulated > 1 μmol cardenolides g^{-1} dry wt. These structures contained mesophyll-like cells with normally developed chloroplasts (Diettrich et al., 1986) bearing considerable quantities of chlorophyll.

The increased cardenolide base level in non--illuminated stage-II-globules signalized the competence of these entities to form and accumulate larger cardenolide amounts. Illumination caused simultaneously the formation of chloroplasts and an increase in cardenolide accumulation. Without the prior development of competence during embryo development, however, cardenolide biosynthesis did not parallel the formation of chloroplasts. Cell strains unable to form shoots or somatic embryos were found to be free of considerable amounts of cardenolides even if turning green upon illumination and containing fully developed chloroplasts (Diettrich et al., 1986, Hagimori et al., 1982; Reinhard et al., 1975).

Embryoid globules treated with light of different wave lengths showed highest cardenolide accumulation upon radiation with blue light (460 - 510 nm) and near red light (630 - 680 nm) in contrast to low cardenolide accumulation after radiation with far red light (680 - 730 nm) (Scheibner et al., in press a). These results indicated that a blue light photoreceptor as well as phytochrome are involved in the regulation of cardenolide biosynthesis and accumulation. Embryoid globules radiated per day with the same amount of white light (equal dayly energy fluence) contained the highest cardenolide contents if the radiation period was $\leq$ 12 h d^{-1} (Scheibner et al., in press a) demonstrating that cardenolide biosynthesis and accumulation in the embryoid globules depended on the photoperiod.

The cardenolides built in the embryoid globules differed structurally from those found in adult plants. Stage-II-globules of D. lanata strain VII contained digitoxigenin derivatives (probably glucodigifucoside and odorobioside G) as main cardenolides (Hering et al., in press) though strain VII was derived from a plant with a high content of digoxigenin derivatives in the leaves of the rosetta. The predominance of digitoxin derivatives agreed, however, with results demonstrating that zygotic D. lanata seedlings in contrast to the adult plants also accumulated either no digoxigenin derivatives or only small amounts (for a discussion cf. Luckner and Diettrich, 1985).

4. SELECTION OF CELL STRAINS WITH SIMPLIFIED DEVELOPMENTAL PROGRAM

Though D. lanata strain VII was stable relative to embryogenesis as well as cardenolide and chloroplast formation for more than 10 years it was not homogeneous. Counting chromosome numbers indicated the existence of a relatively large amount of aneuploid cells (K Scheibner, unpublished results). In agreement regeneration of plants from the somatic embryos was successful only with a small fraction (<1 %) and the regenerated plants frequently were anomalous.

Thorough analysis showed that heterogeneity was also measurable relative to cardenolide accumulation and its connection to other features of embryo development. Analysis of large numbers of parenchyma-like cell colonies and embryoid globules discovered three groups of entities deviating from the bulk (unpublished results):

(a) embryoid globules with normal morphological structure, which formed chloroplasts, but did not accumulate measurable amounts of cardenolides

(b) embryoid globules which accumulated cardenolide contents > 1 μmol g^{-1} dry wt. already in the dark, i.e., without the development of chloroplasts (these globules resembled those of strain S 2 derived from the embryoid D. lanata cell strain S 1, Ohlsson et al., 1983), and

(c) colonies of parenchyma-like cells accumulating about 0.1 μmol cardenolides g^{-1} dry wt., when grown on the surface of an auxin-rich medium without formation of embryos and chloroplasts.

Some cell lines derived from these entities

were stable during several subcultivations. They may be the base for development of new strains with simplified pattern of gene expression.

5. CONCLUSIONS

The results described in section 4 demonstrate that cardenolide biosynthesis and accumulation is only loosely associated with the developmental program "somatic embryogenesis", forming a side branch rather than the backbone. Embryo formation proceeds "normally" (at least in the artificial environment of the laboratory) if cardenolides are absent, and chloroplast formation, as part of the developmental program, may be delated without suppression of cardenolide biosynthesis. This agreed with results obtained with other experimental systems which also showed that secondary metabolism is no essential part of the programs of cell specialization and development (Luckner, 1984; Luckner et al., 1977).

The formation of cardenolides in some lines of rapidly dividing, parenchyma-like cells reflects the fact that secondary metabolism may be expressed fully separated from the differentiation programs in which it is integrated normally. This was shown already with many other types of secondary products formed in cell cultures without morphogenetic activity, though in the mother plants their biosynthesis was part of the programs of organ development, and with individual reactions of cardenolide biosynthesis, like glucosylation (Diettrich et al., in press; Reinhard, 1974; Reinhard et al., 1975) and 12ß-hydroxylation (Heins, 1978; Reinhard, 1974, Reinhard et al., 1975) which were shown to occur also in D. lanata cell cultures without morphogenetic activity.

The usage of cell strains able to express the formation of desired secondary product as part of differentiation programs resulting in the formation of specialized cells and organs, and the subsequent selection of cell lines derived from these strains with more simple pattern of gene expression offers a new rationale for triggering the biosynthesis of secondary products in recalcitrant cell cultures, It may be useful in cases were the direct establishment of cell strains forming secondary products in rapidly dividing parenchyma-like cells could not be achieved. The reason for failures in the classical way of cell strain establishment directly from the explant may base on the fact that only a few types of cells (in most D. lanata explants for example only the xylem parenchyma cells, Diettrich et al., 1986) originate the primary callus. These cells may not be able to form the secondary products in question (the cardenolide content of the vascular bundles of D. lanata leaves is very low, Vogel and Luckner, 1981) and their reprogramming to rapidly dividing parenchyma-like cells of the classical suspension cultures may not include elicitation of the biosynthesis of secondary products. In these cases drastical reprogramming, as triggering of organ development, may demask the genes of secondary metabolism. During long-term cultivation of such strains genome alterations may occur (probably deleterious under the "natural" conditions within the plant body). They may change the regulation of the expression of certain parts of the genetic material and allow the independent elicitation of secondary metabolism. In this indirect way strains forming secondary products may be obtained whose pattern of gene expression is relatively simple. These strains may be useful for studying the principles regulating the formation and accumulation of secondary products and may be the base for the future biotechnological production of valuable compounds.

REFERENCES

Diettrich B, Aster U, Greidziak N, Roos W, Luckner M. Glucosylation of digitoxin and other cardenolides in cell cultures of Digitalis lanata. Biochem Physiol Pflanzen, in press.

Diettrich B, Steup C, Neumann D, Scheibner H, Reinbothe C, Luckner M (1986). Morphogenetic capacity of cell strains derived from filament, leaf and root explants of Digitalis lanata. J Plant Physiol 124:441-453.

Elze H, Pilgrim H, Teuscher E (1974). Die Biotransformation von Cholesterol-26-^{14}C durch Gewebekulturen von _Evonymus europaea_ und _Digitalis purpurea_. Pharmazie 29:727-728.
Graves JMH, Smith WK (1967). Transformation of pregnenolone and progesterone by cultured plant cells. Nature 214:1248-1249.
Hagimori M, Matsumoto T, Mikami Y (1984). Digitoxin biosynthesis in isolated mesophyll cells and cultured cells of _Digitalis_. Plant & Cell Physiol 25:947-953.
Hagimori M, Matsumoto T, Obi Y (1982). Studies on the production of _Digitalis_ cardenolides by plant tissue culture. Plant Physiol 69: 653-656.
Heins M (1978). Screening of Digitalis plants and cell cultures for hydroxylation capacity. In Alfermann AW, Reinhard E (eds): "Production of Natural Compounds by Cell Culture Methods," Proc. Intern. Symp. on plant cell culture, München, pp 39-47.
Hering F, Lehmann T, Luckner M. Glukodigifukosid und Odorobiosid G - Hauptglykoside von somatischen Embryonen und jungen In-vitro-Pflanzen von _Digitalis lanata_. Pharmazie, in press.
Kuberski C, Scheibner H, Steup C, Diettrich B, Luckner M (1984). Embryogenesis and cardenolide formation in tissue cultures of _Digitalis lanata_. Phytochemistry 23:1407-1412.
Luckner M (1984). "Secondary Metabolism in Microorganisms, Plants and Animals," Jena and Berlin: Fischer and Springer.
Luckner M, Diettrich B (1985). Formation of cardenolides in cell and organ cultures of _Digitalis lanata_. In Neumann KH, Barz W, Reinhard E (eds): "Primary and Secondary Metabolism of Plant Cell Cultures", Berlin: Springer, pp 154-163.
Luckner M, Nover L, Böhm H (1977). "Secondary Metabolism and Cell Differentiation", Berlin: Springer.
Nover L, Luckner M, Parthier B (eds) (1982). "Cell Differentiation, Molecular Basis and Problems", Jena and Berlin: Fischer and Springer.
Ohlsson AB, Björk L, Gatenbeck S (1983). Effect of light on cardenolide production by _Digitalis lanata_ tissue cultures. Phytochemistry 22: 2447-

2450.
Pilgrim H (1972). Cholesterol side-chain cleaving enzyme. Aktivität in Keimlingen und in vitro kultivierten Geweben von Digitalis purpurea. Phytochemistry 11:1725-1728.
Reinhard E (1974). Biotransformation by plant tissue cultures. In Street HE (ed): "Tissue Culture and Plant Science," London: Academic Press, pp 433-459.
Reinhard E, Boy M, Kaiser F (1975). Umwandlung von Digitalis-Glycosiden durch Zellsuspensionskulturen. Planta med Suppl pp 163-168.
Scheibner H, Björk L, Schulz U, Diettrich B, Luckner M. Influence of radiation on cardenolide accumulation in somatic embryos of Digitalis lanata. Phytochemistry, in press a.
Scheibner H, Diettrich B, Schulz U, Luckner M. Somatic embryos in Digitalis lanata cell cultures. Synchronization of development and cardenolide biosynthesis. Plant Cell Reports, in press b.
Tschesche R, Lilienweiss G (1964). Cardenolid-Biosynthese aus Pregnenolonglucosid. Z Naturforschg 19b:265-266.
Vogel E, Luckner M (1981). Distribution of cardenolides in Digitalis lanata. Planta medica 41:161-165.
Weiler EW, Zenk MH (1976). Radioimmunoassay for the determination of digoxin and related compounds in Digitalis lanata. Phytochemistry 15: 1537-1545.

Plant Tissue and Cell Culture, pages 199–212

REGULATION OF SECONDARY METABOLISM IN PLANT CELL CULTURES

Hans Becker

Institut für Pharmazeutische Biologie
der Universität Heidelberg
Im Neuenheimer Feld 364, D 69 Heidelberg, FRG

INTRODUCTION

Already more than 20 years ago, Carew and Staba (1965) pointed out that plant cell cultures represent an interesting biotechnical alternative to obtain secondary compounds. The arguments which they presented were essentially repeated in numerous review articles until today. In spite of the potential possibilities, only one product - shikonin - has been known so far to have been manufactured economically (Curtin, 1983). Although many secondary compounds have been found in larger quantities in plant cell cultures, these cell cultures are still competing with the whole plant. In producing microbial products, the biotechnologic production is the only possibility to obtain corresponding secondary products. Therefore, the scientific and the economical treatment of microorganisms for the production of secondary compounds has progressed very much. However, in spite of this, our knowledge about the regulation of the secondary metabolic pathway in microorganisms is still limited (Kleinkauf et al., 1985). In plant cell cultures, the situation is even worse. The present situation of our knowledge about the regulation of the metabolic products in cell cultures is shown in a few examples.

Influence of Medium Composition on Secondary Products

When reading a more recent review of the "Regulation of Secondary Metabolite Formation", one can see that most microbial fermentations are being carried out in a two-step culture (Kleinkauf et al., 1985). The first medium has been optimized for the growth of the cells, the second for the production of secondary compounds. Both media often differ in the production medium, the latter one having a lower content of nitrate or phosphate or of both. Further, it could be shown that a lower sugar content or a lesser-utilizable C-source can have a favorable influence on the secondary compound production. In relation to the influence of sugar, one relates to catabolite-repression (e.g. Martin et al., 1985). The two-step culture - being successful for microorganisms - in the meantime has found use in numerous plant cell cultures (e.g. Zenk et al., 1977; Knobloch and Berlin, 1980; Knobloch and Berlin, 1981; Sasse et al., 1982 a).

Sasse et al. (1982 b) use the term "induction medium" for the second medium in which the production of secondary compounds takes place. They were able to show that in the induction media - none of them contained phytohormones or inorganic phosphate the production of the secondary compounds in _Catharanthus roseus_, _Nicotiana tabacum_ and _Peganum harmala_ was strongly increased compared with the growth media. Parallel to this, a stronger enzymic activity of the phenylalamine ammonia-lyase and tryptophandecarboxylase had taken place which stand at the switching points from primary metabolism to secondary metabolism. When low levels of phosphate were added to the induction medium, the induction of phenylalanine ammonia-lyase and product accumulation in _Catharanthus_ and _Nicotiana_ cells was repressed. However, tryptophan decarboxylase of _Catharanthus_ was not repressed in the induction medium plus Pi while the product levels were decreased. Fujita et al. (1981) did not only look into the influence of macroelements on the production of shikoninderivatives in cell cultures of _Lithospermum erythrorhizon_ but also into the influence of microelements. They thus found that copper has almost no influence on the growth but that in a concentration which is 30 times stronger than the one in White's medium, the shikonin production could be greatly increased.

Since the growth of plant cells depends on the phytohormone content of the corresponding culture media, it does not come as a surprise that numerous research projects are carried out about the dependence of the secondary metabolism of the phytohormone content. There is no unequivocal correlation, though. An increase in the auxin level often leads to a dedifferentiation via a stimulation of the cell division and consequently diminishes the level of secondary metabolites. However, some examples show that high doses of growth factors can increase the content of secondary compounds, too (for reference, see Misawa, 1985).

In one case, a direct involvement of auxin in enzyme activation could be shown. 2,4-D activated in vivo UDP-glucose: scopoletin glucosyltransferase (SGTase) already present in tobacco cells (Hino et al., 1982). The activated SGTase was bound tightly with 2,4-D and molar ratio of bound 2,4 D to SGTase was about one. The specific activity of 2,4-D-bound SGTase was ten times that of 2,4-D free SGTase. No difference in the molecular weight between 2,4-D-bound and free SGTase was observed.

Any other factor of the medium may have a positive or a negative effect on secondary plant metabolism in tissue culture, but except for a few cases, we do not have any clue as to how these effects are mediated.

Key Enzymes for the Regulation of Secondary Products?

An essential precondition for the knowledge of the secondary metabolism is to know the corresponding enzymes which participate in the synthesis of the natural substances. Plant cell cultures are suited to demonstrate this knowledge. A newer review of the enzymatic of the secondary metabolism in plant cell cultures is given by Anderson et al. (1985).

From the above cited observation that secondary product formation was always positively correlated with the activity of the enzymes linking primary and secondary metabolism, Berlin et al. (1985) conclude that these enzymes are a bottleneck for optimal productivity. They started experiments to isolate the respective enzymes, to clone them in bacteria and then to transform the respective cells with Ti-plasmid vectors carrying the cloned genes. These experiments will help to understand the regulatory activity of these enzymes. Hopwood et al. (1985), hovewer, do not think the increased gene dosage using cloned DNA to overcome metabolic bottlenecks to be fruitful. The latter authors think that more tangible benefits of gene cloning in the context of yield improvement are likely to occur through the increased understanding of the normal genetic regulation of secondary metabolites biosynthesis that will follow analysis of the cloned biosynthetic genes.

Another argument that introducing cloned DNA will not lead to a higher production rate is that at least some enzymes (see below: Hrazdina and Wagner, 1985; Amann et al., 1986) are arranged in multienzyme complexes on membranes or integrated in vesicles.

The Enzymes involved in Flavonoid and Isoflavonoid Synthesis

The induction of enzymes catalyzing the synthesis of flavonoids and isoflavonoids is the best studied example of regulation of secondary metabolism in plant cell cultures. Extensive experiments were carried out with cell cultures of *Glycine max* which formed flavonoids spontaneously or light-induced, and with cell cultures of *Petroselinum hortense* which respond to irradiation with ultraviolet light by accumulating large amounts of flavone and flavonol glycosides (for review s. Hahlbrock, 1977; Hahlbrock et al., 1980).

In *Petroselinum hortense*, a selective induction of enzymes related to flavonoid biosyntheses follows after irradiation. The enzymes can be divided in two groups:

1. Enzymes of the general flavonoid metabolism: phenylalanine ammoniumlyase = PAL, cinnamate 4-hydroxylase =C4H, and 4-hydrocinnamate:CoA ligase = 4CL.
2. 12 special enzymes of the flavonoid metabolism.

All three enzyme activities of group I increased drastically after irradiation. After a lag period of 2 - 2.5 h, they increased their activity, then reached maximal values after about 17, 22 and 23 h. Apparent half-lives of the enzyme activities were calculated to be 10, 15 and 17 h. The light-induced activity of enzymes of group 2 had a slightly longer lag period; the peaks in their activities occurred several hours later, and the apparent half-lives, as calculated from the rates of decline subsequent to the peaks, were considerably longer.

In *Petroselinum hortense* cell cultures the amount of flavonoids accumulated with time was in coincidence with expected accumulation, calculated from PAL activity but not of any other enzyme activity involved in the flavonoid pathway. Thus, it can be suggested that PAL is the rate-limiting enzyme for flavonoid glycoside biosynthesis under the conditions used. The author, however, points to the fact that this result is only valid for the chosen system. Although the results of Berlin et al. (1985) suggest the PAL-activity to be a decisive factor in the regulation of the secondary compound synthesis of the phenylpropanoid pathways, there is also some research which consider the substrate supply (phenylalanine) rather than enzymic (PAL) activity as limiting factor in controlling phenylpropanoid accumulation (Margna, 1977).

In a recent study, Hrazdina and Wagner (1985) could show that the coordinated induction of phenylpropanoid and flavonoid pathway enzymes in buckwheat hypocotyls by light resembles that of enzyme group 1 in parsley cell suspension cultures. Evidence was obtained for a multienzyme complex. The authors suggest that the synthesis of phenylpropanoids and flavonoids takes place on a complex consisting of membrane-associated, consecutively assembled enzymes. Such an assembly renders the flavonoid synthesis more effective and makes a coordinated regulation of the enzyme easier in the present case.

In cell cultures of *Glycine max* the synthesis of pterocarpanoids phytoalexins glyceollins is strongly increased by fungal elicitors (for review s. Grisebach, 1985). Large increases and subsequent decreases of a number of enzymes involved in glyceollin biosynthesis occur after infection or elicitor treatment. The increase of PAL and chalcone synthetase activity coinced with high rates of enzyme synthesis and with corresponding changes in mRNA activity for these enzymes. These results suggest that glyceollin synthesis is triggered by enhancement of the transcriptional rate of certain enzymes. Inositoltriphospate could play a role as second messenger in the described course of action elicitor → gylceollin.

When the soybean cells were treated with three different fungal products, the glucan elicitor from *Phytophtora megasperma*, xanthan, an extracellular polysaccharide from *Xanthomonas campestris*, and endopolygalactaronase from *Aspergillus niger*, a similar degree of induction of mRNA activities and of the catalytic activities of PAL and chalcone synthetase were observed (Ebel et al., 1984). The comparable enzymic activities, however, lead to differing amounts of phytoalexins. With the glucan elicitor only the induction of a large amount glyceollin was obtained, endopolygalacturonase induced only low amounts, xanthan did not enhance glyceollin accumulation.

The authors believe that possibly other than up till now examined enzymic systems, such as maybe prenyltransferase, may be involved in the regulation of the phytoalexin synthesis. Most likely, there is no enzyme of the secondary metabolism which is as well examined as PAL. In spite of the extensive research there is still considerably uncertainty concerning the regulation of this enzyme which is so important for many metabolites (Jones, 1984).

Compartimentation in the Regulatory Control

As shown above (Hrazdina and Wagner, 1985), compartimentation may play an important role in secondary product formation. The compartimentation of metabolism intermediate and end products is made more complicated by the fact that for the transport in the corresponding compartments specific carriers must be assumed. Deus-Neumann and Zenk (1986) were able to prove that cell cultures of *Fumaria capreolata* L. energy-dependingly take up (S)-reticuline in vesicles; (R)-reticuline and species-strange alkaloids were not incorporated. These results contradict previously proposed passive ion-trap mechanism for alkaloid accumulation in vacuoles.

Amann et al. (1986) recently were able to prove a localization in vesicles of two out of eight enzymes involved in the biosynthesis of the isoquinoline alkaloid berberine in cell cultures of *Berberis wilsoniae*. The authors discuss that compartimentation of branch point enzymes may be a subtle type of regulation of alkaloid synthesis. It is remarkable that the two respective enzymes do not work in tandem but rather the product of the first vesicular enzyme has to leave the vesicle in order to be modified in the cytoplasm and then serves as substrates for the second vesicular enzyme.

Degradation of Secondary Products

In a review article about regulatory control mechanisms in alkaloid biosynthesis, Floss et al. (1974) made the following statement: Degradation of alkaloids is quite a common phenomenon and certainly in many cases a major factor determining the alkaloid content of a plant. If, as the subsequent discussion will illustrate, our knowledge of the regulation of alkaloid synthesis is scanty, then our knowledge of the control of alkaloid breakdown is virtually nonexistent. Twelve years later, this statement is still fully valid, not only for alkaloids but also for most of the secondary compounds.

Wink and Hartmann (1982) were able to show that quinolizidine alkaloids in cells and medium of *Lupinus polyphyllus* cell suspensions culture as well as in the plant of *Lupinus albus* (Wink and Witte, 1984) are subjected to a fast turnover. The measured secondary metabolism amount is - according to them - always only a momentary recording and is obviously regulated by light.

Inhibitors of Synthesis and Plant Bioregulators

Substantial information about the biosynthesis and regulation of the sterol formation was gained by the use of potential inhibitors of sterol biosynthesis by the team of Benveniste in Strasbourg (e.g. Schmitt et al., 1982). When an enzyme belonging to a biosynthetic pathway was blocked, the compound which accumulated was generally either the substrate of the blocked enzyme or the product of the more or less incomplete action on this substrate of the enzymes situated downstream of the blocked enzyme in biosynthetic sequence. The methology, besides elucidating the biosynthetic pathway and regulation can be used to discover and produce new sterols of pharmacological interest. The authors give as example the gaining of 5 g of (24 R)-24-ethyl-5-d-cholesta-8,14-dien-3ß-ol from one kg (dry wt.) of bramble cells by adding 15-azasterol.

Some plant bioregulators (PBR) excert their effect on plant growth by interfering with gibberellin biosynthesis. 2-chloroethyltrimethyl-ammonium chloride and Amo-1618 are known to inhibit the cyclisation of geranyl-geranyl-pyrophosphate (Graebe and Ropers, 1978). They also show activity on triterpenoid metabolism (Douglas and Paleg, 1974) and carotenoid metabolism (Knypl, 1969). Yokoyama and coworkers (Coggins et al., 1969; Yokoyama, 1984) synthesized a larger number of new compounds and tested them for structure activity relationship on the synthesis of carotenoids and polyisoprenoids in different experimental systems.

Recently, the mechanism of the regulation of the carotinoid biosynthesis by PBR was studied (Benedict et al, 1985). Treatment of lemon pieces with 500 ppm 2-(4-methylphenoxy)triethylamine (MPTA) resulted in the synthesis of 8.4 µg phytofluene and 26.6 µg lycopene / g fresh wt.$^{-1}$ / 72 h^{-1}. The untreated lemons did not accumulate any detectable lycopene during this period. Treatment of lemon pieces prior to MPTA application with amanitin, cordycepin, anisomycin or cycloheximide significantly reduced the MPTA induction of the carotenes. These results support the conclusion that the induction of phytofluene and lycopene by MPTA requires gene expression, polyadenylation of the gene transcript and translation of the poly A^+ RNA on 80 S ribosomes. MPTA may regulate the synthesis of isoprenoids by controlling the expression of genes coding for the enzymes in the carotene pathway.

Using PBR, Staba and Nygaard (1983) found a stimulation of polyisoprenoids in tissue culture of Guayule. A monoterpenoid (iridoid) moiety is involved together with tryptophane in the biosynthesis of indole alkaloids in *Catharanthus roseus*. Lee et al. (1981) demonstrated that alkaloid synthesis in suspension cultures of *C. roseus* can be enhanced by PBR. Since the valepotriate skeleton is also of iridoid nature we found it promising to test PBR for their effect on biosynthesis of valepotriates in cell cultures. For these experiments we selected dimethylmorpholinium-bromide (DMB), dimethylpiperidiniumchloride (DPC, Pix) (Zeeh et al., 1974) as well as 2-(3,4-dichloro-phenoxi)triethylamine (DCPTA) and 2-(3,5-diisopropylphenoxi)-triethylamine (DIPTA).

Studies were conducted with suspension cultures of *Fedia cornucopiae* and *Valeriana wallichii*. For both cell systems the cultivation conditions, growth and production parameters were known. Application of the above mentioned compounds did not alter growth characteristics. When employed in concentrations of 0.01 to 0.04 m mol during the early exponential growth, all PBR increased the production of valepotriates substantially (see tables 1 and 2).

TABLE 1: Effect of equimolar concentration of bioregulator DMB (4ppm) and DPC (3ppm) on total valepotriate yield of V. wallichii and F. cornucopiae. Bioregulators were added to the cell cultures on day 5 and 8 (V. wall.) or 3, 5, and 8 (F. corn.) of the growth cycle. Significant increase in production, absolute and in % of controls (*, confidence interval 95 %).

	valepotriate yield % dry weight	increase in production absolute	%
V. wallichii			
controls	1.61	-	-
4-DMB-5	1.37	-0.24	-15.0
4-DMB-8	1.86	0.26	15.3*
3-DPC-5	1.98	0.32	23.0*
3-DPC-8	1.93	0.32	19.0
F. cornucopiae			
controls	1.50	-	-
4-DMB-3	1.68	0.18	11.4
4-DMB-5	1.74	0.24	15.9*
4-DMB-8	2.10	0.60	40.0*
3-DPC-3	1.75	0.24	16.0*
3-DPC-5	1.84	0.34	22.5*
3-DPC-8	2.26	0.76	50.3*

TABLE 2: Effect of bioregulators DCPTA and DIPTA on total valepotriate production of F. cornucopiae, added on day 8 of the growth cycle. Significant increase in production (*, confidence interval 95 %).

	valepotriate yield % dry weight	increase in production absolute	%
controls	1.38	-	-
3-DCPTA-8	2.35	0.97	70.1*
6-DCPTA-8	1.94	0.56	40.7*
9-DCPTA-8	1.88	0.50	36.2
12-DCPTA-8	0.81	-0.57	-41.6
6.4-DIPTA-8	1.90	0.52	37.7*

Two-Phase Culture

The amount of secondary compounds may be regulated through feedback inhibition or through product-repression. Lack of adequate accumulation in cell cultures could be due to the fact that the corresponding compounds are being synthesized but since the correlating accumulation sites are lacking they remain at the site of the synthesis and inhibit their further synthesis. In some tests with an artificial adsorbens it could be shown that through interfering the balance of the total yield in secondary compounds could be for some parts greatly increased. Knoop and Beiderbeck (1983) isolated phenolic compounds from activated charcoal added to cell cultures of *Matricaria chamomilla* L. The isolated substance, mostly coniferyl aldehyde was 60 times that in controls without charcoal. The authors discuss a balancing

cells ⇄ medium ⇄ activated charcoal.

This hypothesis is being supported by the accumulation of increasing production amounts with increasing amounts of adsorbens. Best suited for the accumulation of lipophilic active ingredients either liquid lipophilic phases - such as defined triglycerides - or solid phases - such as octadecyl silan modified silica gel - (Bisson et al., 1983; Becker and Herold, 1983; Becker et al., 1984). With the latter system we were able to increase the content of valepotriates by the two- till fourfold in cell cultures of *Valeriana wallichii* without influencing the growth.

In the course of one passage (on day 6) the amount of valepotriates bound to the solid phase could be up to the 19fold that present in the cells (Becker and Herold, unpublished results). The mentioned examples allow the conclusion that by changing the balance the regulation of the biosynthesis may be influenced; however, so far studies of the changes in the enzymic activities of such systems are not available.

CONCLUSION

In spite of many activities for a biotechnological production of secondary compounds in plant cell cultures, little is known about the regulation of the secondary substance synthesis. Precondition for any progress in this field must be the knowledge of all enzymes which are involved in the synthesis of the product and also of their compartimentation in the cell. Inhibitors of biosynthesis and plant bioregulators contribute to a elucidation of the biosynthetic pathway and, in some cases, may also lead the way to an increase of the secondary metabolism. In producing cultures, the amount of product may be significantly increased by the use of induction media. An interference with the balance of the product accumulation through an adsorbens has in some cases led to an extensive increase in product.

REFERENCES

Amann G, Wanner G, Zenk MH (1986). Intracellular compartmentation of two enzymes of berberine biosynthesis in plant cell cultures. Planta 167:310-320.

Andersen LA, Phillipson JD, Roberts MD (1985). Biosynthesis of secondary products by cell cultures of higher plants. In Fiechter A (ed.) "Advances in Biochemical Engineering/Biotechnology. 31 Plant Cell Culture", Berlin, Tokyo: Springer, pp 1-36.

Becker H, Herold S (1983). RP-8 als Hilfsphase zur Akkumulation von Valepotriaten aus Zellsuspensionskulturen von Valeriana wallichii. Planta medica 49:191-192.

Becker H, Reichling J, Bisson W, Herold S (1984). Two phase culture - a new method to yield lipophilic secondary products from plant suspension cultures. In Third European Congress on Biotechnology, Vol. I, Weinheim: VCH, pp 209-213.

Benedikt CR, Rosenfield CL, Mahan JR, Madhavan S, Yokoyama H (1985). The chemical regulation of carotenoid biosynthesis in citrus. Plant Science 41: 169-173.

Berlin J, Beier H, Fecker L, Forche E, Noé W, Sasse F, Schiel O, Wray V (1985). Conventional and new approaches to increase the alkaloid production of plant cell cultures. In Neuman et al. (eds): "Primary and Secondary Metabolism of Plant Cell Cultures," Berlin: Springer, pp 272-280.

Bisson W, Beiderbeck R, Reichling J (1983). Die Produktion ätherischer Öle durch Zellsuspensionen der Kamille in einem Zweiphasensystem. Planta medica 47:164-168.

Carew DP, Staba EJ (1965). Plant tissue culture: Its fundamentals, application and relationship to medicinal plant studies. Lloydia 28:1-26.

Coggins C, Henning GL, Yokoyama H (1969). Lycopene

accumulation induced by 2-(4-chlorophenylthio)-triethylamine hydrochloride. Science 168:1589-1590.

Curtin ME (1983). Harvesting profitable products from plant tissue culture. Bio/Technology 649-657.

Deus-Neumann B, Zenk MH (1986). Accumulation of alkaloids in plant vacuoles does not involve an ion-trap mechanism. Planta 167:44-53.

Douglas TJ, Paleg LG (1974). Plant growth retardants as inhibitors of sterol biosynthesis in tobacco seedlings. Plant Physiol 54:238-245.

Ebel J, Schmidt WE, Loyal R (1984). Phytoalexin synthesis in soybean cells: Elicitor induction of phenylalanine ammonia-lyase and chalcone synthetase mRNA's and correlation with phytoalexin accumulation. Arch Biochem Biophys 232:240-248.

Floss HG, Robbers JE, Heinstein PF (1974). Regulatory control mechanisms in alkaloid biosynthesis. In Recent Advances in Phytochemistry 8, pp 141-178.

Fujita Y, Hara Y, Suga C, Morimoto T (1981). Production of shikonin derivatives by cell suspension cultures of Lithospermum erythrorhizon. Plant Cell Reports 1:61-63.

Graebe JE, Ropers HP (1978): Gibberellins. In Letham DS, Goodwin PB, Higgins TJV (eds): Phytohormones and Related Compounds: A Comprehensive Treatise, Vol.I", Amsterdam: Elsevier, pp. 107-203.

Grisebach H (1985). Induction and regulation of phytoalexin synthesis in soybean. In "Regulation of Secondary Metabolite Formation", Workshop Conferences Hoechst, Vol. 16. Weinheim, Dearfield Beach: VCH, pp 355-367.

Hahlbrock K (1977). Regulatory aspects of phenylpropanoid biosynthesis in cell cultures. In Barz W, Reinhard E, Zenk MH (eds): "Plant Tissue Culture and its Biotechnical Application," Berlin: Springer, pp 95-111.

Hahlbrock K, Schröder J, Vieregge J (1980). Enzyme regulation in parsley and soybean cell cultures. In Fiechter A (ed.): "Advances in Biochemical Engineering. 18 Plant Cell Culture II," Berlin: Springer, pp 39-60.

Hino, F, Okazaki M, Miura Y (1982). Effect of 2,4-Dichlorophenoxyacetic acid on glucosylation of scopoletin to scopolin in tobacco tissue culture. Plant Physiol 69:810-812.

Hopwood DA,Malpartida F, Chater KF (1985). Gene cloning to analyse the organization and expression of antibiotic biosynthesis genes in streptomyces. In "Regulation of Secondary Metabolite Formation", Workshop Conferences Hoechst, Vol. 16. Weinheim, Dearfield Beach: VCH, pp 23-33.

Hrazdina G, Wagner GJ (1985). Metabolic pathways as enzyme complexes: Evidence for the synthesis of phenylpropanoids on membrane associated enzyme complexes. Arch Biochem Biophys 226:285-291.

Jones DH (1984). Phenylalanine ammonia-lyase: Regulation of its induction and its role in plant development. Phytochemistry 23:1349-1359.

Kleinkauf H, Döhren v H, Dornauer H, Nesemann G (eds) (1985): "Regulation of Secondary Metabolite Formation", Workshop Conferences Hoechst, Vol. 16. Weinheim, Dearfield Beach: VCH.

Knobloch KH, Berlin J (1980). Influence of medium composition on the formation of secondary compounds in cell suspension culture of _Catharanthus roseus (L) G. Don._ Z Naturforsch 35c:551-556.

Knobloch KH, Berlin J (1981). Phosphate mediated regulation of cinnamoyl putrescine biosynthesis in cell suspensions of _Nicotiana tabacum._ Planta medica 42:167-172.

Knoop B, Beiderbeck R (1983). Adsorbenskultur - ein Weg zur Steigerung der Sekundärstoffproduktion in pflanzlichen Suspensionskulturen. Z Naturforsch 38c:484-486.

Knypl JS (1969). Accumulation of lycopene in detached cotyledones of pumpkin treated with 2-chlorethyltrimylammonium chloride. Naturwissenschaften 56: 572-573.

Lee SL, Cheng KD, Scott AI (1981). Effect of bioregulators on indole alkaloid biosynthesis in _Catharanthus roseus_ cell culture. Phytochemistry 20:1841-1843.

Margna U (1977). Control at the level of substrate supply - an alternative in the regulation of phenylpropanoid accumulation in plant cells. Phytochemistry 16:419-426.

Martin JF et al (1985) Enzymes involved in beta-lactam biosynthesis controlled by carbon and nitrogen regulation. In Kleinkauf H et al (eds) "Regulation of Secondary Metabolite Formation" Workshop Conferences Hoechst, Vol. 16. Weinheim, Dearfield Beach, VCH, pp. 41-75.

Misawa M (1985). Production of useful plant metabolites. In Fiechter A (ed.): "Advances in Biochemical Engineering/Biotechnology. 31 Plant Cell

Culture," Berlin, Tokyo: Springer, pp 59-88

Neumann D, Krauss G, Hicke M, Gröger D (1983). Indole alkaloid formation and storage in cell suspension cultures of Catharanthus roseus. Planta medica 48:20-23.

Sasse F, Heckenberg U, Berlin J (1982 a). Accumulation of ß-carbolin alkaloids and serotonin by cell cultures of Peganum harmala. II. Interrelationship between accumulation of serotonin and activities of related enzymes. Z Pflanzenphysiol 105:315-322.

Sasse F, Knobloch KH, Berlin J (1982 b). Induction of secondary metabolism in cell suspension cultures of Catharanthus roseus, Nicotiana tabacum and Peganum harmala. In Fujiwara A (ed.): "Plant Tissue Culture 1982, Proc 5th Intl Cong Plant Tissue and Cell Culture," Tokyo: Maruzen, pp 343-344.

Schmitt P, Rahier A, Benveniste P (1982). Inhibition of sterol biosynthesis in suspension cultures of bramble cells. Physiol Veg 20:559-571.

Staba EJ, Nygaard BG (1983). In vitro culture of guayule. Z Pflanzenphysiol 109:371-378.

Wink M, Hartmann T (1982). Physiological and biochemical aspects of quinolizidine alkaloid formation in cell suspension cultures. In Fujiwara A (ed.): "Plant Tissue Culture 1982, Proc 5th Intl Cong Plant Tissue and Cell Culture," Tokyo: Maruzen, pp. 333-334.

Wink M, Witte L (1984). Turnover and transport of quinolizidine alkaloids. Diurnal fluctuations of lupanine in the phloem sap, leaves and fruits of Lupinus albus L. Planta 161:519-524.

Yokoyama H, Hsu WH, Hayman E, Poling S (1984). Bioregulation of plant constituents. In Timmermann BN, Steelink C, Loewus FA (eds): "Phytochemical Adaptations to Stress. Recent Advances in Phytochemistry, Vol 18," New York: Plenum Press pp 231-250.

Zeeh B, König KH, Jung J (1974). Development of new plant growth regulators with biological activity related to CCC. Kemia-Kemi 1:621-623.

Zenk MH, El-Shagi H, Arens H, Stöckigt J, Weiler EW, Deus B (1977). Formation of the indole alkaloids serpentine and ajmalicine in cell cultures of Catharanthus roseus. In Barz W, Reinhard E, Zenk MH (eds): "Plant Tissue Culture and its Biotechnical Application," Berlin: Springer, pp 27-43.

Plant Tissue and Cell Culture, pages 213–219

ALKALOID ACCUMULATION IN PLANT CELL CULTURES UPON TREATMENT WITH ELICITORS

Udo Eilert, Wolfgang G.W. Kurz and Friedrich Constabel

Plant Biotechnology Institute, National Research Council of Canada, Saskatoon, S7N OW9

NRCC No. 26102

INTRODUCTION

The development of plant tissue culture as technology for the production of phytochemicals of industrial potential has made 3 substantial advances over the last 10 years: (1) the introduction of production media (Zenk et al., 1977) increased levels of accumulation and widened the spectrum of phytochemicals found; (2) the selection of high-yielding strains due to cell variability demonstrated the feasibility of industrial cell fermentation (Tabata et al., 1978); (3) the employment of elicitors promises to reduce cell fermentation costs and render the process viable (Wolters and Eilert, 1983). Remarkably, advances were made often without proper understanding of the underlying cell biology. This is particularly true for elicitation, a chain of events leading from exposure of cells to elicitor preparations to the induction of product accumulation. Still, considerable insights into the response of cells to elicitation have been obtained (Hahlbrock et al., 1982; Darwill and Albersheim, 1984; Mieth et al., 1986). In fact, the response of cultured cells to elicitation with accumulation of phytochemicals can clearly be accounted for (for reviews see DiCosmo and Towers, 1985; Heinstein, 1985; Eilert, 1987), and will be demonstrated using 2 examples, opium-poppy (Eilert et al., 1985) and periwinkle cell cultures (Eilert et al., 1987).

Opium-poppy (Papaver somniferum)

Material employed was a 2-week-old subculture of a 9-year-old strain of Papaver somniferum L. cv. Marianne, opium poppy, maintained in medium after Gamborg et al. (1968) with 1 mg/l 2,4-D and 1 g/l casein hydrolyzate (1-B5C) with regular transfers once every week. For experiments here, 75 ml fresh 1-B5C medium was inoculated with 25 ml cell suspension. Alkaloid production media were composed as defined for periwinkle by Zenk et al. (1977).

The culture consisted of 2 subpopulations, a small population of big cells which measured up to 1 mm in length, and a bigger population of small cells, 30-60 μm in diameter. While mitoses were seen with the bigger cells, division of the smaller ones clearly provided for maintenance of stock. So far, the subpopulations have not been seen to differently respond to elicitation.

Exposure of poppy cells to elicitor preparations, arachidonic acid, chitosan, nigeran as well as homogenates of cultures of phytopathogenic fungi, resulted in browning of the cultures within a few hours (Eilert et al., 1985). The browning deepened over time. This reaction is thought to be the result of oxidation and polymerization of phenolic compounds. Chromatography of extracts of cells exposed to elicitor preparation for 12 hours and longer and treatment with Dragendorff reagent showed the occurrence of one single alkaloid. Analysis revealed it to be sanguinarine, an antimicrobial compound of some industrial importance. Morphinan alkaloids were not found. The observation would indicate elicitation to activate one single biosynthetic pathway in poppy cell cultures, i.e. the pathway leading to benzophenanthridine alkaloids. So far, poppy cell cultures have not been fractionated according to size in order to determine the degree of participation of subpopulations in sanguinarine accumulation.

Screening of a variety of elicitor preparations for dose response by sanguinarine production resulted in 1% Botrytis culture homogenate to be the most effective preparation, i.e. cells responded with an accumulation of up to 6.6 mg/g dry weight within 18 hours. Colletotrichum, Verticillium, and Alternaria preparations were of substantial, but lesser efficiency. Over a period of 79 hours the accumulation of sanguinarine reached 2.9% of dry weight. Sanguinarine did not only accumulate in the cells, but also in their medium, and here accounted for 10-20% of total sanguinarine.

The occurrence of sanguinarine in the medium could be interpreted as excretion or as leakage due to cell breakdown. An analysis of the fine-structure of poppy cells treated with Botrytis homogenate for 48 h revealed samples containing giant and small cells to be perfectly intact (Eilert and Constabel, 1985). Tonoplast and plasmalemma, in particular, did not show any ruptures, and the cytoplasm contained the regular complement of organelles. The only observation of note was the appearance of stacked endoplasmic reticulum with partially dilated cisternae. The occurrence of sanguinarine may, indeed, be the result of an excretion process.

Stimulation or induction? An answer to this question was found when poppy cells were incubated with cycloheximide or actinomycin D for 1 hour prior to a 24 and 48 hour treatment with 1% Botrytis homogenate and an additional 3 hour incubation with 14C-L-tyrosine (Eilert et al., 1986a). Autoradiographs of thin-layer chromatograms of alkaloid extracts from cells and from their medium clearly showed suppression of sanguinarine formation. It would appear, therefore, that elicitation activates the pathway leading to benzophenanthridine synthesis by induction.

Once treated with elicitor for 48 hours leaves the cells without any structural damage. Re-elicitation, therefore, appeared as a possibility (Eilert et al., 1986b). It would advance the biotechnology of plant cell culture since elicitation products can be obtained by extracting the medium of the cells, while permitting to recycle the cells. Preliminary experiments showed that poppy cells which had been elicited once for 48 hours, then cultured in fresh medium for 10 days produced 0.3 mg sanguinarine per 100 ml culture after 48 hours of re-elicitation, 17 mg/100 ml culture in the first, 6 mg/100 ml in the second exchange medium. A similar performance was observed upon another and a fourth elicitation. Remarkably, re-elicitation changed the texture of the cultures, i.e. they became clumpy, and the chemistry, i.e. in addition to sanguinarine dihydrosanguinarine was found. A number of questions regarding the response of poppy cell cultures to elicitor preparation remain to be answered:

(1) which of the subpopulations is participating in accumulation of sanguinarine preferentially;
(2) would the supply of immediate precursors elicit the formation of morphinan alkaloids;
(3) how does extracellular sanguinarine affect poppy cells cultured;
(4) will any cell line of the opium poppy respond as seen for material employed here;
(5) would treatment with elicitor preparations of cells in alkaloid production medium lead to further increases in sanguinarine accumulation?

Work with periwinkle cell cultures allowed to make some predictions.

Periwinkle (*Catharanthus roseus*)

Seven to 10-day-old subcultures of periwinkle (*Catharanthus roseus* (L.) G. Don) initiated in 1979-81, grown in 1-B5 or 1-B5C medium and under conditions as described previously (Kurz et al., 1981), and characterized with respect to their alkaloid production were now subjected to treatment with homogenate of various fungi (Eilert et al., 1987). Homogenates of *Pythium aphanidermatum* was the most effective elicitor when used at a concentration of 5%. The response by the cultures to a 24 and 48 hour treatment with elicitor was not the same as seen before with poppy cell cultures, i.e. one periwinkle cell line did not turn brown and did not accumulate any alkaloids (non-alkaloid producing cell line #916); 4 cell lines responded by accumulation of tryptamine, precursor of indole alkaloids (cell lines #916, #943, #953, #200); only one cell line, i.e. #615 responded with accumulation of several alkaloids, i.e. strictosidine lactam, ajmalicine, tabersonine, lochnericine, and catharanthine, as found earlier when culturing #615 cells in alkaloid production medium. So, periwinkle cell cultures responded preferentially to Pythium, not Botrytis homogenate as did poppy cells, and showed differences in response to elicitation with cell lines chosen. Again, elicitation did not expand the spectrum of alkaloids found in cell cultures as determined previously when using alkaloid production medium. Noticeably occurrence of vindoline and dimeric alkaloids did not result from elicitation. Also, upon treatment, indole alkaloids accumulated in the medium of periwinkle cells. Furthermore, five days after transfer of cells of line #615 to alkaloid production medium, elicitor treatment resulted in substantial accumulation of ajmalicine and catharanthine in the cells and the medium, not however exceeding levels observed before with cells in growth medium, i.e. 1-B5C medium plus elicitor.

DISCUSSION

In general, upon treatment of poppy and periwinkle cell cultures with elicitors unspecific browning reaction and alkaloid accumulation always occurred coincidentally. A direct correlation between intensity of the pigmentation and amount of alkaloids accumulated could not be seen. Predictability of optimal elicitor-quality and quantity would be convenient and time saving. However, the diversity and variation in elicitor activity as shown in the case of sanguinarine accumulation in Papaver (Eilert et al., 1985) and indole alkaloid accumulation in Catharanthus material (Eilert et al., 1987) clearly demonstrated the necessity to screen a wide variety of compounds for elicitor activity. The course of alkaloid accumulation after elicitation was comparable to data obtained with induction by elicitors of other metabolites. Most compounds appeared according to their sequence in the biosynthetic pathway, with the exception of catharanthine. Its early accumulation could not be explained. Elicitation experiments with both growth and alkaloid production medium demonstrated that metabolite synthesis was not restricted to a particular medium. The production medium thus was not required for alkaloid formation and a comparison of media indicated that similar yields of products could be obtained in the growth medium through the elicitor stimulation. Accumulation by culturing cells in the alkaloid production medium and simultaneously treating them with elicitor did not result in a higher yield, only in a reduction of the time required for alkaloid accumulation. Elicitor stimulation of accumulation of secondary metabolites as observed here with alkaloids is expected to grealy impact plant cell technology.

REFERENCES

Darvill AG and Albersheim P (1984). Phytoalexins and their elicitors-a defense against microbial infections in plants. Annu Rev Plant Physiol 35: 243-275.

DiCosmo F, Towers GHN (1984). Stress and secondary metabolism in cultured plant cells. In Timmerman BN, Steelink C and Loewus FA (eds): "Phytochemical Adaptation to Stress," Loewus Plenum Publishing Corporation pp 97-175.

Eilert U, Kurz WGW, Constabel F (1985). Stimulation of sanguinarine accumulation in *Papaver somniferum* cell cultures by fungal elicitors. J Plant Physiol 119: 65-76.

Eilert U, Constabel F (1985) Ultrastructure of *Papaver somniferum* cells cultured in vitro and treated with fungal homogenate eliciting alkaloid production. Protoplasma 128: 38-42.

Eilert U, Kurz WGW, Constabel F (1986a). Elicitation of sanguinarine accumulation in *Papaver somniferum* cells by fungal homogenates-an induction process. J Plant Physiol (in press).

Eilert U, Kurz WGW, Constabel F (1986b). Elicitor-induction of sanguinarine formation in *Papaver somniferum* cell cultures and semicontinuous sanguinarine production by re-elicitation. 34th Ann Congr Med Plant Res, Hamburg, 22 to 27-09-1986.

Eilert U, Kurz WGW, Constabel F (1987). Elicitor stimulation of monoterpene indole alkaloid formation in suspension cultures of *Catharanthus roseus*. J Plant Physiol (in press).

Eilert U (1987). Elicitation of product formation. In Vasil IK, Constabel F (eds.) "Cell Culture and Somatic Cell Genetics of Plants," Academic Press, N.Y. (accepted).

Gamborg OL, Miller RA, Ojima K (1968). Nutrient requirements of suspension cultures of soybean root cells. Exp Cell Res 50: 151-158.

Hahlbrock H, Kreuzaler F, Rogg H, Fautz E, Kuhn DN (1982). Regulation of flavonoid and phytoalexin accumulation through mRNA and enzyme induction in cultured plant cells. 33. "Colloquium-Mosbach," Springer Verlag, Berlin. pp 34-43.

Heinstein PF (1985). Future approaches to the formation of secondary natural products in plant cell suspension cultures. J Nat Prod 48: 1-9.

Kurz WGW, Chatson KB, Constabel F, Kutney JP, Choi LSL, Kolodziejczyk P, Sleigh SK, Stuart KL, Worth BR. (1981).

Alkaloid production in Catharanthus roseus cell cultures VIII. Characterization of the PRL #200 cell line. Planta Medica 42: 22-31.

Mieth H, Speth V, Ebel J. (1986) Phytoalexin production by isolated soybean protoplasts. Zeitschr f Naturf 41c: 193-201.

Tabata M, Ogino T, Yoshioka K, Yoshikawa N, Hiraoka N (1978). Selection of cell lines with higher yield of secondary products. In "Frontiers of Plant Tissue Culture 1978", Thorpe TA (ed.), University of Calgary Press, Calgary pp 213-222.

Wolters S, Eilert U (1983). Elicitoren-Ausloser der Akkumulation von Pflanzenstoffen. Ihre Anwendung zur Produktionssteigerung in Zellkulturen. Dtsch Apoth Zeitg 123: 659-667.

Zenk MH, El-Shagi H, Arens H, Stockigt J, Weiler EW, Deus B. (1977). Formation of the indole alkaloids serpentine and ajmalicine in cell suspension cultures of Catharanthus roseus. In "Plant Tissue Culture and its Bio-technological Application," Barz W, Reinhard E, Zenk MH (eds.). Springer Verlag, Berlin, Heidelberg, New York pp 27-43.

Section IV.
Genetics

Plant Tissue and Cell Culture, pages 223–241

HAPLOIDS FROM GAMETOPHYTIC CELLS - RECENT DEVELOPMENTS AND FUTURE PROSPECTS

Wilfred A. Keller[1], Paul G. Arnison[2] and Brian J. Cardy[3]
[1]Plant Research Centre, Agriculture Canada, Ottawa, Ontario, Canada K1A 0C6
[2]Paladin Hybrids Inc., Brampton, Ontario, Canada L6V 2L2
[3]BioTechnica International of Canada Inc. Calgary, Alberta, Canada T2E 7H7

INTRODUCTION

Numerous studies on the *in vitro* culture of gametophytic cells with the aim of producing haploid plants have been reported during the last two decades. A large number of reviews, book chapters and symposia proceedings have been published as well (Chu, 1982; Hu and Zeng, 1985; Heberle-Bors, 1985; Hu and Yang, 1986). Haploids can be regenerated from both male and female gametophytic cells through the culture of anthers, microspores, ovaries and ovules. A positive *in vitro* response will lead to the development of embryos and/or callus from which plants can be regenerated. Early events during *in vitro* culture have been characterized at the cytological, ultrastructural and biochemical level (Chen et al., 1984; Raghavan, 1984; Huang, 1986) and will not be considered in detail here. The purpose of this presentation is to review progress in the development of techniques for haploid production, identification of factors influencing embryogenesis/callus proliferation from gametophytic cells, characterization of regenerants and utilization of haploid cells/plants in plant breeding and genetics.

EVALUATION OF HAPLOID PRODUCTION TECHNIQUES

1. Anther/Microspore Culture

Anther culture has been employed to obtain microspore-derived callus, embryos and plants in well over 200 species (Chu, 1982; Maheshwari et al., 1982; Hu and

Zeng, 1985). However, active and sustained research in the development of anther culture techniques has been confined to a few species which include tobacco, barley, wheat, triticale, rice, maize, potato, rapeseed and cole crops. A modification of the technique involves the culture of the inflorescence under conditions which selectively permit proliferation of the microspores into callus or embryos (Toriyama and Hinata, 1985). This approach may be useful in cases where small anther size makes isolation technically difficult.

During the last four years, major advances have been made in the development of microspore culture procedures. The 'shed pollen' technique originally developed in tobacco involves flotation of anthers on liquid medium followed by dehiscence and release of microspores and/or their division products into the medium where they continue to develop as callus or embryos. This approach has been successfully adapted to barley (Sunderland and Xu, 1982; Kohler and Wenzel, 1985) and wheat (Wei, 1982). Microspores can also be mechanically isolated from fresh or preconditioned anthers. Preconditioning including preliminary culture of anthers on medium may be essential for any *in vitro* response or it may significantly improve the efficiency of callus/embryo formation (Chen, 1986a). The culture of microspores directly isolated from fresh anthers has only been successful in a few species including tobacco (Rashid and Reinert, 1983; Aruga et al., 1985), *Nicotiana rustica* (Kyo and Harada, 1985), barley (Wei et al., 1986), *B. napus* (Lichter, 1982; Keller and Armstrong, 1983b; Chuong and Beversdorf, 1985) and *B. carinata* (Chuong and Beversdorf, 1985). In most cases efficiency of callus formation or embryogenesis in mechanically isolated microspores is reduced in comparison to anther culture systems. An exception to this generalization is *B. napus* where high frequencies of embryogenesis are now consistently obtained.

2. Ovary/Ovule Culture

Reports on haploid production via ovary and ovule culture have been available for approximately a decade but during the last four years there has been growing interest in this approach. In 1982, Yang and Zhou listed species in which haploids had been produced and identified key factors affecting the process. More recent reports have described the development of new techniques or improvement of established protocols for haploid production in rice (Zhou et al., 1986), maize (Truong-Andre and Demarly, 1984), sugarbeet (Bossoutrot and Hosemans, 1985), *Gerbera* (Meynet and Sibi, 1984), sunflower (Yang et al., 1986), *Petunia* (Raquin, 1985) and poplar (Wu and Xu, 1984). Ultrastructural studies have revealed that proliferation can be induced in synergid cells (Zhou et al., 1986) or the egg cell (Yang et al., 1986). A major disadvantage of the ovule culture approach relates to its inbuilt limitation of one embryo sac per ovule. Even if all the haploid cells in the embryo sac could give rise to callus/embryos, total yield would still be low in comparison to the potential numbers of haploids that could be recovered from anther or microspore cultures. *In vitro* response of ovary/ovule cultures is usually less than 15%; this is not significantly different from the response that can be obtained in anther culture.

FACTORS INFLUENCING EMBRYO/CALLUS FORMATION FROM GAMETOPHYTIC CELLS

1. Donor Plant Physiology

The physiological condition of the donor plant is a key factor determining the success of haploid production. Donor growth temperature is critical in determining subsequent embryogenic response in isolated microspores of *B. napus*. Research at Agriculture Canada has shown that high frequencies of embryogensis are consistently obtained in microspore cultures derived from plants grown at a 10°C day/5°C night cycle in comparison to higher growth temperatures (Table 1). A beneficial effect of lower growth temperature has also been reported for the establishment of isolated microspore cultures of *Nicotiana* (Rashid and Reinert, 1983). The effect of light on donor plants has been studied in some species. In *Brassica campestris*, anther response increased as light intensity received by donors was increased (Keller et al., 1982). Photoperiod may also be important. Increased frequencies of microspore embryogenesis were reported in tobacco anther cultures if donor plants were grown under short days (Heberle-Bors, 1982).

Table 1. Effect of Donor Plant Growth Temperature on Embryogenesis in Isolated Microspores of _Brassica napus_ cv. Topas.*

Donor temp (^{o}C) day/night (16 h photo-period)	No. plants tested	No. plants responding	% Response	Ave. No. embryos/1000 anthers in responding plants**
20/15	18	0	0	0
15/10	55	28	51	6,000
10/5	8	8	100	193,000

* Data from Keller et al. (in prep.)
**Microspores cultured at 32.5^{o}C for 3-4 d prior to transfer to 25°C.

2. Developmental Stage of Gametophytic Cells

Numerous studies have demonstrated that microspores will respond only at a specific development stage which ranges from the early uninucleate to the early binucleate stage for virtually all species analyzed. There are few reported exceptions such as some wheat genotypes where plantlets were obtained from anthers at the pollen mother cell to the trinucleate stage (He and Ouyang, 1984). Visual selection of buds or anthers at the optimal development stage is feasible but may require modification if changes in donor genotype or donor growth conditions are made. Information on optimal stage of development for ovule culture is limited but it appears that ovules from a broad range of developmental stages are responsive (Yang and Zhou, 1982).

3. Pretreatments of Gametophytic Cells

Pretreatments of inflorescence, buds, or anthers at reduced temperatures prior to culture has long been claimed to increase anther response _in vitro_ and appears to be particularly effective in graminaceous species including rice (Chen, 1986a), wheat (Lazar et al., 1985), barley

(Huang and Sunderland, 1982) and sugar cane (Fitch and Moore, 1984). However, in other studies a positive effect was not found (Marsolais et al., 1984; Ouyang, 1986). Low temperature pretreatments have not been shown to be beneficial for ovule cultures of several species in which they were tested (Yang and Zhou, 1982). Other pretreatments including high temperature shocks (Keller and Armstrong, 1983a), subjection to reduced atmosphere pressure (Klimaszewska and Keller, 1983), and osmotic shock (Ouyang, 1986) have been reported to improve anther culture response but remain unconfirmed with regard to general applicability.

4. Culture Medium Composition

Within recent years there has been a general switch to the use of liquid anther culture media as they have been shown to increase microspore-derived callus/embryo yield in several species including rapeseed (Cardy, 1986), potato (Uhrig, 1985), wheat (Lazar et al., 1985) and rice (Chen, 1986a). A double layer system of liquid over agar has been effectively employed for some species (Johansson et al., 1982).

The role of sucrose in embryogenesis or callus proliferation has been widely studied with differing effects reported for different species. In barley (Wei et al., 1986) and *Nicotiana* spp (Aruga et al., 1985; Kyo and Harada, 1985) it was essential to subject isolated microspores to sucrose starvation for several days in order to achieve embryogenesis or callus formation. Other studies which have focussed on effect of sucrose concentration have shown that in *Brassicas* and most cereals elevated sucrose levels of at least 6% are necessary for establishing anther/microspore cultures. In *B. napus*, microspore viability during the early stages of anther culture was increased in the presence of 16% sucrose compared to 8% (Dunwell and Thurling, 1985). Significant embryo yield increases have been obtained by sequentially reducing sucrose concentration in *B. napus* anther culture media during the first two weeks of culture (Chen and Chen, 1983).

The influence of growth regulators on haploid callus or embryo development has been studied in many species. Several studies have revealed that auxins are essential or beneficial (see Chu, 1982, Maheshwari et al., 1982; Yang and Zhou, 1982). Cytokinins have also been shown to be beneficial in some cases (Chien and Kao, 1983; Ye et al., 1985) and a few studies have reported positive _in vitro_ response in the presence of gibberellic acid (P. Arnison, unpublished) and abscisic acid (Torrizo and Zapata, 1986). However, in many cases a definite requirement for growth regulators has not been demonstrated and in some instances their addition to the medium has been shown to be inhibitory. For example, recent experiments with isolated microspores of _B. napus_ have revealed that exogenous growth regulators are _not essential_ and may in fact be inhibitory to embryogenesis (Keller, unpublished).

Numerous basal media modifications have been made to maximize _in vitro_ response of gametophytic cells. These include _manipulation_ of source and concentration of inorganic nutrients, particularlity nitrogen (see Chen, 1986a) as well as supplementation with organic nutrients such as amino acids (Kyo and Harada, 1985; Chen, 1986a). Activated charcoal, which presumably acts by adsorbing deleterious compounds is beneficial in some instances (Johansson et al., 1983). Undefined factors such as potato extract have consistently been shown to enhance anther response (Chen 1986a; Ouyang, 1986). Undefined, beneficial substances are also produced through conditioning of media by preculturing with ovaries and anthers (Kohler and Wenzel, 1985).

5. Physical Factors

Anther/microspore density affects subsequent _in vitro_ response. In barley, very high densities of more _than 60_ anthers/ml are optimal (Xu and Sunderland, 1983), whereas in _B. napus_ the optimal appears to be nearer to 6/ml (_Keller and_ Armstrong, 1983b; Cardy, 1986). In _B. napus_ microspore cultures, optimal density is in the _order of_ 2-5X10^4 (W. Keller, unpublished). Density effects are most likely related to medium conditioning. Another factor shown to be important is anther orientation where studies with barley have shown that anthers plated "up" with locules in contact with the medium yielded large numbers of calli from those locules not in direct contact with the medium (Shannon et al., 1985).

6. Culture Environment

An initial period of culture of anthers and isolated microspores at elevated temperatures (30°C or more) prior to maintenance at a lower temperature has been shown to be highly beneficial for Brassica spp. (Keller and Armstrong, 1983a, b; Dunwell et al., 1985; Sharma and Bhojwani, 1985). A positive effect of an initial high temperature culture period has also been shown in pepper (Dumas de Vaulx et al., 1981) and eggplant (Dumas de Vaulx and Chambonnet, 1982). The biochemical changes induced by these trestments have remained largely undefined. Research carried out thus far with ovule cultures has not revealed a specific requirement for high temperature (Yang and Zhou, 1982). The effects of other environmental parameters on haploid callus/embryo formation have been evaluated in some studies. For example, Johansson and Eriksson (1984) reported that the presence of 2% CO_2 stimulated embryo formation in anther cultures of several unrelated species.

7. Donor Genotype

An overiding factor determining the degree of success in haploid production relates to donor plant genotype (Uhrig 1985; Chen, 1986a; Ouyang, 1986). A survey of B. napus genotypes has revealed that winter types generally have a higher response than spring types. A greatly enhanced response was also observed in spring x winter F_1 hybrids (Table 2). The ability to undergo callus proliferation or embryogenesis is heritable and has generally been shown to be controlled by a limited number of nuclear genes (Henry and deBuyser, 1985; Uhrig, 1985; Chen, 1986a) however, cytoplasmic effects have been observed in at least one case (Charmet and Bernard, 1984). Genotype-environmental interactions must be taken into consideration in the development of haploid production protocols. Different genotypes may have different optimal growth conditions, pretreatments, culture medium compositions, and culture environments (Keller and Armstrong, 1983b).

PLANT REGENERATION

Microspore-derived callus is generally transfered to medium with elevated cytokinin/auxin ratios to induce shoot-regeneration. Several studies have demonstrated the importance of early transfer of the callus to regeneration medium in order to maximize regeneration efficiency and reduce genetic abberations (Chen, 1986a). In cases where embryos have formed, transfer to an embryo culture medium which usually consists of basal medium with low sucrose

Table 2. Frequency of Embryogenesis in Anther Cultures of Different _Brassica napus_ Genotypes.*

Category	No. of Genotypes	Frequency of Embryogenesis per 1000 Anthers**	
		Range	Mean
Spring	19	20-5780	1010
Winter	7	330-6430	2170
SpringXSpring F_1's	10	20-1090	400
WinterXWinter F_1's	9	150-4510	1250
SpringXWinter F_1's	5	1500-5300	3460
Double haploids***	9	90-9260	1470

* Data from Cardy (1986).
** Based upon a sample of 108 anthers per donor plant with a minimum of 3 plants surveyed per genotype; anthers cultured at 35° for 2 days prior to maintenance at 25°C.
***Derived from a single winter X winter F_1 plant which had a mean embryo yield of 4510 embryos per 1000 anthers.

(<3%) is usually required. In some species, particularly in _Brassicas_, microspore-derived embryos are abnormal and the majority do not develop directly into plantlets. Often multiple transfers onto cytokinin-containing medium are necessary to achieve regeneration (Keller, 1984). The composition of the original anther or microspore culture medium can influence the capability of _Brassica_ embryos to develop into plantlets. Cardy (1986) has shown that embryos obtained from anthers cultured on solid medium are superior to those derived from liquid medium in terms of their ability to develop into plantlets.

CHARACTERIZATION OF PLANTS DERIVED FROM GAMETOPHYTIC CELLS

In addition to haploids, diploids and polyploids have also been detected in regenerated populations. Diplo- and polyploidization mechanisms include nuclear fusion and endomitosis during the early stage of culture. These processes have been well documented in a number of cytological studies (Chen et al., 1984; Huang, 1986). Aneuploids have also been detected at low frequencies and are usually found in cases where intermediate callus proliferation has occurred (Chu et al., 1985; Gu, 1986; Hu, 1986). However, a detailed study with triticale revealed that most cytological variations detected in anther-derived plants probably existed in the microspores *in vivo* (Charmet et al., 1986).

Cytoplasmic changes can occur in plants regenerated from gametophytic cells. Revertants of a plastid-related mutation have been detected in anther-derived tobacco haploids (Matzinger and Burk, 1984). Albinism which occurs in both microspore and ovule-derived cereal haploids appears to be due to cytoplasmic DNA modifications. Organelle DNA restriction analyses revealed deletions in the plastid DNA of anther-derived albinos of wheat (Day and Ellis, 1986). In independent studies on green anther-derived regenerants of wheat and triticale, detectable changes in cytoplasmic DNA were not found (Charmet et al., 1985; Rode et al., 1985).

An important question relevant to the usefulness of haploids in genetic research relates to how well they represent the genetic array found within the *in vivo* microspore/ovule population. Contradictory results have thus far been reported. In *Brassica oleracea*, isozyme analysis of anther-derived plants has revealed a higher than expected proportion of fast-migrating types (Orton and Bowers, 1985). However, in rice, genotypic analysis has revealed that anther-derived plants are representative of the theoretical genotypic ratios that should be found in the microspore population (Chen, 1986a). It is possible that suboptimal medium composition and culture environment favor the *in vitro* selection of certain genotypes and that through further methodology improvements such selection processes will be eliminated or reduced.

Another potential problem that may relate to diploid regenerants is the occurrence of heterozygous plants. Such plants presumably originate from diploid heterozygous, gametic cells. Isozyme analyses of anther-derived B. napus lines revealed that up to 7% of the lines may be heterozygous (Cardy, 1986). Up to 10% of anther-derived diploid rice plants were shown to be heterozygous (Chen, 1986b). However, studies in barley failed to detect heterozygous regenerants (Franzone et al., 1984).

Several reports have claimed that homozygous lines obtained from anther-derived haploids are significantly depressed in vigor in comparison to highly inbred lines or homozygous lines derived from haploids produced by other methods (Huang et al., 1984; Kumashira and Oinuma, 1985; Schnell and Wernsman, 1986). In Nicotiana spp the degree of depression was intensified with successive cycles of anther culture (Brown and Wernsman, 1983; de Paepe et al., 1983). Measurement of nuclear DNA in doubled haploids of tobacco has revealed significant increases in DNA (Dhillon et al., 1983) suggesting that such changes are responsible for depressed vigor. It has also been suggested that high frequences of mutations occur in microspores during culture (Kumashira and Oinuma, 1985). However, in other studies no evidence for depressed yield or vigor of anther- or microspore-derived lines has been found (Charmet and Branlard, 1985; de Buyser et al., 1985). It is possible that observed reduction in vigor may be associated with specific donor genotype, inadvertant selective advantage of certain genotypes in vitro, and suboptimal media and culture environments.

UTILIZATION OF GAMETE-DERIVED PLANTS IN GENETICS

1. Plant Breeding

Production of haploids from F_1 donors permits the breeder to effectively select desirable genetic recombinants. Homozygous lines are established through spontaneous chromosome doubling during early stages of in vitro culture or through colchicine-induced chromosome doubling of haploids. Because homozygous lines can be made rapidly available, a saving of time of up to 50% can be achieved in developing new cultivars. In China in vitro - produced haploids have been used in breeding programs for more than a decade and superior cultivars of rice and wheat have been developed from them (Chen, 1986b; Hu, 1986b). Progress has also been made in applying haploidy in the improvement of woody species such as poplar and rubbertree (Chen, 1986c). In Europe, and North America, field evaluation of potato, barley, wheat and triticale anther-derived lines has revealed their potential usefulness in breeding (Foroughi-Wehr and Friedt, 1984; Charmet and Branlard, 1985; de Buyser et al., 1985; Uhrig, 1985). New cultivars may soon be released from some of these lines. Rapid progress should also be expected in rapeseed breeding through the use of doubled haploids.

In highly heterozygous, cross-pollinating crops, haploidy creates a rapid method of producing pure breeding lines which can serve as parents in hybrid cultivar development. Chinese scientists have released a number of superior 'inbred lines' of maize derived through anther culture (Wu, 1986). The cole crops are particularly suited to this approach as they are readily amenable to anther culture (Table 3). Broccoli hybrids produced through crosses of doubled haploids compared favorably in many traits and were superior to commercially-available hybrids which were developed through conventional inbreeding methods (Arnison et al. in prep.). Anther culture has now been widely adopted by a number of European and American establishments involved in cole crop breeding.

Another potential application of haploidy in breeding is in the production of desirable genetic translocations, substitution and addition lines through the culture of anthers of interspecific and intergeneric hybrids. This approach is being employed on triticale X wheat hybrids (Wang and Hu, 1985).

Breeders are still faced with numerous problems in employing haploidy techniques. Firstly, there are still few species in which haploid production can occur at frequencies which makes routine use feasible. These include rice, barley, wheat, maize, potato, tobacco and *Brassicas*. Even in these species, problems with genotype specificity and reliability of production still exist. Other potential problems already discussed include the production of heterozygous diploids, *in vitro* selection of certain genotypes, high frequency albino production and possible reduced vigor of doubled haploids.

Table 3. Anther Culture Response in *Brassica oleracea* Varieties.*

Variety	Common name	Cultivars responsive / No. tested	Embryogenic capacity (per 1000 anthers) under optimal conditions
acephala	- thousand head kale	1/1	444
	- marrowstem kale	1/1	333
	- curley kale	2/3	3,000
botrytis	- cauliflower	3/3	241
	- calabrese	5/11	1,998
capitata	- red/white cabbage	18/19	3,797
gemmifera	- brussels sprouts	11/12	475
gongylodes	- kohlrabi	5/6	2,130
italica	- broccoli	41/51	5-12,000**
sabauda	- savoy cabbage	2/2	26,056

* Data from Arnison et al. (in prep.)
**Rates between 25-50,000 can be routinely achieved with certain doubled haploid lines.

2. Mutant Isolation

Microspores can be used directly in mutant isolation providing that frequencies of embryogenesis or callus formation are sufficiently high. This approach has been applied with some success in barley and rice (Wenzel 1985; Chen 1986b). In *B. napus* where frequencies of embryogenesis of 5% can be obtained and where more than 20 M microspores can be plated, mutant isolation studies are also feasible and have recently been undertaken in several Canadian and European laboratories. It is also possible that anther/microspore cultures can be used to identify agronomically-desirable variants without the use of mutagens and specific *in vitro* selection strategies (Schaeffer et al., 1984; Parisi and Picard, 1986).

3. Recipient Systems for Gene Transfer

Highly embryogenic isolated microspore cultures offer great potential as recipient cells for introduction of foreign genes. Through careful bud selection, *B. napus* microspores can be cultured at efficiencies of >20% and will also undergo embryogenesis when cultured as individual cells in microdrops (Pechan and Keller, 1986). Such microspore cultures are now being utilized for intranuclear microinjection of plasmid vectors. Microspores may also be effective in cocultivation and electroporation approaches for achieving gene transfer and they may be particularly valuable as a cell system suitable for use in cereal transformation.

REFERENCES

Aruga K, Nakajima T, Yamamoto K (1985). Embryogenic induction in pollen grains of *Nicotiana tabacum*. Japan J Breed 35: 50-58.

Bossoutrot D, Hosemans D (1985). Gynogenesis in *Beta vulgaris* L.: From *in vitro* culture of unpollinated ovules to the production of doubled haploid plants in soil. Plant Cell Rep 4: 300-303.

Brown JS, Wernsman EA, Schnell RJ (1983). Effect of a second cycle of anther culture of flue-cured lines of tobacco. Crop Sci 23: 729-733.

Cardy BJ (1986). Production of anther-derived doubled haploids for breeding oilseed rape (*Brassica napus* L.). Ph.D. Thesis, University of Guelph, Guelph, Ont. Canada.

Charmet G, Bernard S (1984). Diallel analysis of androgenetic plant production in hexaploid Triticale (X. triticosecale, Wittmack). Theor Appl Genet 69: 55-61.

Charmet G, Bernard S, Bernard M (1986). Origin of aneuploid plants obtained by anther culture in triticale. Can J Genet Cytol 28: 444-452.

Charmet G, Branlard G (1985). A comparison of androgenetic doubled-haploid, and single seed descent lines in Triticale. Theor Appl Genet 71: 193-200.

Charmet G, Vedel F, Bernard M, Bernard S, Mathieu C (1985). Cytoplasmic variability in androgenetic doubled haploid lines of triticale. Agronomie 5: 709-717.

Chen Y (1986a). Anther and pollen culture of rice. In Hu H, Yang H (eds): "Haploids of Higher Plants in Vitro", Berlin, Heidelberg: Springer-Verlag, pp 3-25.

Chen Y (1986b). The inheritance of rice pollen plant and its application in crop improvement. In Hu H, Yang H (eds): "Haploids of Higher Plants in vitro", Berlin, Heidelberg: Springer-Verlag, pp 118-136.

Chen Z (1986c). Induction of androgenesis in woody plants. In Hu H, Yang H (eds): "Haploids of Higher Plants in Vitro", Berlin, Heidelberg: Springer-Verlag, pp. 42-66.

Chen CC, Kasha KJ, Marsolais A (1984). Segmentation patterns and mechanisms of genome multiplication in cultured microspores of barley. Can J Genet Cytol 26: 475-483.

Chen Z, Chen Z (1983). High frequency induction of pollen-derived embryoids from anther cultures of rape (*Brassica napus* L.) Kexue Tongbao 28: 1690-1694.

Chien YC, Kao KN (1983). Effects of osmolality, cytokinin, and organic acids on pollen callus formation in Triticale anthers. Can J Bot 61: 639-641.

Chu C-C (1982). Haploids in plant improvement. In Vasil IK, Scowcroft WR, Frey KJ (eds): "Plant Improvement and Somatic Cell Genetics", New York: Academic Press, pp 129-158.

Chu Q, Zhang Z, Gao Y (1985). Cytogenetical analysis on aneuploids obtained from pollenclones of rice (*Oryza sativa* L.) Theor Appl Genet 71: 506-512.

Chuong PV, Beversdorf WD (1985). High frequency embryogenesis through isolated microspore culture in *Brassica napus* L. and *B. carinata*. Braun. Plant Science 39: 219-226.

Day A, Ellis THN (1984). Chloroplast DNA deletions associated with wheat plants regenerated from pollen: possible basis for maternal inheritance of chloroplasts. Cell 39: 359-368.

deBuyser J, Henry Y, Taleb G (1985). Wheat androgenesis: cytogenetical analysis and agronomic performance of doubled haploids. Z. Pflanzenzuchtg 95: 23-34.

de Paepe R, Prat D, Knight J (1983). Effects of consecutive androgenesis on morphology and fertility in *Nicotiana sylvestris*. Can J Bot 61: 2038-2046.

Dhillon SS, Wernsman EA, Miksche JP (1983). Evaluation of nuclear DNA content and heterochromatin changes in anther-derived dihaploids of tobacco (*Nicotiana tabacum*) cv Coker 139. Can J Genet Cytol 25: 169-173.

Dumas de Vaulx, Chambonnet D (1982). Culture in vitro d'anthères d'aubergine (Solanum melongena L.): stimulation de la production de plantes au moyen de traitements à +35°C associés à de faibles teneurs en substances de croissance. Agronomie 2: 983-988.

Dumas de Vaulx R, Chambonnet D, Pochard E (1981). Culture in vitro d'anthères de piment (Capsicum annum L.): Amélioration des taux l'obtention de plantes des différents génotypes par des traitement à +35°C. Agronomie 1: 859-864.

Dunwell JM, Cornish M, de Courcel AGL (1985). Influence of genotype, plant growth temperature and anther incubation temperature on microspore embryo production in Brassica napus spp oleifera. J Exp Bot 36: 679-689.

Dunwell JM, Thurling N (1985). Role of sucrose in microspore embryo production in Brassica napus spp oleifera. J Exp Bot 36: 1478-1491.

Fitch MM, Moore PH (1984). Production of haploid Saccharum spontaneum L - Comparison of media for cold incubation of panicle branches and for float culture of anthers. J Plant Physiol 117: 169-178.

Foroughi-Wehr B, Friedt W (1984). Rapid production of recombinant barley yellow mosaic virus resistant Hordeum vulgare lines by anther culture. Theor Appl Genet 67: 377-382.

Franzone PM, Fischbeck G, Foroughi-Wehr B, Friedt, W (1984). Analysis of origin of androgenetic plants of Hordeum vulgare by the use of balanced lethal systems. Z Pflanzenzuchtg 93: 320-330.

Gu M (1986). Cytogenetic stability and variability of calli and cell clones originated from maize pollen and their regenerated plants. In Hu H, Yang H (Eds): "Haploids of Higher Plants in Vitro", Berlin, Heidelberg: Springer-Verlag pp 79-90.

He D-G, Ouyang J-W (1984). Callus and plantlets formation from cultured wheat anthers at different developmental stages. Plant Sci Lett 33: 71-79.

Heberle-Bors E (1982). In vitro pollen embryogenesis in Nicotiana tabacum L. and its relation to pollen sterility, sex balance, and floral induction of the pollen donor plants. Planta 156: 396-401.

Heberle-Hors E (1985). In vitro haploid formation from pollen: a critical review. Theor Appl Genet 71: 361-374.

Henry Y, de Buyser J (1985). Effect of the IB/IR translocation on anther culture ability in wheat. Plant Cell Rep 4: 307-310.

Hu H (1986a). Variability and gametic expression in pollen-drived plants in wheat. In Hu H, Yang H (eds): "Haploids of Higher Plants in Vitro", Berlin, Heidelberg: Academic Press 67-78.

Hu D (1986b). Jinghua No. 1, a winter wheat variety derived from pollen sporophyte. In Hu H, Yang Y (eds): "Haploids of Higher Plants in Vitro", Berlin, Heidelberg: Springer-Verlag, pp 137-148.

Hu H, Yang H (1986). "Haploids of Higher Plants in Vitro". Berlin, Heidelberg, Springer-Verlag.

Hu H, Zeng JZ (1985). Development of new varieties via anther culture. In Evans DA, Sharp WR, Ammirato PV (eds.):"Handbook of Plant Cell Culture Vol. 3, Crop Species", New York: Macmillan, pp 65-90

Huang B (1986). Ultrastructural aspects of pollen embryogenesis in Hordeum, Triticum and Paeonia. In Hu H, Yang H (eds): "Haploids of Higher Plants in Vitro", Berlin, Heidelberg: Springer-Verlag pp 91-117.

Huang B, Dunwell JM, Powell W, Hayter AM, Wood W (1984). The relative efficiency of microspore culture and chromosome elimination as methods of haploid production in Hordeum vulgare L. Z Pflanzenzuchtg 92: 22-29.

Huang B, Sunderland N (1982). Temperature-stress pretreatment in barley anther culture. Ann Bot 49L 77-88.

Johansson L, Andersson B, Eriksson T (1982). Improvement of anther culture technique: Activated charcoal bound in agar medium in combination with liquid medium and elevated CO_2 concentration. Physiol Plant 54: 24-30.

Johansson L, Eriksson T (1984). Effects of carbon dioxide on anther cultures. Physiol Plant 60: 26-30.

Keller WA (1984). Anther culture of Brassica. In Vasil IK (ed): "Cell Culture and Somatic Cell Genetics of Plants Vol I", Orlando: Academic Press, pp 302-310.

Keller WA, Armstrong KC (1983a). Production of haploids via anther culture in Brassica oleracea van italica. Euphytia 32: 151-159.

Keller WA, Armstrong KC (1983b). Production of Brassica napus haploids through anther and microspore culture. In Proc. Sixth Int Rapeseed Cong. Paris, pp 239-245.

Keller WA, Armstrong KC, de la Roche AI (1982). The production and utilization of microspore-derived haploids in Brassica crops. In Giles KL, Sen SK (eds):

"Plant Cell Culture in Crop Improvement," New York: Plenum Pub Corp, pp 169-183.

Klimaszewska K, Keller WA (1983). The production of haploids from Brassica hirta Moench (Sinapis alba L.) anther cultures: Z Pflanzenphysiol 109: 235-241.

Kohler F, Wenzel G (1985). Regeneration of isolated barley microspores in conditioned media and trials to characterize the responsible factor. J. Plant Physiol 121: 181-191.

Kumashiro T, Oinuma T (1985). Comparison of genetic variability among anther-derived and ovule-derived doubled haploid lines of tobacco. Japan J Breed 35: 301-310.

Kyo M, Harada H (1985). Studies on conditions for cell division and embryogenesis in isolated pollen culture of Nicotiana rustica Plant Physiol 79: 90-94.

Lazar MD, Schaeffer GW, Baenziger PS (1985). The physical environment in relation to high frequency callus and plantlet development in anther cultures of wheat (Triticum aestivum L.) cv Chris. J Plant Physiol 121: 103-109.

Lichter R (1982). Induction of haploid plants from isolated pollen of Brassica napus. Z Pflanzenphysiol 105: 427-434.

Maheshwari SC, Rashid A, Tyagi AK (1982). Haploids from pollen grains-retrospect and prospect. Amer J Bot 69: 865-8798.

Matzinger DF, Burk LG (1984). Cytoplasmic modification by anther culture in Nicotiana tabacum L. J Hered 75: 167-170.

Meynet J, Sibi M (1984). Haploid plants from in vitro culture of unfertilized ovules in Gerbera jamesonii. Z. Pflanzenzuchtg 93: 78-85.

Marsolais AA, Séguin-Swartz, Kasha KJ (1984). The influence of anther cold pretreatments and donor genotypes on in vitro androgenesis in wheat (Triticum astivum L.) Plant Cell Tissue Organ Culture 3: 69-79.

Orton TJ, Bowers MA (1985). Segregation of genetic markers among plants regenerated from cultured anthers of broccoli (Brassica oleracea var. italica). Theor Appl Genet 69: 637-643.

Ouyang J (1986). Induction of pollen plants in Triticum aestivum. In Hu H, Yang H (eds): "Haploids of Higher Plants in vitro", Berlin, Heidelberg; Springer-Verlag, pp 26-41.

Parisi L, Picard E (1986). Disease response of doubled haploid lines and their original cultivars in wheat (Triticum aestivum L.). Z. Pflanzenzuchtg 96: 63-78.
Pechan P, Keller WA (1986). Developmental study of microspore embryogensis of Brasica napus L. Proc. Sixth Int Cong Plant Tissue and Cell Culture, Abstr.
Raghavan V (1984). Protein synthetic activity during normal pollen development and during induced pollen embryogenesis in Hyoscyamus niger. Can J Bot 62: 2493-2513.
Raquin C (1985). Induction of haploid plants by in vitro culture of Petunia ovaries pollinated with irradiated pollen. Z. Pflanzenzuchtg 94: 166-169.
Rashid A, Reinert J (1983). Factors affecting high-frequency embryo formation in Ab initio pollen cultures of Nicotiana Protoplasma 116: 155-160.
Rode A, Hartmann C, Dron M, Picard E, Outier F (1985). organelle genome stability in anther-derived doubled haploids of wheat (Triticum aestivum L cv "Moisson"). Theor Appl Genet 71: 320-324.
Shaeffer GW, Sharpe FT, Cregan PB (1984). Variation for improved protein and yield from rice anther culture. Theor Appl Genet 67: 383-389.
Schnell RJ, Wernsman EA (1986). Androgenic somaclonal variation in tobacco and estimation of its value as a source of novel genetic variability. Crop Sci 26: 84-88.
Shannon PRM, Nicholson AE, Dunwell JM, Davies DR (1985). Effect of orientation on microspore callus production in barley (Hordeum vulgare L.) Plant Cell Tissue Organ Culture 4: 271-280.
Sharma KK, Bhojwani SS (1985). Microspore embryogenesis in anther cultures of two Indian cultivars of Brassica juncea (L) Czern Plant Cell Tissue Organ Culture 4: 235-239.
Sunderland N, XU ZH (1982) Shed pollen culture in Hordeum vulgare. J Exp Bot 33: 1086-1095.
Toriyama K, Hinata K (1985). Panicle culture in liquid media for obtaining anther calli and protoplasts in rice. Japan J Breed 35: 449-452.
Torrizo LB, Zapata FJ (1986). Anther culture in rice: IV. The effect of abscisic acid on plant regeneration. Plant Cell Rep. 5: 136-139.
Truong-Andre I, Demarly Y (1984). Obtaining plants by in vitro culture of unfertilized maize ovaries (Zea mays L) and preliminary studies on the progeny of a

gynogenetic plant. Z. Pflanzenzuchtg 92: 309-320.
Uhrig H (1985). Genetic selection and liquid medium conditions improve the yield of androgenetic plants from diploid potatoes. Theor Appl Genet 71: 455-460.
Wang X, Hu H (1985). The chromosome constitution of plants derived from pollen of hexaploid triticale x common wheat F_1 hybrids. Theor Appl Genet 70: 92-96.
Wei ZM (1982). Pollen callus culture in *Triticum aestivum*. Theor Appl Genet. 63: 71-73.
Wei ZM, Kyo M, Harada H (1986). Callus formation and plant regeneration through direct culture of isolated pollen of *Hordeum vulgare* cv. "Sabarlis", Theor Appl Genet 72: 252-255.
Wenzel G (1985). Strategies in unconventional breeding for disease resistance. Ann Rev Phytopathol 23: 149-172
Wu J (1986). Breeding corn by anther culture. In Hu H, Yang H (eds): " Haploids of Higher Plants in Vitro", Berlin, Heidelberg: Springer-Verlag, pp. 149-164.
Wu K, Xu M (1984). Induction of maternal haploid plants from unpollinated ovaries of poplar *in vitro* Acta Genet Sinica 11: 47-51.
Xu Z, Sunderland N (1983). Inoculation density in the culture of barley anthers. Scientia Sinica 25B: 961-968.
Yang HY, Zhou C (1982). In vitro induction of haploid plants from unpollinated ovaries and ovules. Theor Appl Genet 63: 97-104.
Yang H, Zhou C, Cai D, Yan H, Wu Y, Chen (1986). in vitro culture of unfertilized ovules in *Helianthus annuus*. In Hu H, Yang H (eds): "Haploids of Higher Plants in Vitro", Berlin, Heidelberg, Springer-Verlag, pp 182-191.
Ye JM, Harvey BL, Kao KN (1985). Effects of 2,4-D and zeatin riboside on pollen callus induction in barley anther culture. Can J Plant Sci 65: 29-32.
Zhou C, Yang H, Tian H, Liu Z, Yan H (1986). In vitro culture of pollinated ovaries in *Oryza sativa* L. In Hu H, Yang H (eds): "Haploids of Higher Plants in Vitro", Berlin, Heidelberg: Springer-Verlag, pp. 165-181.

Plant Tissue and Cell Culture, pages 243–264

SELECTION PROGRAMMES FOR ISOLATION AND ANALYSIS OF MUTANTS IN PLANT CELL CULTURES

Michel Jacobs, Ioan Negrutiu, Rob Dirks and Danielle Cammaerts

Laboratory of Plant Genetics, Vrije Universiteit Brussel
B - 1640 Sint-Genesius-Rode, Belgium

INTRODUCTION

The possibility of extrapolating methodology and basic knowledge from microbial and mammalian cell genetics to somatic plant cell cultures led during the last decade to the establishment of reliable in vitro model systems and to the development of appropriate selection strategies. Despite the restrictions imposed by the life cycle characteristics and the genome complexity of a higher plant, a relatively wide range of mutants affecting biochemical pathways and developmental processes are available. Various reviews have given a comprehensive coverage of existing data (Berlin and Sasse 1985, Chaleff 1983, Flick 1983, King 1984, Maliga 1984, Negrutiu *et al* 1983, Widholm 1984, Bright *et al* 1985).

There are nowadays many reasons to believe that mutants will play an important role in the development of plant genetics and physiology. They represent suitable tools for the analysis of biochemical pathways. They are essential to test and develop somatic cell genetic techniques such as somatic hybridization and transformation. They also represent convenient material to study gene expression and to isolate key genes governing essential metabolic functions. Some of them may concern functions important in crop production and protection.

The purpose of this paper is to summarize the main results of the research which has involved the isolation and use of mutants, and more specifically those in which biochemical lessons have been identified.We like also to stress how the design of successful methods for mutant isolation is dependent on the knowledge of the biochemistry and physiology of the process under investigation. By discussing the properties of a few well characterized selectable biochemical mutants, we will also emphasize their potential value in the development of genetic manipulations in higher plants.

SELECTION OF BIOCHEMICAL MUTANTS: IN VITRO VS WHOLE PLANT.

A first aspect concerns the comparison of categories of biochemical mutants at the whole plant level with those issued of *in vitro* technology. Screening at the level of the whole plant has resulted in obtaining mutants affected in biosynthetic pathways and developmental processes as well as numerous polymorphic isozyme systems (Nelson and Burr, 1973, Marx 1983) (TABLE 1) The cell culture approach allowed efficient selective techniques and large scale screening methods to extend the spectrum of mutants to traits expressed at the cellular level.

TABLE 1. Categories of biochemical mutants isolated at whole plant (I) and cell level (II)

I	II
. Carbon metabolism (photosynthesis, photorespiration)	. Carbon assimilation . Nitrogen assimilation
. Starch biosynthesis	. Amino acid metabolism
. Carotenoid biosynthesis	. Purine and pyrimidine metabolism
. Storage proteins	
. Nitrogen metabolism	. Antibiotic resistance
. Growth factors	. Herbicide resistance
. Incompatibility factors	. Stress resistance
. Isoenzymes	. Auxotrophs
. Developmental processes	. Autotrophs
	. Developmental systems

VARIANT AND MUTANT

An important remark concerns the term *mutant* which is used at different levels of stringency in reports of their obtention in plant cell cultures. TABLE 2 indicates the criteria which must be fulfilled by a novel phenotype to be considered as a mutant. The term *mutant* should be reserved to phenotypes for which genetic or at least biochemical evidence has been provided. Besides point mutations, genetic events may include deletions, chromosomal rearrangements, gene amplification or transposition of control elements. The term variant is used for cell lines identified by a phenotypic change in culture and eventually regenerated plants therefrom, but without a clear evidence of the cause of the variation.

TABLE 2. Criteria used to assess that a novel phenotype deserves the term mutant. They are listed by order of relative importance and validity.

- Sexual transmission of the altered trait to the progeny
- Mode of inheritance (linkage data with defined genetic markers if possible)
- Genetic stability
- Molecular or biochemical evidence for the alteration (from change in nucleotide sequence to altered or deficient gene product)
- When fused with appropriate cell partners, the mutant shows a pattern of complementation consistent with that of single gene mutation (Fusion-complementation test)
- Expression of the altered trait in regenerated plants or/and in cell culture initiated therefrom
- Stability of variant phenotype in the absence of the selective pressure
- Spontaneous frequencies of forward and reverse mutation are low and significantly increased by mutagenesis

The limitations of the in vitro mutant selection due to an insufficient knowledge of the molecular basis of the studied processes and of the relationship between cellular and whole plant phenotypes are reflected by the number of reports in which only variant phenotypes are defined without accurate biochemical and in most cases genetical characterization (TABLE 3). The discrepancy between the number of mutants and the amount of variants can be also due to the type of applied selection and to the category to which belongs the cell line phenotype, as exemplified by the comparison between auxotrophs and types with potential agronomic value. For somatic cell variants obtained after mutagenesis and which have been regenerated and give a progeny, enough evidence exists demonstrating underlying genetic rather than epigenetic alteration. But a lot of variants have also resulted of the variation due to the *in vitro* culture phase.

TABLE 3. Reported characteristics of cell lines isolated in cell cultures (based on Flick 1983, and mentioned literature).

Phenotype	Induced by mutagenesis	Biochemically defined	Genetically defined	Total number reported
Amino acid and amino acid analog resistance	18	34	6	52
Antibiotic resistance	4	9	6	16
Nucleic acid base analog resistance	9	6	2	16
Agronomical value	9	7	13	31
Nitrate reductase deficiency	17	17	6	18
Auxotrophs	15	12	1	19
Miscellaneous	5	3	3	11
All types	77	88	37	163

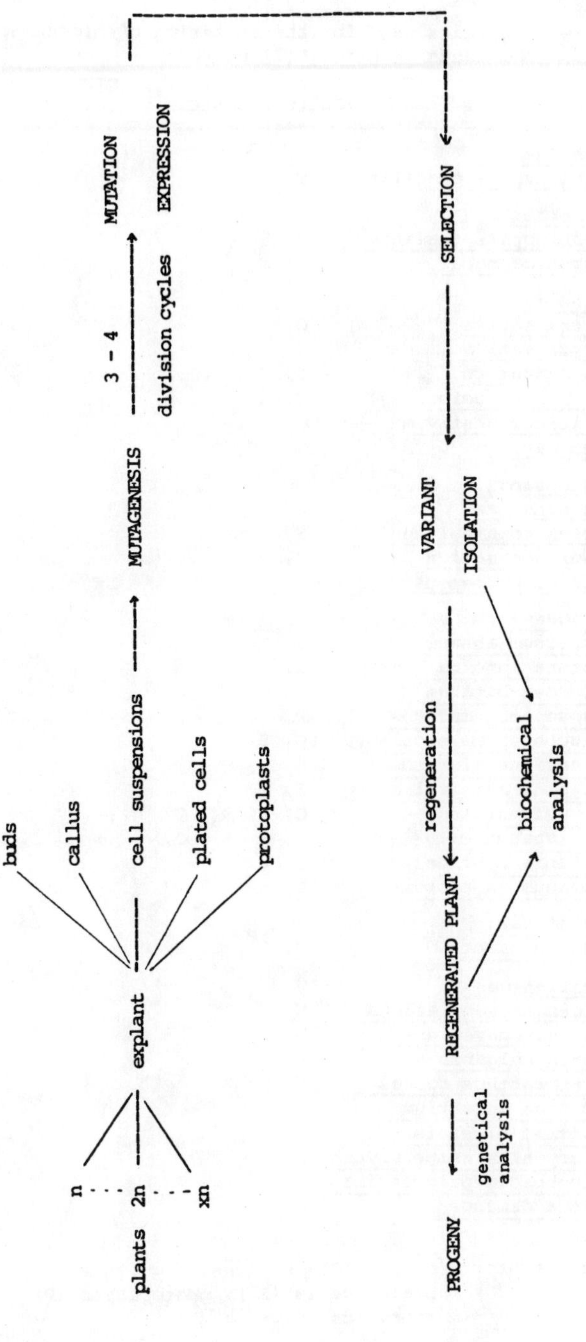

FIG. 1 . CULTURE SYSTEMS IN WHICH MUTAGENESIS-SELECTION PROGRAMMES CAN BE APPLIED.

TABLE 4 . Species used for the isolation of biochemical mutants in plant cell culture.

Species	Culture systems*
Compositae:	
Haplopappus gracilis	S^P
Cruciferae:	
Arabidopsis thaliana	S, S^P
Brassica napus	C, S
Gramineae:	
Avena sativa	C
Oryza sativa	C
Saccharum sp.	S, S^P
Sorghum bicolor	C, S
Triticum aestivum	C
Zea mays	C^E, S
Leguminosae:	
Glycine max	S
Lotus corniculatus	S
Medicago sativa	C, S
Trifolium repens	S
Solanaceae:	
Capsicum annuum	S
Datura innoxia	S, S^P
Ipomoea batatas	S
Hyoscyamus muticus	P
Lycopersicum esculentum	C, S, S^P
Nicotiana glutinosa	S
N. plumbaginifolia	S, P
N. sylvestris	C, S, S^P, C
N. tabacum	C, S, S^P, C
Petunia hybrida	S, S^P
Solanum tuberosum	C, S
Umbelliferae:	
Daucus carota	S, S^P, C, E, P
Miscellaneous:	
Acer pseudoplatanus	S, S^P
Apium graveolens	S^P
Beta vulgaris	S
Cataranthus roseus	S
Citrus aurantium	S
Citrus sinensis	C
Corydalis sempervivens	S
Euphorbia pulcherrima	S
Rosa damascena	S

*Culture systems: callus (C), suspension cultures (S), plated cells (S^P), protoplasts (P), embryoids (E).

CULTURE SYSTEMS FOR SELECTION

FIGURE 1 describes the various culture systems used to screen and select cell lines variants and possibly mutants. The ploidy level of the original explant is an essential parameter with regard to the purpose of the programme and the chance of success. Selection and screening for desired traits can be performed with meristems, callus or suspension cultures, plated cells or protoplasts. Each type of culture may have its advantages or disadvantages. With callus, exposure to the selective agent is not uniform and the possibility of mixed population of selected and normal cells can lead to the obtention of chimeral callus and plants. Suspension cultures and plated cells have to be obtained from already established cell cultures with the risks of increased chromosome instability. Protoplasts allow to manipulate uniform cell populations, eventually issued directly from differentiated tissues and to obtain microcolonies from individual protoplasts (clones). However, protoplasts of many plant species have not been regenerated into whole plants.

Whatever, culture systems are used, it is clear that the number of species utilized for the isolation of biochemical mutants in plant cell culture is still very limited and practically restricted to members of Solanaceae and a few other families (TABLE 4.)

IN VITRO SELECTION METHODS

That the particular selection method employed can markedly influence the nature of the variants obtained is reflected by their classification according to the type of selection strategy.

Most of the mutants described until now are the results of positive selection methods which allow the isolation of resistants to a selective pressure which inhibits or kills wild type cells. Such resistant cell lines are thus the most represented and are characterized by an increased tolerance to toxic products such as amino acids, amino acid analogues, antibiotics, nucleic acid base analogues, herbicides, heavy metals, toxins and pathogens, phytohormones, physical or chemical stress.

Negative selection in which growing cells are preferentially killed while those lacking a metabolic function survive is the method of choice to isolate auxotrophic mutants for amino acids, vitamins, nucleic acid bases or other nutrients. Such types have been only obtained by manipulating haploid protoplast cultures. However, sensu stricto negative selection procedures have proven difficult and most auxotrophs have been identified by total isolation methods or via enrichment procedures.

Non-selective screening remains the only possible approach when a selection strategy cannot be applied for in vitro manipulation. Analytical screening devices based on visual detection, staining methods, use of probes, radioimmuno assays, electrophoretical patterns can then be helpful to identify the few cell lines with wanted traits.

The selection schemes vary also according to how the selective pressure is applied. In most of the cases, of positive selection, the concentration of the toxic product adjoined in the culture medium is the one which stops completely the growth of wild type cells. As mentioned by Meredith, 1983,step-wise method characterised by a gradually increasing selection pressure may favor the obtention of two types of genetic change, gene amplification and mutations in organelle genome. In mammalian cells, amplification of genes have been obtained by stepwise selection in progressively increasing concentrations of the toxic agent (Arfin and Gantt,1983).

In all cases of selection programmes mentioned until now, it is obvious that the desired trait must be expressed in culture. This limits severely the phenotypes which can be identified in cell culture systems. However, the cellular components of even complex phenotype as drought tolerance remain accessible to selection at the cellular level and their modification can significantly contribute to a change in the behaviour of the whole plant (Meredith, 1983).

CLASSIFICATION OF MUTANTS OBTAINED IN CELL CULTURES.

Besides the type of selection procedure which led to their isolation, biochemical mutants can be classified in various categories which illustrate the diversity of the collection of phenotypes and genotypes available. (TABLE 5). We have considered sometimes tentatively, the following characteristics : the type of biochemical lesions, the category of genetic events associated with the mutant phenotype and the agronomical interest of the modified trait.

TABLE 5 : Classification of mutants.

CRITERIA	OBSERVED TYPES
BIOCHEMICAL LESION	. decrease or loss of activity of a metabolic enzyme . alteration of regulatory properties of enzyme . increase in enzyme activity . alteration of ribosomal protein . synthesis of a new protein

GENETIC BASIS	. nuclear events
	  . gene mutation (dominant-recessive, mono-,di- or multigenic)
	  . gene amplification
	  . transposition events
	  . chromosomal rearrangements
	. cytoplasmic events
	  . organelle genome modification

AGRONOMICAL VALUE	. nutritional quality . resistance to stress (temperature, drought, flooding, salt, toxic metals and pollutants, herbicides, pathogens and pests) . efficiency of carbon utilization . efficiency of uptake and utilization of nutrients.

This illustrates the variety of mechanisms involved in the origin of plant biochemical mutants and their value in plant biochemistry, molecular biology and breeding.

We shall now consider more specifically first a few biochemical alterations which have been identified in mutants resistant against analogs and secondly mechanisms leading to the expression of herbicide tolerance.

RESISTANCE TO ANALOGS.

Resistance can arise through modification in transport systems with as consequence no uptake or a decreased uptake of the antimetabolite in the cell. A barley mutant resistant to S-aminoethylcysteine through a recessive mutation affecting the membrane permeability for basic amino acids of the root cells has been analysed by Bright et al, 1983. Mutants of N. tabacum resistant to valine obtained from UV-mutagenized haploid mesophyll protoplasts were shown to be under the control of two recessive genes (Bourgin et al, 1985). According to the results of uptake experiments, the high level of resistance could be accounted for by a reduced uptake of neutral and acidic amino acids (Borstlap et al, 1985).

In only a few instances has resistance to analogues been associated with a modified enzyme which is involved in the metabolisation of the analogue. 6-azauracil-resistant variants of Haplopappus gracilis and Datura innoxia lack activity of uracilphosphoribosyltransferase, an enzyme involved in the conversion of uracil- and here of its analog - into a form that can be used for pyrimidine synthesis (Jones, 1984).

In other instances overproduction of amino acids has been reported as cause for the resistance. That increase has been related to altered regulation mechanisms, such as a decrease or lack of the feedback control normally exerted by the amino acid or its analog at the level of key enzymes of the pathway. For example, aspartokinase activity in lysine + threonine resistant callus of Zea mays is resistant to the feedback inhibition by lysine (Hibberd et al, 1980). Datura innoxia cell lines and regenerated plant resistant to inhibitory concentration of 5-methyltryptophan possess an altered anthranilate synthase less sensitive to feedback inhibition by tryptophan (Ranch et al , 1983).

Increase of the activity of the regulatory enzyme considered to be one of the target for analog molecules has been in a few cases correlated with the resistance expressed by variant cell lines. Tobacco cell lines selected tor resistance to the methionine analog ethionine show a 16 times increase of lysine-sensitive aspartokinase activity compared to that found in the sensitive line (Gonzales et al, 1984). However the molecular basis for the elevated aspartokinase activity in the ethionine-resistant cells has yet not been elucidates.

RESISTANCE TO HERBICIDES

The isolation and characterisation of herbicide resistant cell lines and plants illustrate quite well the potentialities offered by selecting at cell level (TABLE 6). Herbicide tolerance may be due to mutation as in the case of picloram, sulphonyl ureas, paraquat and the hormone naphtalene acetic acid, or gene amplification as is the case for phosphinithricin and probably for glyphosate. Alfalfa cell lines showing tolerance to phosphinithricin an inhibitor of glutamine synthase (GS), are characterized by an increase in GS enzyme level due to a five to ten-fold amplification of one of the alfalfa GS genes. (Doms et al, 1984 , Fischer et al, 1985). Properties shown by sulphonylureas and glyphosate tolerant plants for which detailed knowledge of the enzymic site of action exists demonstrate that crucial pathways in the metabolism of plant cells are targets for herbicide action.

Such selected resistant lines provide also a source from which resistance genes can be cloned and reinserted in the appropriate crop by transformation.

EXAMPLES OF SELECTABLE MARKERS : ANALYSIS AT THE BIOCHEMICAL AND GENETIC LEVEL.

Biochemical mutants isolated by positive selection methods include dominant as well as recessive markers (TABLE 7). Plant species and material, mutagen treatment (when applied) and selection procedures were chosen with the goal to ensure the obtention of mutants plants allowing genetic analysis. Inhibitory concentration of lysine plus threonine, S-aminoethylcysteine (a lysine analogue), chlo-

Table 6 : Examples of herbicide resistance obtained in in vitro cultures.

Herbicide	Species	Culture System	Possible cause for resistance	Inheritance	References
Picloram	*N. tabacum*	S^P	-	Mendelian, D, SD	Chaleff and Parsons, 1978
Sulphonyl ureas	*N. tabacum*		insensitive form of acetolactate synthase	Mendelian, SD	Chaleff and Mauvais, 1984
Glyphosate	*N. tabacum*	C, S^P	insensitive form of enolpyruvyl shikimate synthase	-	Singer and Mac Daniel 1985
Triazine	*N. plumbaginifolia*	P^*	altered chloroplastic binding protein	maternal	Cseplo *et al*, 1985
1-phosphinitricin	*M. sativa*	S	glutamine synthetase overproduction	gene amplification (35 kb)	Donn *et al*, 1984 Ticher *et al*, 1986
Naphtalene acetic acid	*N. tabacum*	P	-	Mendelian, D	Muller *et al*, 1985
Bentacone Phenmedipharm	*N. tabacum*	leaf and C	-	Mendelian, R	Robin and Carlson,1978
Paraquat	*L. esculentum*	S^P	-	Mendelian, D	Thomas and Pratt, 1982

C : callus, S : suspension cultures, S^P : plates cells, P: protoplasts, P^*: photomixotrophic cultures.

TABLE 7. Phenotypical, biochemical and genetic characteristics of biochemical mutants isolated in this laboratory from protoplast cultures, embryogenic cell suspensions, and embryos and seed cultures. The number of isolated mutants are into brackets.

Obtention procedure	Marker phenotype	Altered function	Inheritance	Expression: Regenerants	Expression: Progeny	Species
	I. Resistant mutants.					
Mutagenesis - selection	Lysine + threonine	Thr overproduction:	MGD	E	E	N. sylvestris (2)
		Altered aspartate	?	E	?	D. carota (1)
		kinase.	MGD	-	E	H. sativum (3)
		Decreased uptake, or THR or LYS overproduction	DGR, MGD or (?)	-	E	A. thaliana (numercus)
	s-aminoethylcysteine	Lys overproduction altered DHPS,	MGD	E	E	N. sylvestris (2)
		Decreased uptake or not established	MGR or D or (?)	-	E	A. thaliana (numerous)
Transformation	kanamycin	neomycin phosphotransferase (NPT II)	D	E	E	N. plumbaginifolia (numerous)
	hygromycin	hygromycin B phosphotransferase	D	E	E	N. plumbaginifolia (numerous)
	II. Deficient (resistant) mutants.					
Mutagenesis - selection	chlorate	defective nitrate reductase	MGR	E	E	N. plumbaginifolia (30)
	allylalcohol	defective alcohol dehydrogenase	MGR	-	E	A. thaliana (32)
	III. Deficient mutants.					
Mutagenesis - enrichment	Aminoacid requiring	His^-, Ile^-, Trp^-, Met^-, Lys^-, Ile^- + Val^-, Leu^-; Defective enzymes where assayed	R	E	(?)	N. plumbaginifolia (15)
	IV. Ti transformed plants.					
Transformation	opine	opine biosynthesis and opine dehydrogenases	D	E	E	N. sylvestris (7)
						N. plumbaginifolia (1)

Abbreviations: MG: monogenic, DG: digenic; D: dominant, R: recessive; E: expressed, nE: not expressed; (?): not established.

rate and allyl alcohol were applied to select for resistance. A type of enrichment-selection was used to get recessive, deficient amino acid requiring cell lines isolated from populations of haploid *N. plumbaginifolia* protoplasts. Opine synthetising *N. sylvestris* and *N. plumbaginifolia* plants were obtained via tumor induction or root induction.

RESISTANCE AGAINST S-AMINOETHYLCYSTEINE (AEC)

Selection in protoplast cultures of diploid *N. sylvestris* resulted in the isolation at a low frequency (2.10^{-7}) of two AEC-resistant cell lines characterized by lysine overproduction (10 to 20 fold) and by an altered dihydropicolinate synthase (DHPS), the main control enzyme of own lysine biosynthesis (Negrutiu *et al*, 1983). Lysine overproduction, resistance to AEC and the feedback insensitive form of DHPS were also expressed in regenerated plants and their progeny. Genetic analysis indicated that the resistance trait was inherited as a single dominant nuclear gene. Through a series of back-crosses, one of the mutant(RAEC 1) has been obtained as a homozygous type characterized by a DHPS form completely insensitive to lysine. This allows to study the properties of the modified enzyme and to attempt the cloning of the mutated DHPS gene for using it to transfer the lysine overproduction capacity to crop species.

CHLORATE-RESISTANT, NITRATE-REDUCTASE LESS MUTANTS.

Protoplast-derived colonies of haploid *N. plumbaginifolia* plants were selected for chlorate resistance in media supplemented with casamino acids (Negrutiu *et al*,1983). Among fifty resistant lines which could be regenerated into plants, 30 were characterized biochemically. They were classified as defective in the apoenzyme (nia type, 26 lines) or the cofactor (cnx type, 4 lines). By means of genetic crosses and fusion-complementation analysis, two of the cnx types, cnx 20 and cnx 82, were identified as belonging to the complementation group cnx A (molybednum restorable group) and cnx 27 to the molybednum non-restorable complementation group B. A new complementation group, cnx D, was revealed as the isolated type cnx 103 showed complementation with mutants belonging to the already reported categories cnx A, B and C. Very recently, a fifth cnx mutant type (cnx E) has been described (Gabard *et al*, 1986). A biochemical analysis of the cnx D mutant showed that the cnx D gene is still able to produce a 7.65 NR apoproteine indicating that the cofactor, although deficient, can still keep the two apoprotein subunits together (de Vries *et al*, 1986).

Conditional lethal mutants such as nitrate reductase deficient types can thus complement by sexual crosses. The genetic analysis demonstrated that the deficiency was inherited as a single recessive gene in both *nia* en *cnx* mutants. Crosses confirmed also the genetic complementation between apoenzyme and cofactor deficient types as well as between molybednum restorable and non-restorable *cnx* mutants (Dirks *et al*, in press). Spontaneous revertants of nitrate reductase less mutants were isolated by screening for nitrate utilization in diploid mutant protoplast culture. The reversion frequency for the *nia* mutants varies from 10^{-6} to 5.10^{-7}. For *cnx 20* about $1.9.10^{-7}$ colonies were screened on nitrate medium without finding revertants. Revertants of the *nia* type proved to be tetraploid and segregation analysis showed that only one of the four *nia* gene had reverted to a functional allele. (Dirks *et al*, 1986).

AUXOTROPHIC MUTANTS

The isolation of a wide spectrum of nutritional mutants is perhaps the most successful example of the use of haploid cell and mainly protoplast cultures in plants. With the excemtion of thiamine less mutants in *Arabidopsis* and *Lycopersicon*, such auxotrophs were not obtained by rescuing non-growing seedlings in M_2 generations.

Non-selective procedures as well as enrichment/selection schemes have been applied to obtain auxotrophs. Although they have led to the recovery of auxotrophs, neither arsenate or bromodeoxyuridine enrichment selection technique has proved to be ideal and reproducible methods. This is most probably due to the limitation of the plant cell culture system in itself, particularly the lack of synchrony in the first divisions after protoplast isolation and the requirement for high densities (at least 10^4 protoplasts/ml) of cells.

The spectrum of isolated auxotrophs comprises requirements for pantothenate, adenine, isoleucine plus valine, threonine and histidine in *Datura innoxia* (King, 1985) for histidine, tryptophan, nicotinamide and leucine in *Hyoscyamus muticus* (King, 1984). In *N. plumbaginifolia* isoleucine, leucine and uracil have been reported (Sidorov and Maliga, 1982). In the same species, altogether 15 auxotrophs were isolated in our laboratory (see TABLE 8). All the lines are stable both as callus after subcultures and as regenera-

TABLE 8. Properties of auxotrophic lines isolated in haploid protoplast culture of *N. plumbaginifolia*.

Auxotroph type and number ( )	Amino acid concentration (mM) for callus and regenerants	Precursors and amino acids supporting growth	Defective enzyme	Fusion - complementation +	Fusion - complementation -	Regeneration and meiotic transmission
his^-I (3)	0.3 - 0.5	histidinol (H-ol), H-ol phosphate, his	IAP-amino-transferase	NR^-, trp^- (H.m.)	his^-I	abortive embryos
his^-II (1)	0.3 - 0.5	histidinol (H-ol), H-ol phosphate, his	not determined	his^-I		plant (transmission to be tested)
met^- (1)	0.3	homocysteine, met	β-cystathionase	$ileu^-$, his^-I	-	seeds requiring met for germination
$ileu^-$ (3)	0.2 - 0.3	2-oxobutyrate 2-aminobutyrate, ileu	threonine deaminase	met^-	$ileu^-$	only rosettes
val^- + $ileu^-$ (3)		val + ileu val+ileu+leu	dihydroxy-acid dehydratase	not tested		callus
leu^- (2)	0.1 - 0.3	ketoisocaproic acid, leu	2. OH-3 carboxy-caproate isomerase or dehydrogenase	not tested		no regenerants
trp^- (1)		only trp	tryptophan synthase	NR^-, his^-	trp^- (H.m.)	only rosettes
lys^- (1)	0.15	lysine	not determined	not tested		plant, to be tested.

(trp^-(H.m.) = trp^- line of *Hyoscyamus muticus*

ted plants. They were characterized partly by precursor feeding tests, by enzyme assays, by protoplast fusion-complementation experiments and by their transmission when fertile plants were available.

In the case of the histidine-requiring types (his^-), three lines are characterized by a very low activity of the enzyme imidazole acetol phosphate (IAP) aminotransferase which controls the second step of the histidine biosynthesis. Somatic hybridization between a fourth his^- type and lines deficient in IAP aminotransferase resulted in complementation, indicating the existence of a second mutation. The methionine- requiring auxotroph lacks β-cystathionase activity. The biochemical lesions in six other amino acid-requiring auxotrophic lines have been deeply investigated by means of feeding experiments with ^{14}C, and unlabelled substrates and enzyme analysis. (Wallsgrove et al, 1986). Three of the lines require isoleucine for growth, are unable to synthetise 2-oxobutyrate in vivo and have no detectable threonine dehydratase in vitro. The other three lines requiring isoleucine + valine accumulate radioactive dihydroxymethylvalerate when fed with ^{14}C-L-threonine and contain no detectable dihydroxyacid dehydratase, the third enzyme of the common pathway of the branched-chain amino acids, leucine, valine and isoleucine.

For some of the amino acid auxotrophs it was possible to obtain regeneration but the transmission of the mutant trait through meiosis was only obtained in selfed methionine-requiring auxotroph. White, non-growing seedlings were observed when methionine was not supplemented in the medium. Histidine deficiency is not sexually transmitted. Revertants obtained from his^- lines do not give rise to his^- segregants by selfing while nitrate reductase revertants segregate 1 deficient to 3 wild type as theoretically expected. While a normal development of auxotrophic plants can be obtained by supplementation of the lacking nutrient, on the contrary normal fertilisation and/or seed formation appear to be severely hampered. Such data may explain why amino acid auxotrophs have so far not been isolated in systems based on seed and seedling screening.

CONCLUSIONS AND PROSPECTS

The development of the cellular systems envisaged in this article allowed to obtain an impressive collection of variant and mutant lines in still a limited amount of species. By disrupting physiological and developmental functions, new tools were created for investigating and modifying genetical and biochemical processes, as illustrated for mutants in the nitrate reduction system (TABLE 9)

TABLE 9 : Possible uses of nitrate reductase-deficient mutants.

- behaviour of auxotrophs in cell cultures
- biochemical markers in fusion experiments
- biochemical markers in genetic crosses
- effects of mutagenic treatments in cell cultures
- reversion studies and behaviour of revertants
- transformation experiments taking profit of the selection for NR^+
- knowledge of nitrogen metabolism and synthesis of the NR enzyme
- basis for a more efficient uptake and utilization of nitrate and ammonium

The availability of selectable markers is a prerequisite to assess methods for transferring limited amounts of genetic information into plant cells. Methods to improve the transformation conditions of plant protoplasts by DNA- and vector-mediated transfer are dependent on the use of convenient receptor systems as available in N. plumbaginifolia. High transformation rates will allow to develop gene isolation procedures based on the rescue of mutants using genomic clones harbouring the corresponding wild type genes. Protoplast fusion by using irradiated donor partners carrying "rescue" genes for correcting deficient receptor cells represents another approach for partial chromosome transfer.

In a more applied perspective devising better selection programmes makes possible to get mutants providing valid information on the molecular basis of important agricultural traits. In that respect herbicide resistance mutants appear already as a first meaningful example of the practical importance of in vitro selection.

ACKNOWLEDGMENTS

The authors like to thank J. Calcoen and J. Zwaenepoel for typing the manuscript.

REFERENCES

1. Arfin SM, Pratt D (1982). Amino acid analogs as tools for the study of amino acid metabolism in mammalian cells. Trends in Bioc Science 8: 163-164.
2. Berlin J, Sasse F (1985). Selection and screening techniques for plant cell cultures. Advances in Biochemical engineering/ Biotechnology 31 : 99-132.
3. Borstlap AC, Schuurmans, J. Bourgin JP (1985). Amino-acid transport mutant of Nicotiana tabacum. Planta :140-144.
4. Bourgin JP, Goujaud J, Missonier C, Pethe C (1985). Valine-resistance, a potential marker in plant cell genetics, I Distinction between two types of valine-resistant tobacco mutants isolated from protoplast-derived cells. Genetics 109 : 393-407.
5. Bright SWJ, Kueh JSH, Rognes SE (1983) Lysine transport in two barley mutants with altered uptake of basic amino acids in the root. Plant Physiol 72: 821-24.
6. Bright SWJ, Ooms G, Foulger D, Evans N, Karp A (1985). Mutation and tissue culture, in press.
7. Chaleff RS (1983). Isolation of agronomically useful mutants from plant cell cultures. Science 219 : 676-682.
8. Chaleff RS (1980) Further characterization of picloram-tolerant mutants of Nicotiana tabacum . Theor. Appl. Genet. 58 : 591-95.
9. Chaleff RS, Mauvais, CJ (1984). Acetolactate synthase in the site of action of two sulfonylurea herbicides in higher plants. Science 224 : 1443-4.
10. Cseplo A, Medgyesty P, Hideg E, Demeter S, Marton L, Maliga P (1985). Triazine-resistant Nicotiana mutants from photomixotrophic cell cultures. Mol Gen Genet,200 : 508-10
11. Devries SE, Dirks R, Mendel RR, Srhaart JG, Feenstra WJ (1986). Biochemical characterization of some nitrate reductase-deficient mutants of Nicotiana plumbaginifolia Plant Science, in press.
12. Dirks R, Negrutiu I, Sidorov V, Jacobs M (1985). Complementation analysis by somatic hybridization and genetic crosses of nitrate reductase-deficient mutants of Nicotiana plumbaginifolia. Evidence for a new category

of cnx mutants. Mol Gen Genet 201 : 339-41.

13. Dirks R, Negrutiu I, Heinderyckx M, Jacobs M (1986) Genetic analysis of revertants for the nitrate reductase function of Nicotiana plumbaginifolia . Mol Gen Genet 202 : 309-11.
14. Dirks, R, Negrutiu I, Jacobs M (1986). Genetic analysis of nitrate reductase deficient mutants of Nicotiana plumbaginifolia plants regenerated from haploid protoplast cultures, in preparation.
15. Donn G, Fischer E, Smith A, Goodman M (1984). Herbicide resistant alfalfa cells : an example of gene amplification in plants. J Mol Appl Genet 2, 621-35.
16. Flick CE (1983). Isolation of mutants from cell culture In : Handbook of Plant Cell Culture Vol I (Evans DA, Shays WR, Ammirato, PV, Yamada Y Eds). Macmillan Publ. C° New York 393-442.
17. Gabard J, Marion-Poll A, Goujaud J, Missonier C, Grafe R, Caboche M (1986). Mutant selection from Nicotiana plumbaginifolia mesophyll protoplasts. In Genetic Engineering of Plants and Micro-organisms important for agriculture, C.E.E. meeting, Wageningen, 71-72.
18. Gonzales RA, Das PK, Widholm JM (1984).Characterization of cultured tobacco cell lines selected for resistance to a methionine analog, ethionine. Plant Physiol 74: 640-44.
19. Hibberd KA, Walter T, Green CE, Gengenbach BG (1980). Selection and characterization of feedback-insensitive tissue culture of maize. Planta 145 : 183-87.
20. Jones GE (1984) 6-azaurucil-resistant variants of cultured plant cells lack uracil phosphoribosyl transferase activity. Plant Physiol 75 : 161-5.
21. King J (1985) The response of Datura innoxia wild-type and auxotrophic cells to several enrichment agents. Plant Cell Reports 4 : 123-125.
22. King PJ (1984). From single cells to mutant plants. Oxford Surveys of Plant Molecular and Cell Biology 1:7-32.
23. Maliga P (1984). Isolation and characterization of mutants in plant cell culture. Ann Rev Plant Physiol 35: 519 -542.
24. Marton L, Dung TM, Mendel R, Maliga P (1982) Nitrate reductase deficient cell lines from haploid protoplast cultures of Nicotiana plumbaginifolia .Mol Gen Genet 182: 301-4.
25. Marx GA (1982) Developmental mutants in some annual seed plants. Ann Rev Plant Physiol 34 : 389-417.

26. Meredith CP (1983) On being selective : mutants from cultured cells. Plant Mol Biol Report 1 : 105-11.
27. Muller JF, Goujaud J, Caboche M (1985) Isolation in vitro of naphtalene acetic acid-tolerant mutants of Nicotiana tabacum which are impaired in root morphogenesis. Mol Gen Genet 189 : 194-200.
28. Negrutiu I, Jacobs M, Caboche M (1983) Advances in somatic cell genetics of higher plants. The protoplast approach in basic studies on mutagenesis and isolation of biochemical mutants. Theor Appl Genet 67 : 289-304.
29. Negrutiu I, Dirks R, Jacobs M (1983). Regeneration of fully nitrate reductase-deficient mutants from protoplast culture of N. plumbaginifolia.Theor Appl Genet 66 : 341-347.
30. Negrutiu I, Cattoir-Reynaerts A. Verbruggen I, Jacobs M (1984). Lysine overproducer mutants with an altered dihydrodipicolinate synthase from protoplasts of N. sylvestris. Theor Appl Genet 68 : 11-20.
31. Negrutiu I, De Brouwer D, Dirks R, Jacobs M (1985). Amino acid auxotrophs from protoplast cultures of Nicotiana plumbaginifolia I BUDR enrichment selection, plant regeneration and general characterization. Mol Gen Genet. 199: 330-337.
32. Nelson OE, Burr B (1973) Biochemical genetics of higher plants. Ann Rev Plant Physiol 24 : 493-518
33. Radin DN, Carlson PS (1978) Herbicide-tolerant tobacco mutants selected in situ and recovered via regeneration from cell culture. Genetic Research 32 : 85-89.
34. Ranch JP, Ricks, Brotherton JE, Widholm J (1983) Expression of 5-methyltryptophan resistance in plants regenerated from resistant cell lines of Datura innoxia . Plant Physiol 71 : 136-40.
35. Sidorov VA, Menczel L, Maliga P (1981) Isoleucine-requiring Nicotiana plant deficient in threonine deaminase. Nature 294 : 87.
36. Sidorov V, Maliga P (1982) Fusion-complementation analysis of auxotrophic and chlorophyll-deficient lines isolated in haploid Nicotiana plumbaginifolia protoplast cultures. Mol Gen Genet 186: 328 - 332.
37. Singer SR, Mc Daniel CN (1985) Selection of glyphosate-tolerant tobacco calli and the expression of this tolerance in regenerated plants. Plant Physiol 78 : 411-16
38. Thomas BR, Pratt D (1982) Isolation of paraquat-tolerant mutants from tomato cell cultures. Theor Appl Genet 63 : 169-76.

39. Fischer E, Das Sarma S, Goodman HM (1986) Nucleotide sequence of an alfalfa glutamine synthetase gene. Mol Gen Genet 203 : 221-9.
40. Wallsgrove RM, Risiott R, Negrutiu I, Bright SWJ (1986) Biochemical characterization of Nicotiana plumbaginifolia auxotroph that require branched-chain amino acids Plant Cell Reports, in press.
41. Widholm JM (1984) Selection and characterization of plant cell mutants for molecular biology studies. IAPTC Newsletter 44 : 2 - 8.

Plant Tissue and Cell Culture, pages 265–274

RECOMBINATION OF CHLOROPLASTS IN CULTURED PLANT CELLS

Pal Maliga, Erzsebet Fejes, Zora Svab
and Dean Engler

Advanced Genetic Sciences, Inc.,
Oakland, CA 94608 USA

INTRODUCTION

An almost universal feature of the circular chloroplast genome (cpDNA) in flowering plants is a large inverted repeat sequence, some 10-25 kilobase pairs (kb) in size, which separates the remainder of the molecule into single copy regions of about 80 kb and 20 kb (Palmer, 1985; Fig. 1). Several lines of evidence suggest that genetic recombination occurs in chloroplasts of flowering plants. Formation of circular head-to-head dimers observed by electron microscopy was explained by intermolecular recombination between two circular monomers through the repeat (Kolodner and Tewari, 1979). Presence of two types of chloroplast genomes in a single plant which differ in the relative orientation of the single copy regions was attributed to interconversion of the circular genome (flip-flop recombination) through the inverted repeats (Palmer, 1983; Mubumbila et al. 1983). Intramolecular recombination was the suggested mechanism by which deleted forms of plastid DNA were obtained in albino wheat plants (Day and Ellis, 1984; 1985) and the same mechanism was implicated in the evolution of the chloroplast genome (reviewed in Palmer, 1985).

Attempts to demonstrate recombination between genetically marked chloroplast genomes of *Oenothera* species have failed (Kutzelnigg and Stubbe, 1974; Chiu and Sears, 1985). In *Oenothera*, chloroplasts are transmitted from both parents to the offspring. Screening for chloroplast recombinants was by complementation of chloroplast-encoded pigment mutations in seedlings. Lack of recovering

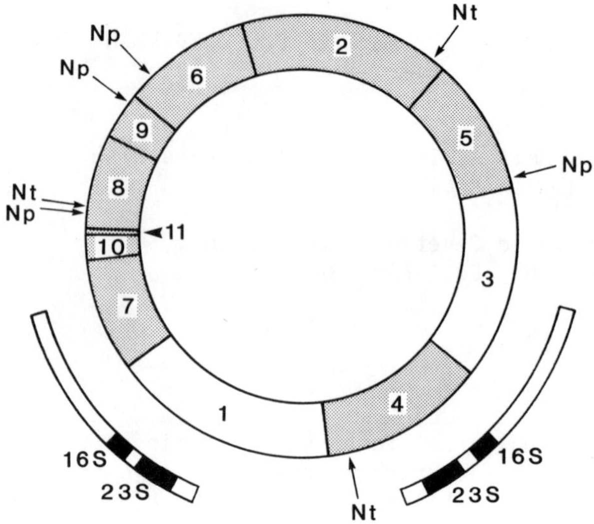

Figure 1. SalI restriction map of the pt14 chloroplast genome. The inverted repeat regions are denoted by concentric bars, showing the location of the 16 S and 23 S rRNA genes. The SalI maps of the parental species, *N. tabacum* and *N. plumbaginifolia* and their somatic hybrid pt14 are identical. Arrows point to parent-specific restriction sites that were tested by hybridizing cloned cpDNA to total cellular DNA (Nt, *N. tabacum*; Np, *N. plumbaginifolia*; data from Medgyesy et al., 1985). Cloned SalI fragments tested for recombination sites are shaded.

chloroplast recombinants in a population of 7500 plants was attributed to recombination being a rare event (Chiu and Sears, 1985).

Somatic cell fusion techniques made it feasible to obtain heteroplastidic cells in species in which chloroplasts are maternally inherited. Formation of cytoplasmic hybrids has been reviewed by Galun and Aviv (1983), Gleba and Evans (1983), Maliga and Menczel (1986) and Maliga (1986). No chloroplast recombination has been detected in fusion-derived clones, as judged by analysis of restriction fragment length polymorphism in *Nicotiana*, *Petunia* and *Brassica* lines, and in *Solanum* and *Lycopersicon* somatic hybrids (reviewed in Maliga and Menczel, 1986). The only chloroplast recombinant described has been identified in a somatic hybrid of

Nicotiana tabacum and Nicotiana plumbaginifolia, pt14, reported by Medgyesy et al. (1985). In this paper intergenomic chloroplast recombination will be discussed using the pt14 line as an example.

Selection of Chloroplast Recombinants by Genetic Markers

A genetic screen utilizing chloroplast-encoded drug resistance was designed to facilitate the recovery of the chloroplast recombinants. Streptomycin resistance carried by the N. tabacum line SR1 allows greening of cells in culture medium in the presence of streptomycin (Maliga et al. 1975). Sensitive cells, however, are bleached on a selective medium. In a derivative of the SR1 mutant, SR1-A15, greening is blocked by a second cytoplasmic mutation (Svab and Maliga, 1986). Thus in SR1-A15 streptomycin resistance is present but is not expressed phenotypically (i.e. greening). Chloroplast recombination events were screened and recovered following fusion of the SR1-A15 mutant with the LR400 N. plumbaginifolia line that is streptomycin sensitive, normal green, and carries a maternal lincomycin resistance mutation. Selection of recombinant chloroplast genomes was based on the expression of streptomycin resistance from the SR1-A15 chloroplast genome as the result of chloroplast recombination (Medgyesy et al., 1985).

Study of chloroplast recombination should also be feasible using other chloroplast markers, including pigment mutations and resistance to streptomycin, lincomycin, spectinomycin, chloramphenicol, triazine herbicides and tentoxin (reviewed in Maliga, 1984; Maliga and Menczel, 1986).

Frequency of Chloroplast Recombination

Of the 1.9×10^5 colonies screened one chloroplast recombinant, line pt14, was recovered. According to a tentative estimate, this population contained 2×10^3 to 2×10^4 fused cells containing a mixed chloroplast population (heteroplastidic cells; Medgyesy et al., 1985). Since recombination between the chloroplast genomes of this line was very extensive (see below) we assume that the rate limiting step of chloroplast recombination is the physical availability of the different chloroplast genomes for

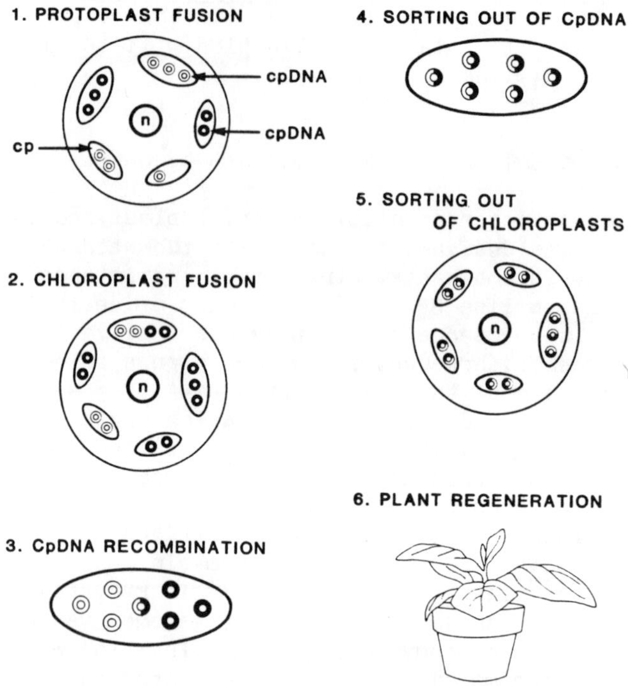

Figure 2. Plants with recombinant cpDNA from fused protoplasts.

recombination. Recombination takes place within chloroplasts that contain cpDNA from the two parental lines. The mechanism by which the heterozygous chloroplasts are formed is not known. It could be chloroplast fusion, or a transformation-like process. Since organelle fusion has been reported on the basis of microscopic observations (Wildman et al. 1962) at this time we assume that heterozygous chloroplasts are formed by organelle fusion. This assumption has been incorporated into Fig. 2 which shows the stages of chloroplast recombinant formation from heteroplastidic cells to a regenerated cybrid plant.

Sorting Out of the Recombinant Chloroplast Genome

Sorting out of the parental cpDNA molecules had been complete by the time of regenerating the pt14 plant. This conclusion is based on probing total cellular DNA isolated from regenerated plants with cloned cpDNA fragments that were suitable for differentiation of the parental cpDNA molecules by restriction fragment length polymorphism. Incomplete segregation would have been detected by the presence of characteristic fragments from both parents. No heterogeneity in hybridization patterns was seen for the seven parent-specific restriction sites that are randomly distributed around the chloroplast genome (Fig. 1, data from Medgyesy et al., 1985; Maliga and Fejes, 1985). Accordingly, the stable cybrid carrying the recombinant cpDNA was obtained by sorting out of cpDNA molecules within a chloroplast yielding a pure population of recombinant cpDNA. Sorting out of cpDNA at the chloroplast level was followed by the elimination of parental chloroplasts (Fig. 2). Since the green, resistant cells proliferate faster than the sensitive cells (Maliga et al., 1975) application of the selective medium not only allowed identification of the clone carrying the recombinant cpDNA but probably also accelerated the sorting out process.

Physical Evidence for CpDNA Recombination

The *SalI* maps of the parental lines SR1-A15 and LR400, and of the chloroplast recombinant pt14 are identical (Svab and Maliga, 1986; Medgyesy et al., 1985). *SalI* fragments with identical size contain the corresponding DNA sequences and therefore could be conveniently compared for the presence of recombination sites. Cloned *Sal* I fragments were digested with a number of restriction endonucleases, and separated by size on acrylamide gels. Lanes containing DNA from the recombinant line were inspected for the presence of parent-specific fragments and novel, non-parental fragments. Presence of parent-specific fragments from each parent was evidence that somewhere within the *SalI* clone recombination had taken place. Novel fragments, if generated by restriction at conserved parent-specific sites, were indicative of recombination within the non-parental fragment. Such evidence for chloroplast recombination sites has been provided in *SalI* fragments No. 2, 4, 5, 6, 7, 8, 9, and 10 (Maliga and Fejes, 1985; unpublished; shaded area on Fig. 1) that in total represent about 70 % of the chloroplast genome (Seyer et al., 1981). No recombination site was found in the smallest (0.65 kb) *SalI* fragment No. 11. *SalI* fragments No. 1 and 3 have not been tested.

Evidence discussed above has indicated that recombination occured within the studied region. The number of recombination sites could not be estimated from these data because the same recombination site may have been identified with different enzymes. In order to estimate the number of recombination sites within the pt14 cpDNA a detailed physical map of the SalI fragment No. 8 was prepared for the parental species and line pt14 (Fig. 3A). In the 11.5 kb fragment six recombination sites have been identified. The borders of the region within which the recombination took place are marked by arrowheads in Fig. 3C. Positions of the recombination sites have been determined by combining three kinds of data. One was the presence or absence of parent-specific sites, such as the PstI sites around recombination border #6. Parent-specific distances between conserved sites, such as the SalI site and the HindIII sites on the left in Fig. 3A, were also used. Length polymorphism was probably generated by a series of small insertions and deletions that is characteristic of chloroplast genome evolution (Palmer, 1985). The third method to identify parent-specific regions and recombination sites was hybridization of isolated, labelled parent-specific fragments to restriction fragments with known map positions (Fig. 3B). If the entire 160 kb chloroplast genome (Seyer et al., 1973) of line pt14 was involved in the recombination process, and the recombination sites are as frequent in the rest of the chloroplast genome as in SalI fragment No. 8, line pt14 should contain about 80 recombination sites.

Mechanism of CpDNA recombination

The high number of recombination sites identified in the pt14 cpDNA indicates that recombination in chloroplasts should be a frequent event. An active recombination system is also suggested by finding equimolar amounts of the products of the chloroplast flip-flop recombination through the inverted repeat (Palmer 1983, Mubumbila et al., 1983) and by the frequent formation of dimers (Kolodner and Tewari, 1979). Since the chloroplasts have an active recombination system, the limiting step in the formation of chloroplast recombinants is probably the step that generates the heterozygous chloroplasts, that is the frequency of chloroplast fusion.

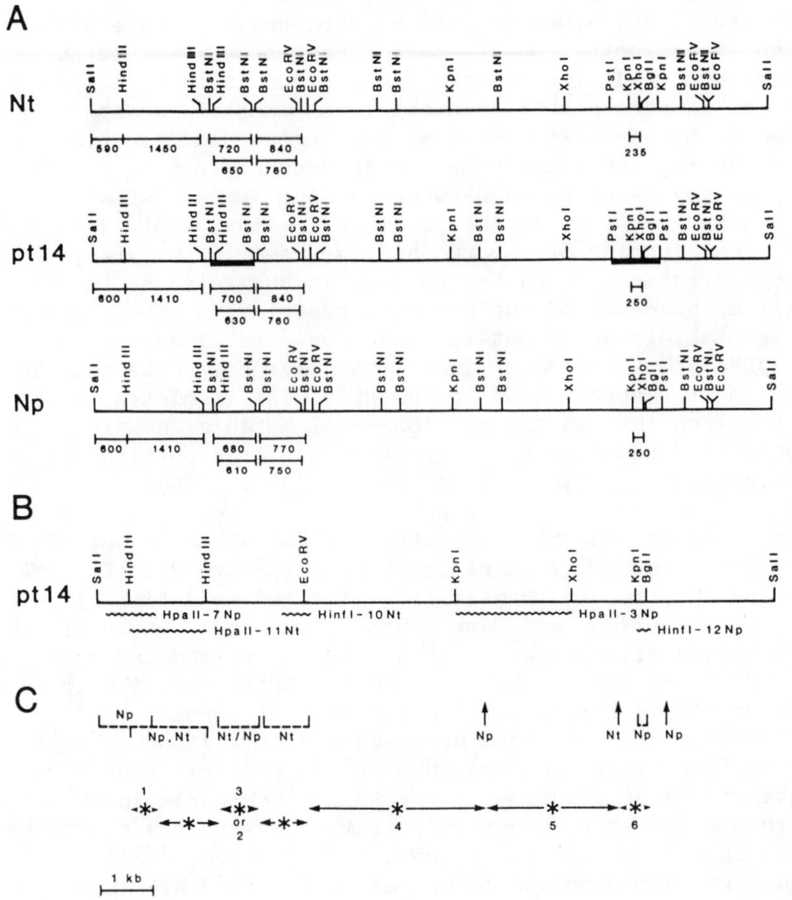

Figure 3. Identification of recombination sites in the 11.5 kb Sal I fragment isolated from pt14 cpDNA. Physical maps of the corresponding regions of N. tabacum (Nt), pt14 and N. plumbaginifolia (Np) cpDNA are shown in A. Numbers below the bars give distances in base pairs between conserved restriction sites. Hybridization of isolated parent-specific fragments to fragments with known map position is shown in B. Deduced position of recombination sites (*), and data used to identify their position is summarized in C. Brackets mark regions within which a parent-specific or recombination site should be present, vertical arrows indicate the position of parent-specific restriction sites.

Frequent recombination events, the flip-flop recombination and dimer formation, were shown to take place through the inverted repeat. *SalI* fragment No. 8 is in the large unique region. Finding recombination in this region indicates therefore that frequent recombination is not limited to the inverted repeats. Homologous recombination in the unique regions would remain undetected unless recombination takes place between cpDNA molecules with distinct markers such as the polymorphism that evolved in the chloroplast genomes of the two species used in our study. Homologous recombination in our case is suggested by the overall maintenance of the parental restriction sites, and of the parental distances between the conserved sites. In the pt14 cpDNA no deletions or inversions have been detected. The mechanism of recombination in the pt14 line is clearly distinct from that in the albino cereal mutants in which recombination resulted in elimination of large portions of the chloroplast genome (Day and Ellis, 1984; 1985).

The precise mechanism of cpDNA recombination is not known. One possibility is recombination through specific DNA sequences, a proposition made for cpDNA recombination by Palmer et al. (1985) and Howe (1985). The large number of recombination sites suggests that this may not be the case, or if there are such recombinogenic sequences they should be widely scattered throughout the chloroplast genome. Alternatively, chloroplasts may have a generalized homologous recombination system and are involved in multiple rounds of recombinations as described for yeast mitochondria in which genetic linkage is apparent only if the markers are separated by less than 1000 base pairs (reviewed by Dujon, 1981). Branch migrating junction formed as an intermediate in a single reciprocal recombination event, coupled with heteroduplex repair would also be a suitable mechanism to generate a large number of recombination sites if the repair is according to an alternating template. This mechanism was suggested by Bedbrook et al. (1977) to maintain sequence indentity by copy correction within the chloroplast repeats that was found in plastome mutants (Gordon et al., 1982; Myers et al., 1982) and in the sequenced repeats in *Nicotiana tabacum* (Shinozaki et al., 1986).

REFERENCES

Bedbrook JR, Kolodner R, Bogorad L (1977). Zea mays chloroplast ribosomal RNA genes are part of a 22,000 base pair inverted repeat. Cell 11:739-749.

Chiu WL,Sears BB (1985). Recombination between chloroplast DNAs does not occur in sexual crosses of Oenothera. Mol Gen Genet 198:525-528.

Day A, Ellis THN (1984). Chloroplast DNA deletions associated with wheat plants regenerated from pollen: possible basis for maternal inheritance of chloroplasts. Cell 39:359-368.

Day A, Ellis THN (1985). Deleted forms of plastid DNA in albino plants from cereal anther culture. Curr Genet 9:671-678.

Dujon B (1981). Mitochondrial genetics and functions. In Strathern JN, Jones EW, Broach JR (eds): "The Molecular Biology of the Yeast Saccharomyces: Life Cycle and Inheritance", New York: Cold Spring Harbor Laboratory, pp 505-635.

Galun E, Aviv D (1983). Cytoplasmic hybridization: genetic and breeding applicatins. In Evans DA, Sharp WR, Ammirato PV, Yamada Y (eds): "Handbook of Plant Cell Culture Vol I. Techniques for Propagation and Breeding," New York: Macmillan, pp 358-392.

Gleba YY, Evans DA (1983). Genetic analysis of somatic hybrid plants. In Evans DA, Sharp WR, Ammirato PV, Yamada Y (eds): "Handbook of Plant Cell Culture Vol I. Techniques for Propagation and breeding," New York: Macmillan, pp 322-357.

Gordon KHJ, Crouse EJ, Bohnert HJ, Herrmann RG (1982). Physical mapping of differences in chloroplast DNA of five wild-type plastomes in Oenothera subsection Euoenothera. Theor Appl Genet 61:373-384.

Howe CJ (1985). The endpoints of an inversion in wheat chloroplast DNA are associated with short repeat sequences containing homology to att-lambda. Curr Genet 10:139-145

Kolodner R, Tewari KK (1979). Inverted repeats in chloroplast DNA from higher plants. Proc Natl Acad Sci USA 76:41-45.

Kutzelnigg H, Stubbe W (1974). Investigations on plastome mutants of Oenothera. I. General considerations. Sub Cell Biochem 3:73-89.

Maliga P (1984). Isolation and characterization of mutants in cell culture. Annu Rev Plant Physiol 35:519-542.

Maliga P (1986). Cell fusion to introduce genetic information coded by chloroplasts and mitochondria in flowering plants. In Bogorad L (ed): "Molecular Developmental Biology," New York: Alan R. Liss, pp 45-53.

Maliga P, Fejes E (1985). Physical evidence for recombination of chloroplast DNA in a somatic hybrid of _Nicotiana tabacum_ and _Nicotiana plumbaginifolia_. In Freeing M (ed): "Plant Genetics," New York: Alan R. Liss, pp 641-650.

Maliga P, Menczel L (1986). Chloroplast transfer and recombination through protoplast fusion. In Vasil IK (ed): "Cell culture and Somatic Cell Genetics Vol III.," Orlando, Academic Press, in press.

Maliga P, Breznovits A, Marton L (1975). Non-Mendelian streptomycin-resistant tobacco with latered chloroplasts and mitochondria. Nature 255:401-402.

Megyesy P, Fejes E, Maliga P (1985). Interspecific chloroplast recombination in a _Nicotiana_ somatic hybrid. Proc Natl Acad Sci USA 82:6960-6964.

Mubumbila M, Gordon KHJ, Crouse EJ, Burkard G, Weil JH (1983). Gene 21:257-266

Myers AM, Grant DM, Rabert DK, Harris EH, Boynton JE, Gillham NW (1982). Plasmid 7:133-151.

Palmer DJ (1983). Chloroplast DNA exists in two orientations. Nature 301:92-93.

Palmer DJ (1985). Comparative organization of chloroplast genomes. Ann Rev Genet 19:325-354.

Palmer JD, Boynton JE, Gillham NW, Harris EH (1985). Evolution and recombination of the large inverted repeat in _Chlamydomonas_ chloroplast DNA. In Steinback KE, Bonitz S, Arntzen CJ, Bogorad L (eds): "Molecular Biology of the Photosynthetic Apparatus," New York: Cold Spring Harbor Laboratory, pp 269-278.

Seyer P, Kowalik KV, Herrmann RG (1981). A physical map of _Nicotiana tabacum_ plastid DNA including the location of structural genes for ribosomal RNAs and the large subunit of ribulose bisphosphate carboxylase/oxygenase. Curr Genet 3:189-204.

Shinozaki K, Ohme M, Tanaka M, Wakasugi T, Hayashida N, Matsubayasji T, Zaita N, Chunwongse J, Obokata J, Yamaguchi-Shinozaki K, Ohto C, Torazawa K, Meng BY, Sugita M, Deno H, Kamogashira T, Yamada K, Kusuda J, Takaiwa F, Kato A, Tohdoh N, Shimada H, Sugiura M (1986). The complete nucleotide sequence of tobacco chloroplast genome: its gene organization and expression. EMBO J, in press.

Svab Z, Maliga P (1986). _Nicotiana tabacum_ mutants with chloroplast encoded streptomycin resistance and pigment deficiency. Theor Appl Genet, in press.

Wildman SG, Hongladarom T, Honda SI (1962). Chloroplasts and mitochondria in living plant cells: cinephotomicrographic studies. Science 138:434-436.

Plant Tissue and Cell Culture, pages 275–286

SOMACLONAL VARIATION AND GENOMIC FLUX

W.R. Scowcroft, R.I.S. Brettell, S.A. Ryan,
P.A. Davies, M.A. Pallotta

BioTechnica Canada, Calgary, Alberta (W.R.S.),
CSIRO Division of Plant Industry, Canberra, ACT,
Australia (R.I.S.B., S.A.R., P.A.D., M.A.P.)

INTRODUCTION

The displacement of eukaryotic cells from their normal course of organized development leads to perturbation of supposedly stable genomes. Nowhere is this more apparent than when plant tissues have been subjected to a cell and/or tissue culture regime.

Cytological analyses of cells from tissue culture have shown that chromosomal instability appears to be the rule rather than the exception (Bayliss, 1980). Apart from chromosome number changes, structural alterations such as deletions, duplications, translocations and centric fusions have been observed at the morphological level. More refined structural discontinuities in the chromosomes of tissue cultured cells have been uncovered by the use of techniques such as Giemsa-banding (Gould, 1982).

The capacity to regenerate mature, fertile plants from tissue cultures provided a new dimension to the analysis of variation arising during tissue culture. The analysis of plants regenerated from tissue culture enabled the application of a genetic, and particularly a Mendelian, perspective to tissue culture variation. This led to the surprising conclusion that clonal uniformity through tissue culture propagation was a myth. Rather, it appeared, tissue culture "gave rise" to mutations at an unexpectedly high level. Larkin and Scowcroft (1981) argued that this phenomenon was sufficiently widespread that it could no longer be ignored as artifactual. These authors proposed the term somaclonal variation to describe enhanced genetic variability arising during tissue culture.

SOMACLONAL VARIATION

A number of recent reviews have treated the subject of somaclonal variation from the standpoint of its ubiquity, the likely causes and origins and its potential impact on plant improvement (Larkin and Scowcroft, 1981; 1983; Orton, 1984; Evans *et al.*, 1984; Larkin *et al.*, 1985; Scowcroft *et al.*, 1985; Karp and Bright, 1985; Scowcroft, 1985). A relatively comprehensive list of species displaying somaclonal variation is provided by Scowcroft (1985).

Various analyses have revealed a wide spectrum of mutational events which occur during tissue culture. These include mutants which behave as classical Mendelian mutants, genetic changes which affect characters under the control of multigene families, those which affect polygenically determined traits and those which give the appearance of resulting from activation of transposable elements. Polyploid and aneuploid variants occur frequently, as do chromosomal rearrangements such as duplications, deficiencies, translocations and other interchanges.

In this discussion we will provide details of several recently analyzed somaclonal variants which provide some insight into the nature of the genetic changes which occur during tissue culture.

Mendelian Somaclonal Mutants

The genetic analysis of plants from tissue cultures of diverse species such as rice, wheat, maize, lettuce, tobacco, tomato, celery, and rapeseed demonstrate the occurrence of classical point mutations. Many of these mutants are allelic to other known mutants. In many cases there is substantial evidence that these mutations occurred during the tissue culture phase because sister plants regenerated from the same callus were non-mutant. One such mutant, which has been analyzed at the nucleotide level, arose in regenerated maize plants which were screened for isozyme variants at the alcohol dehydrogenase (*Adh1*) locus (Brettell *et al.*, 1986a).

Maize ADH Somaclonal Mutant

Cultures were initiated from immature embryos of an F1 hybrid between *Adh1-F* (A188) and *Adh1-S* (Berkley Slow) and plants regenerated over time, the earliest being 3 months after culture initiation. Roots which developed on individual regenerants could be electrophoretically assayed for ADH1 and ADH2 without specific anaerobic induction.

Among 645 regenerants screened, 18 had perturbed isozyme patterns. One of these, which was lacking the expected F/S ADH heterodimer in the regenerant, proved to have a unique mutation which affected electrophoretic mobility of one of the ADH1 isozymes. The homozygous variant, called *Adh1-Usv*, elaborated an ADH1 homodimer which had a slower migration rate in starch gel electrophoresis than the ADH1-S homodimer (Figure 1).

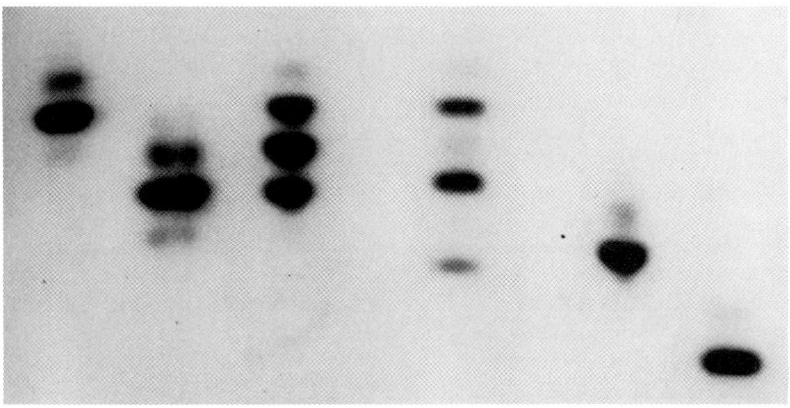

Figure 1. Zymograms of maize ADH1. Left to right: F/F, S/S, F/S, F/Usv, U725/U725, W/W.

The segregation of the mutant allele among seed progeny from the original variant plant confirmed the Mendelian expectation. Four other plants, regenerated from the same callus which gave rise to the mutant, had an ADH1 isozyme pattern which was identical to that of the parent.

The mutant *Adh1-Usv* was cloned by E.S. Dennis using an ADH probe derived from the *Adh1-S* allele. Restriction endonuclease analysis of the mutant allele revealed that it derived from the parental *Adh1-S* allele rather than from the *Adh1-F* allele. Subsequent sequencing of the entire gene revealed one base change relative to the sequence of the *Adh1-S* parental allele (see Dennis *et al.*, 1984). This nucleotide substitution was found in the last codon triplet of exon 6 (Figure 2).

```
                 EXON 6     ↓  INTRON 6
Adh1-S       -G/GTG/CAG/GAG/GTCTGTTTCTTT-
               Val Gln [Glu]

Adh1-Usv     -G/GTG/CAG/GTG/GTCTGTTTCTTT-
               Val Gln [Val]
```

Figure 2. Comparison of the sequence of *Adh1-S* with that of the mutant *Adh1-Usv* at the boundary of exon 6 with intron 6 where a single base change was found in the DNA coding sequence of the gene.

The adenine base of this codon was changed to a thymine which resulted in a substitution of valine for glutamic acid in the ADH1-Usv monomer. Such an amino acid substitution is expected to cause loss of a single negative charge on the altered ADH molecule and hence its slower electrophoretic mobility.

Modification of gene copy number

Plant genomes are large with haploid genome sizes of 10^9–10^{10} nucleotide pairs. Though variable from one species to another, about 60% of the genome of a "typical" plant is composed of repeated DNA. The reiteration frequency per sequence can be as high as 10^5–10^6, and highly repeated DNA can comprise up to 20% of the DNA. Repetitive DNA is of two basic types. The repeat units can be clustered as tandem arrays with specific genome location, or the repeat families can be dispersed throughout the genome often in different repeat unit permutations. Such repeated DNA provides the opportunity for extensive somaclonal variation.

Analyses of dihaploid plants of Nicotiana sylvestris derived by androgenesis has identified both qualitative and quantitative changes in DNA (De Paepe *et al.*, 1981). In potato, some plants regenerated from a small number of protoplast derived somaclones were characterized by deficiencies for specific rDNA repeat sequences (Landsmann *et al.*, 1985).

In collaboration with R. Appels, one of us (R.I.S.B.) has conducted an extensive molecular analysis of possible variation in the rRNA genes of triticale following passage through tissue culture (Brettell *et al.*, 1986b).

From six triticale cultivars, cultures were initiated from immature embryos, maintained for periods of up to six months and then regenerated. DNA which was isolated from 193 regenerants, was digested with the restriction endonuclease, Taq 1. Southern hybridizations were carried out with a Taq 1 rDNA spacer probe. This probe encompasses most of the rDNA spacer region of wheat. Southern hybridization of triticale with this probe elaborates 4 bands with molecular weights of 3.2, 3.1, 2.8 and, 2.5 kb. The first three are from the wheat genome and the smallest fragment is from chromosome 1R of rye.

Among the 192 somaclones assayed, several showed perturbation in the relative concentrations of the 4 bands as evidenced by densitometry. One variant, of the cultivar Currency, had an 80% reduction in the 2.5 kb fragment, i.e. the 1R rDNA spacer sequence. This depleted 1R rDNA was transmitted through meiosis. Mitotic C-banding of this variant also revealed a significantly reduced level of staining in the interstitial C-band proximal to the secondary constriction of 1R. C-banding correlates with heterochromatic regions which in cereals are known to contain highly reiterated DNA sequences. The depleted 1R site is known to be the location of the nucleolar organizer.

This change was clearly a consequence of tissue culture because six other plants regenerated from the same culture, as well as the original parent, did not carry the mutation.

Among the triticale somaclones several were found to be

aneuploid for some of the wheat chromosomes. Using N-banding for wheat chromosome identification, these somaclone aneuploids have enabled the assignment of the 3.2 and 2.8 kb fragment to chromosome 6B and the 3.1 kb fragment to chromosome 1B.

This was an unexpected benefit of somaclonal variation in that tissue culture variation can give rise to specific aneuploid chromosome variants. These isogenic monosomic lines provide added precision in the chromosomal assignment of mutants.

Chromosomal Rearrangements and Somaclonal Variation

Larkin *et al.*, (1985) cite 21 reports of structural modification to the chromosomes of plants regenerated from 14 different species. Most of these modifications were discerned from meiotic analysis of either the regenerated plants and/or their progeny. These structural alterations included deletions, interchanges, inversions, DNA amplification, isochromosome formation and subchromatid exchange.

From genetic analysis of somaclonal variants affecting constitutive alcohol dehydrogenase synthesis in wheat we have been able to identify specific chromosomal rearrangements as a consequence of tissue culture (Davis *et al.*, 1986).

The progeny of 551 regenerants of the hexaploid wheat cultivar "Millewa" were analyzed for qualitative and quantitative changes in the ADH zymogram pattern. Constitutive ADH1 isozymes in wheat are complicated by the fact that the locus is triplicated, one locus from each genome. Analysis of monosomic and nullisomic aneuploids for the chromosomes which carry these loci indicate that the active ADH isozymes are dimers formed by random association of each of the three different monomers. Evidence supports the assumption of the model used for this analysis that the three monomers are produced in equal quantities and that the six different dimers have essentially equal activity.

Among the 551 regenerants analyzed, 17 were variant. Thirteen of these were attributed to aneuploidy for a chromosome of the homeologous group 4 which carry the *Adh1* loci. The remaining four variant families were euploid. These variants were interpreted as resulting from either isochromosome formation, a translocation involving a 7BS/4Aα translocation or a 3BS/4Aα translocation, respectively. A brief description of the latter two variants will serve to illustrate the complexity of the events which give rise to somaclonal variation.

7BS/4Aα Translocation (SV1 Mutant)

Adh-A1 duplication. This mutant, called *SV1*, appears to result from a translocation involving 4Aα and probably chromosome 7B as identified by N-banding. Mitotic counts on germinated embryos of seed with variant ADH patterns were normal, but meiosis of homozygous mutant plants revealed a ring quadrivalent along with 19 bivalents. Seedlings of *SV1* were much less responsive to gibberellic acid induced shoot elongation than the Millewa parent which carries the *Rht1* dwarfing allele. In addition mature plant height of *SV1* was much shorter than that of Millewa.

The *Adh-A1* and *Rht1* loci are located on the same arm of chromosome 4A and it can be concluded that *SV1* is a duplication resulting from a translocation. The duplication of *Adh-A1* accounts for the variant ADH pattern. The altered GA_3 response and reduced mature plant height could be due either to duplication of the *Rht1* locus or an altered *Rht1* gene as a result of the translocation.

3BS/4Aα Translocation (SV5 Mutant)

This mutant is a deletion-like mutation as a result of a tranlocation of a segment of chromosome 3BS to 4Aα. The seed of this variant segregated for an ADH pattern which indicated loss of the *Adh-A1* locus. Monosomic segregants for the lost *Adh-A1* locus were uniformly taller than either the parent or the wild type *Adh-A1* segregants, which suggested inactivation of the dwarfing allele, *Rht1*. Cytology of the heterozygous plants indicated the regular occurrence of a rod quadrivalent, and N-banding implicated chromosome arms 3BS and 4Aα. All homozygous mutants were sterile. Chromosome arm 4Aα carries a gene (*Mslc*) which regulates male fertility, the loss of which renders plants sterile. Thus the translocation of 3BS to 4Aα and probable subsequent loss of the reciprocal translocation leads to a variant with loss of function of the *Adh-A1*, *Rht1* and *Mslc* loci.

It is highly likely that the genetic events which gave rise to the mutants occurred during culture. For each mutant, most of the sister regenerants derived from the same embryo explant displayed a wild type ADH phenotype. There were 65 sister plants to SV1, none of which possessed the SV1 translocation. Similarly the 7 sister plants of SV5 were all wild type.

Multigene Families

Multigene families of more or less complexity are frequent in plants. The synthesis of seed storage proteins in most plants, particularly crop species, result from multigene families. The analysis of somaclonal variation in multigene complexes has the advantage that a given somaclone, or a single progeny of it, has a number of target sites at which genetic changes can occur.

A class of wheat seed storage proteins, the gliadins, are coded by multigene loci located principally on two chromosomes. We have described heritable variation in the electrophoretic pattern of the gliadins for a number of somaclonal families. Heritable changes included loss of specific bands, appearance of new bands and changes in band intensity. Though varying in the frequency of changes, somaclonal variation for wheat gliadins has been confirmed by Cooper *et al.*, (1986) and Maddock *et al.*, (1986).

In a separate analysis of the multigene determined trait, β-amylase, Ryan and Scowcroft (1986) recovered one seemingly unique variant from among the progeny of 149 wheat somaclones. This variant was characterized by at least five new isozyme bands as well as enhanced intensity of two other preexisting bands. These multiple changes were under the control of a single gene and the changes could not be accounted for by electrophoretic shift of preexisting bands. The variant pattern did not occur among more than 100 diverse genotypes screened, nor among more than 200 single seed of the parent cultivar. The variant was cytologically euploid and meiotically normal. Several sister regenerants derived from the same embryo callus as the variant did not display the variant pattern.

The origin of this unique variant is not clear. It is highly unlikely that the mutation has arisen as a result of *de novo* creation of a new β-amylase locus. The variant gene could represent a previously silent locus which has been activated by mutational derepression of regulator gene(s) or excision of a transposable element from a structural gene. DNA sequence transposition in wheat has been observed but there is no conclusive evidence for the existence of transposable elements, or the mutational events resulting from their transposition, as has been described for maize.

Origin of Somaclonal Variation

There has been a number of assessments of the possible origins of somaclonal variation. An understanding of the mechanisms which give rise to tissue culture instability and somaclonal variation is necessary for several practical reasons. First, where somaclonal variation is to be utilized to generate variability for plant improvement it could be desirable to enhance its occurrence. Equally important where uniformity of tissue culture plants is essential, as in rapid micropropagation or for *in vitro* germplasm conservation, it is necessary to be able to control the mechanisms which generate variation.

Apart from chromosomal aberrations there is insufficient evidence to support, let alone favour, any one of several possible mechanisms to account for somaclonal variation. In addition to DNA amplification and transposable elements which are topical primarily because they reflect the direction of current research, other possibilities include, somatic rearrangement of genes particularly those in multigene families, somatic crossing over, sister chromatid exchange, altered nucleotide methylation patterns, perturbation of DNA replication by altered nucleotide pools, and silencing or activation of genes by mutations in associated non-coding regions.

Benefits and Problems of Somaclonal Variation

The major likely benefit of somaclonal variation is in plant improvement. One of the major potential benefits of somaclonal variation is the creation of additional genetic variability in co-adapted, agronomically useful cultivars, without the need to resort to hybridization. This takes on added appeal if *in vitro* selection is possible or if rapid plant screening procedures are available. Characteristics for which somaclonal mutants can be enriched during *in vitro* culture include resistance to disease pathotoxins, herbicides and tolerance to environmental or chemical stress. Somaclonal variation for other characters for which there is no adequately defined *in vitro* response, such as seed protein quality, photosynthetic efficiency and yield, can be of benefit provided effective whole plant screening protocols are available.

A serious disadvantage of somaclonal variation occurs in operations which require clonal uniformity, as in the horticulture and forestry industries where tissue culture is employed for rapid propagation of elite genotypes. Similarly, the genetic integrity of germplasm conserved by *in vitro* techniques is at risk because of somaclonal variation. Where clonal uniformity is desired, an understanding of the mechanisms which cause variation might help establish protocols to reduce its incidence.

A new option to introgress alien genes from wild relatives, into crop species has emerged from the observation that chromosome rearrangements occur in plants regenerated from tissue culture. The occurrence of chromosomal rearrangements during the tissue culture of interspecific hybrids and synthetic crop species, such as triticale, is particularly compelling. Armstrong *et al.*, (1983) found that karyotypes of triticale were quite unstable during culture. Aneuploidy was frequent and among 51 plants regenerated from 6 month old cultures more than half the plants possessed altered chromosomes. These included telocentrics, acrocentrics, deletions and euchromatic additions. C-banding enabled identification of rye-wheat recombinant chromosomes.

Larkin and Scowcroft (1981), McCoy *et al.*, (1982) and Ahloowalia (1983) recognized that the enhanced level of chromosome exchange in tissue culture might be employed to effect alien gene transfer. In crops such as wheat, wide crossing programs have produced alien chromosome addition lines following hybridization with relatives such as *Aegilops*, *Agropyron*, *Elymus*, *Haynaldia* and *Hordeum*. These alien species carry genes for resistance to wheat disease pathogens and tolerance to environmental stress such as heat, salinity and metal toxicity. Research has been initiated to utilize tissue culture in an attempt to enhance the rate of transfer of specific alien genes from these wheat alien addition lines. This could emerge as one of the most salutary consequences of somaclonal variation for plant improvement.

REFERENCES

Ahloowalia BS (1983). Spectrum of variation in somaclones of triploid ryegrass. Crop Sci 23:1141-1147.

Armstrong KC, Nakamura C, Keller WA (1983). Karyotype instability in tissue culture regenerants of triticale (X Triticosecale Wittmack) cv. 'Welsh' from 6-month-old callus cultures. Z Pflanzenzuchtg 91:233-245.

Bayliss MW (1980). Chromosomal variation in plant tissue culture. Int Rev Cytol 11A:113-143.

Brettell RIS, Dennis ES, Scowcroft WR, Peacock WJ (1986a). Molecular analysis of a somaclonal mutant of maize alcohol dehydrogenase. Molec Gen Genet 202:235-239.

Brettell RIS, Pallotta MA, Gustafson JF, Appels R (1986b). Variation at the Nor loci in triticale derived from tissue culture. Theor Appl Genet 71:637-643.

Cooper DB, Sears RG, Lookhart GL, Jones BL (1986). Heritable somaclonal variation in gliadin proteins of wheat plants derived from embryo callus culture. Theor Appl Genet 71:784-790.

Davies PA, Pallotta MA, Ryan SA, Scowcroft WR, Larkin PJ (1986). Somaclonal variation in wheat: genetic and cytogenetic characterization of alcohol dehydrogenase 1 mutants. Theor Appl Genet (in press).

Dennis ES, Gerlach WL, Pryor AJ, Bennetzen JL, Inglis A, Llewellyn D, Sachs MM, Ferl RJ, Peacock WJ (1984). Molecular analysis of the alcohol dehydrogenase (Adh1) gene of maize. Nuc Acid Res 12:3983-4000.

De Paepe R, Bleton D, Gnangbe F (1981). Basis and extent of genetic variability among doubled haploid plants obtained by pollen culture in Nicotiana sylvestris. Theor Appl Genet 59:177-184.

Evans DA, Sharp WR, Medina-Filho HP (1984). Somaclonal and gametoclonal variation. Amer J Bot 71:759-744.

Gould AR (1982). Chromosome instability in plant tissue cultures, studied with banding techniques. In Fujiwara A (ed): "Plant Tissue Culture 1982," Tokyo: Jap Assoc Plant Tiss Culture, pp 431-432.

Karp A, Bright SWJ (1985). On the causes and origins of somaclonal variation. Oxford Survey Plant Mol Cell Biol 2:199-234.

Landsmann J, Uhrig H, Saedler H (1985). Somaclonal variation in Solanum tuberosum detected at the molecular level. Theor Appl Genet 71:500-505.

Larkin PJ, Brettell RIS, Ryan SA, Davies PA, Pallotta MA, Scowcroft WR (1985). Somaclonal variation: impact on plant biology and breeding strategies. In Day P, Zaitlin M, Hollaender A (eds): "Biotechnology in Plant Science" New York: Academic Press, pp 83-100.

Larkin PJ, Scowcroft WR (1981). Somaclonal variation - a novel source of variability from cell culture for plant improvement. Theor Appl genet 60:197-214.

Larkin PJ, Scowcroft WR (1983). Somaclonal variation and crop improvement. In Kosuge T, Meredith CP, Hollaender A (eds): "Genetic Engineering of Plants," New York: Plenum, pp 289-314.

Maddock SE, Rissiott R, Parmar S, Jones MGK, and Shewry PR (1986). Somaclonal variation in the gliadin patterns of grains of regenerated wheat plants. J Exp Bot (in press).

McCoy TJ, Phillips RL, Rines HW (1982). Cytogenetic analysis of plants regenerated from oat (Avena sativa) tissue cultures: high frequency of partial chromosome loss. Can J Genet Cytol 24:37-50.

Orton TJ (1984). Somaclonal variation: theoretical and practical considerations. In Gustafson JP (ed): "Gene Manipulation in Plant Improvement," New York: Plenum Press, pp 427.

Ryan SA, Scowcroft WR (1986). A somaclonal variant of wheat with additional B-amylase isozymes. Theor Appl Genet (in press).

Scowcroft WR (1985). Somaclonal variation: The myth of clonal uniformity. In Hohn B, Dennis ES (eds): "Genetic Flux in Plants," Vienna: Springer - Verlag, pp 217-245.

Scowcroft WR, Davies P, Ryan SA, Brettell RIS, Pallotta MA, Larkin PJ (1985). The analysis of somaclonal mutants. In Freeling M (ed): "Plant Genetics," New York: Alan R Liss, pp 799-815.

Section V.
Molecular Biology

Plant Tissue and Cell Culture, pages 289–302

DIRECT GENE TRANSFER TO PLANTS: FACTS AND FUTURE

Ingo Potrykus, Jerzy Paszkowski, Michael W. Saul,
Ioan Negrutiu and Raymond D. Shillito

Friedrich Miescher-Institut, P.O.Box 2543,
CH-4002 Basel, Switzerland

INTRODUCTION

Gene transfer to plants is of basic and applied interest. It enables to study the structure and function of isolated genes via modification, reintroduction and subsequent analysis of their function in transgenic plants, cell lines or even cell populations. Gene transfer also enables to introduce foreign genes, or even specifically designed hybrid genes to plant genomes, thus creating novel varieties with specifically designed characters. Gene transfer in plants was, until recently, possible only by use of the natural gene transfer mechanism of *Agrobacterium tumefaciens*. Though developed to very efficient methods (e.g. Gheysen et al., 1985), *A. tumefaciens*-based gene transfer is hampered by hostrange limitations, which are especially disturbing when aiming at gene transfer into cereals or other graminaceous monocots. The hostrange poses an even more severe problem in the only alternative biological vector system, plant DNA viruses, for which it has been shown that they can be used to carry and systemically spread a foreign gene in a host plant (Brisson et al., 1984). Such limitations are not to be expected if it were possible to transfer genes "directly", without the use of biological vectors, into plant cells. As it could be foreseen that the cell wall would constitute an effective barrier for DNA molecules to pass, naked plant cells, the "protoplasts", have been considred the ideal experimental system to study vectorless gene transfer. Based on the positive experience from earlier experiments (e.g. Lurquin and Kado, 1977; Davey et al.,1980; Draper et al., 1982, and Krens et al., 1982) we developed a routine and efficient protocol for gene transfer into protoplasts. The basic prerequisites for success were a) a selectable marker gene which could be expected to yield a clear phenotype in protoplast-derived cell clones, b) expression signals which could be expected to be active independent of the physiological state of the host cell, c) precise linkage of the protein-coding region of the selectable gene to the expression signals guaranteeing expression of the hybrid gene in the host genome, d) DNA stretches on both sides of the hybrid gene which would allow for integration into the host genome without disruption of the gene construct itself, e) a protoplast system with routine high plating efficiency and plant regeneration capacity, f) an absolutely tight selection protocol for presumptive transformants.

THE BASIC EXPERIMENT

This is described in detail in Paszkowski et al.(1984). The selectable marker gene was the protein-coding region of the bacterial gene for aminoglycoside-phosphotransferase (APH(3')II) from transposon Tn5, causing resistance to aminoglycosydic antibiotics like kanamycin. The "plant expression signals" were taken from the small (19S) transcript of Cauliflower Mosaic Virus. The hybrid marker gene contained finally the bacterial gene under control of the 5'/3' CaMV expression signals. A control construct contained the gene in inverse orientation versus the expression signals. The acceptor protoplasts were from *Nicotiana tabacum* "Petit Havana SR1", which routinely proliferate with plating efficiencies around 10% and have a plant regeneration capacity of better than 90% from protoplast-derived clones. An absolutely tight selection protocol was based on growth in agarose segments in large volumes of liquid culture medium containing 50 mg/l kanamycin sulfate. Kanamycin-resistant clones were recovered from treatments, where the hybrid marker gene with the correct orientation was incubated with polyethylene glycol according to Krens et al. (1982). Kanamycin-resistant plants were grown to maturity. Southern blot hybridization with radioactively labelled probe covering the protein-coding region of the gene as well as enzyme analysis for kanamycin-dependent phosphorylation proved that the novel phenotype of the clones surviving selection was indeed based on the integration and activity of functional copies of the foreign gene into the host genome. It was also shown that the transgenic plants were neither chimeras nor mosaics and that the gene was transmitted to offspring.

PROTOCOL FOR ROUTINE HIGH TRANSFORMATION FREQUENCIES

In the course of the optimization of the protocol it turned out that the following parameters contribute to an increase in transformation frequency: a) form, size, and concentration of the transforming DNA, b) size and concentration of carrier DNA, c) a heatshock treatment, d) concentration and molecular weight of polyethyleneglycol, e) electroporation treatment, f) the sequence of the addition of the transforming DNA and PEG to the protoplasts, g) type and quality of the protoplasts, h) handling and culture of the protoplasts after the transformation treatment. The following protocol gives routinely high frequencies of transgenic plants with leaf protoplasts of <u>N. tabacum</u> (see also Shillito et al. (1985). Protoplasts are isolated as appropriate for subsequent culture (e.g. Potrykus and Shillito, 1985). They are suspended in 0.4 M mannitol containing 6 mM $MgCl_2$ and 0.1% MES buffer (morpholinoethansulfonic acid) at pH 5.6 to a population density of 1.5×10^6/ml. $MgCl_2$ is added to bring the electric resistance of the suspension into the range of 1.2 Kohm. This is followed by incubation at 45°C for 5 minutes and cooling to room temperature on ice. The gene is added to a final concentration of 10 µg/ml and carrier DNA to 40 µg/ml, followed by polyethyleneglycol (MW 6000) to a final concentration of 8%. For electroporation aliquots are transferred to the chamber of an electroporator and three high voltage pulses applied at 10 second intervals to give 50% survival. Thereafter the protoplast suspension is transferred into Petri dishes, a 10-fold volume of culture medium containing 0.6% agarose is added without prior washing of the protoplasts. The protoplasts are carefully distributed and the gel allowed to solidify. After incubation under optimal growing conditions for 7 days the protoplast-agarose gel is cut into segments and transferred into a 10-fold volume of liquid culture medium containing, if desired, a selective agent.

HOSTRANGE

One of the major reasons to develop the biological vector-independent gene transfer was the hope that this would circumvent any hostrange limitation and would thus enable gene transfer to such interesting plants as e.g. the cereals. This is indeed the case as has been demonstrated for *Lolium multiflorum* (Potrykus et al., 1985), *Triticum monococcum* (Loerz et al., 1985), and *Zea mays* (Fromm et al., 1985). The remaining problem with genetic engineering of cereals is however in plant regeneration from protoplasts. There will be, probably, several presentations on plant regeneration from rice (*Oryza sativa*) protoplasts on this Conference and there is good reason to hope that the final proof for plant regeneration from cereal protoplasts will be presented in the near future. It will, however, still be a difficult task to repeat with other cereals what might be possible with rice (Yamada et al., 1986). With protoplasts from herbaceous dicots we have not seen any limitation, so far, with regard to plant species or protoplast type, and we would therefore assume that "direct gene transfer" is possible with every protoplast.

FREQUENCY OF STABLE INTEGRATIVE TRANSFORMATION

We relate the transformation frequency to the number of clones growing after the transformation treatment. The calculation is based on either screening without prior selection of a few hundred randomly picked protoplast-derived clones, or by relating the number of clones surviving selection to the number of clones growing in an unselected aliquot of the treatment. This gives precise figures on the frequency of stable integration of at least one functional copy of the foreign gene (which is only a fraction of the frequency of integration of the foreign DNA). With this method of calculation surprisingly high frequencies of routinely 1-2% are achieved with *N. tabacum* "Petit Havana SR" (Shillito et al., 1985) and with other partially optimized protoplast systems (e.g. *Petunia hybrids*, *Hyoscyamus muticus*, *N. plumbaginifolia*, *N. tabacum* cell lines) frequencies around 0.1% are obtained. Screening for integration without function of the foreign gene yielded frequencies around 15% and for the very specific rDNA of *Physarum polycephalum* of higher than 50% (P. Kuenzler, FMI, personal communication). From these data one might conclude that almost every protoplast has the competence to take up and to integrate foreign DNA.

REPRODUCIBILITY

Direct gene transfer has produced transgenic clones and plants in our laboratory in more than 270 experiments at frequencies which are predictably in the range stated above. It also works efficiently in other laboratories,e.g. Max-Planck-Institut, Cologne (Loerz et al., 1985; Hain et al., 1985), Cornell University, Ithaca (Langridge et al., 1985), Stanford (Fromm et al., 1985), Pavia (F. Sala, personal communication), and, probably, numerous others.

TRANSFORMING DNA

The effect of the transforming DNA on transformation frequency is, at least, 4-fold: a) there is a clear concentration effect with a linear range between 1 to 50 µg/ml; b) linearized molecules transform about 2-fold better than circular ones; c) the size plays a role in as far as small and large molecules transform better than medium-sized ones; d) stronger expression signals produce higher transformation frequencies under selective conditions than weaker ones.

CARRIER DNA

The presence of carrier DNA is not a prerequisite for even high transformation frequencies; however, it improves and stabilizes transformation frequencies. There is again a concentration effect which levels off at ca. 100 µg/ml. Also the size plays an important role, yielding optimal effects with small molecules and toxic effects with large ones. There is not much difference with regard to the type of carrier DNA when comparing plant DNA, calf thymus DNA, salmon sperm DNA, and lambda DNA (Shillito et al., 1985).

HEATSHOCK TREATMENT

Heatshock treatment, optimally 5 minutes at 45°C for tobacco leaf protoplasts, may improve transformation frequency by 10-fold. It also may improve the integration pattern of the transforming DNA by reducing rearrangement and inactivation. Both effects are, however, not entirely consistent in different experiments, and they may vary considerably with plant species and protoplast type.

ELECTROPORATION

Electroporation describes the phenomenon of pore formation in membranes under the influence of short discharge pulses of high voltage. This is conveniently done with protoplasts by subjecting them, in a cylindrical chamber between parallel electrodes, to controlled discharge pulses from a capacitor. High initial field strength (1400 volts) and very short decay constant (10 μsec) in several repetitions balanced to 50% survival, yield the highest effect on stable integrative transformation, which is up to ca. 10-fold with _N. tabacum_ leaf protoplasts. The initial field strength has to be adjusted to the size of the protoplast type used. Electroporation is no prerequisite for even high frequency direct gene transfer and its effect on transformation frequency may vary with plant species and protoplast type. For more details see Shillito et al. (1985).

INTEGRATION INTO THE HOST GENOME

The foreign genes integrate, obviously via illegitimate recombination, and therefore at random sites, into chromosomal DNA. There is no clear case of integration into either chlDNA or mtDNA known. In 77% of 220 primary transformants studied integration was at one single locus. The number of integrated functional copies can range from 1 to 30 or more. Also in the cases of multiple copies these are preferentially at one locus. There is, however, also simultaneous integration at independent loci. The integration pattern as expressed in Southern blot analysis can be very simple (only one functional copy integrated) or relatively complex (numerous bands representing additional non-functional copies inactivated by integration within the gene itself). There is again an effect of the plant species and the cell type on the integration pattern which is, however, not consistent; and even different transgenic clones from the same protoplast population may, from the same treatment, arrive at simple as well as complex integration patterns.

LOCALISATION OF THE FOREIGN GENE IN THE HOST GENOME

The precise localisation of a foreign gene in transgenic plants is of interest under various aspects. Most of the transgenic plants produced are with _N. tabacum_, a species for which genetic localisation via linkage studies with marker genes is rather complicated and laborious. We have, therefore, attempted to develop *in-situ* hybridization to metaphase chromosomes as an alternative method. Homozygous transformants carrying one genetic locus for kanamycin resistance on different chromosomes and selected from the offspring of independent transformants, were induced to produce roots on cuttings. Protoplasts isolated from the root meristems were subjected to *in-situ* hybridization with a radioactively labelled probe of the foreign gene. Following exposure for 12 weeks the position of the foreign gene was visible on homologous chromosomes which were different in independent transformants (Mouras et al., 1986). It can be envisaged that the same procedure might allow for the localisation of original plant genes for which DNA probes are available.

EXPRESSION

The level of expression of the foreign gene is, of course, primarily influenced by the expression signals linked to the protein-coding region. From work with *A. tumefaciens*-based gene transfer a wealth of knowledge on regulated promoter sequences is available. Our own experience is limited to the constitutive expression signals from the 19S and 35S transcript of CaMV and the 35S promoter is probably one of the strongest known plant promoters. The effect of the expression signal on the transformation frequency under selective conditions has been mentioned earlier; the present hypothesis to explain this phenomenon is that single copy gene-transformants might have a better chance to survive selection when under a stronger promoter. There are, however, also other factors besides the expression signals which influence the level of expression, as transgenic clones carrying one copy may have higher levels of expression than others carrying multiple copies of the same gene.

INHERITANCE

The bacterial gene for neomycinphosphotransferase under the constitutive CaMV expression signals constitutes a very simple, clear, and dominant phenotype on kanamycin: seedlings (and plants) containing the gene stay green, those lacking the gene bleach to complete white. With seedlings this phenotype is very clear within 7 days. The large number of seeds produced per flower in *N. tabacum* (up to 3000) in the majority of the transgenic plants being morphologically normal and fully fertile, in combination with the simple phenotype of the gene, facilitated a detailed genetic analysis. In the great majority of the transgenic plants analysed (more than 220) the inheritance of the phenotype followed strictly Mendelian rules. In all cases the inheritance was strictly correlated with the transmission of the foreign DNA as well as with the gene activity (Potrykus et al., 1985b). In the majority (77%) the gene was inherited as a single dominant Mendelian factor. The remaining fraction of the population contained two, three, and more, linked or unlinked genes. Of special interest is a group which showed clear reciprocal differences and maternal inheritance of the phenotype in crosses with the untransformed wildtype (Shillito and Potrykus, in preparation).

STABILITY OF THE FOREIGN GENE

The fate of the foreign gene has been studied, so far, through seven sexual generations and for 30 months without selective pressure, and both stability and instability have been found. Stability predominates: a) subpopulations from one experiment yield the same transformation frequency when grown directly under selective conditions as when grown for 5-6 weeks without selection and thereafter transferred to selective conditions; b) of 60 subclones regenerated from two primary transformants, 57 maintained the gene under nonselective conditions ; c) transgenic plants are neither chimeras nor mosaics; d) the gene is regularly transmitted to sexual offspring; e) the inheritance of the gene follows strictly Mendelian rules; f) the gene has been maintained over seven sexual generations. These data demonstrate a high degree of mitotic and meiotic stability. These genetic data are supported by the inheritance of the transformant-specific hybridisation pattern (Potrykus et al., 1985c and unpublished data).

INSTABILITY OF THE FOREIGN GENE

Besides those clones with a high degree of stability there were others which showed considerable instability: A selfcross of a homozygous transformant yielded an R2 population of 422 resistant and 4 sensitive seedlings. Sixty further R2 populations of the same plant, produced by selfing 60 flowers from primary, secondary, and tertiary inflorescences all produced seedling populations containing ca. 0.5% sensitive individuals. One resistant R2 plant tested further produced R3 populations with again ca. 0.5% sensitive individuals. One resistant R2 plant tested further produced R3 populations with again ca. 0.5% sensitive seedlings. Three resistant R3 plants were taken to produce the R4 generation. Two of those again had 0.5% sensitives, one, however, produced a population segregating 840:98 for resistant versus sensitive . There are also data to demonstrate instability at the molecular level: Among numerous cases where the transformant-specific hybridisation pattern was maintained and inherited unaltered, one case was found, where all bands had been lost except for one of 1.2 kb representing the functional copy of the gene, and this plant was resistant. Another case was also detected, where all bands except the 1.2 kb fragment were maintained, and this plant was sensitive (Potrykus et al., 1985c).

CO-TRANSFORMATION

Recovery of transformants with a gene not producing a selectable phenotype would be considerably facilitated if the phenomenon of "co-transformation" known from work with animal cells (Wigler et al., 1979) also existed for plant cells. To study this question the plasmid pMSI carrying the Zein gene (Wienand et al., 1981) was mixed with the plasmid pABDI (Paszkowski et al., 1984) carrying the gene for kanamycin-resistance, in ratios of 1 + 1 and 3 + 1. Kanamycin-resistant clones were recovered and randomly picked clones from both treatments analysed by Southern blot analysis for simultaneous presence of Zein DNA. Seventytwo % of the kanamycin-resistant clones recovered from the 1 + 1 mixture, and 88% of the resistant clones from the 3 + 1 mixture contained DNA hybridizing to the Zein probe. There was no cross-hybridization to DNA from control clones. A full sized fragment containing the entire Zein gene was found in 28% of the clones. In co-transformation experiments with the nopaline synthase gene up to 47% of the kanamycin-resistant clones contained a full copy of the non-selected gene and all those clones were shown to express the gene (Schocher et al.,1986). This method of co-transformation is now in routine use in our laboratory for transfer of non-selectable genes.

GENE TRANSFER FROM TOTAL GENOMES

Successful transformation in reconstruction experiments down to ratios of 10 copies of the kanamycin-resistance gene per tobacco genome prompted us to attempt transfer of a single copy gene from a tobacco genome. To this end DNA was isolated from plants containing one copy of the kanamycin-resistance gene stably integrated into the genome, partially digested and used to treat tobacco protoplasts with the standard direct gene transfer protocol. Transformants with this gene were recovered at the expected low frequency of ca. one per 3 million protoplast-derived clones (Mueller et al., in preparation). This result indicates that it should be possible to transfer plant genes from one species to another without prior isolation. The low frequency will make this feasible only for genes producing a selectable phenotype.

TRANSIENT EXPRESSION

It is possible to express foreign genes, following their uptake into protoplast populations by electroporation (or polyethyleneglycol treatment) over a series of days, independent of their stable integration into the host genome (Fromm et al., 1985). This is an important technique for rapid testing of the effects of modifications in gene constructs. However, the expression of the gene in the whole plant may bear no relation to the expression seen in such systems. In addition, in our experience the conditions for optimal transient expression (low initial field strength and long decay constant) and stable integrative transformation (high initial field strength and short decay constant) are so different that transient expression cannot be used to optimize an experimental system for stable integrative transformation.

LIMITATIONS

There is probably no limitation with regard to any kind of "hostrange". So far protoplasts from all plant species tested have also been transformed. There is probably also no limitation in respect to the protoplast type, which might be of special importance when attempting study of tissue-specific expression. There is probably little limitation concerning uptake and size of the DNA molecules (DNA larger than 50 kb is readily taken up), but there may be some limitation with regard to an unrearranged integration into the genome for molecules larger than 20 kb. There may be other limitations which have not been identified yet, however, the fact that "direct gene transfer" requires, to date, isolated protoplasts is a limitation which may cause problems in cases where transgenic plants are the aim of the experiment. Although plant regeneration from protoplasts has been described for over a hundred plant species it is not routine for many and it has not been achieved at all for rather important ones.

OTHER VECTORLESS GENE TRANSFER SYSTEMS

Liposome fusion has been used to convincingly demonstrate uptake, integration, expression, and inheritance of foreign genes in tobacco (e.g. Deshayes et al., 1985). Also bacterial spheroplasts harboring plasmids with a marker gene have been employed to transfer genes into protoplasts via spheroplast fusion (Hain et al., 1984). Finally has microinjection of DNA into nuclei of protoplasts been developed recently to an efficient method for gene transfer (Miki et al., 1985; Reich et al., 1985; Crossway et al., 1986). This technique may open the way for genes into plants which are neither accessible to Agrobacterium-mediated transformation nor regenerable from protoplasts.

FUTURE

Direct uptake of DNA molecules into protoplast populations, subsequent transient expression or highly efficient stable integrative transformation, as well as clonal development and plant regeneration from genetically modified protoplasts and general progress in plant regeneration from protoplasts offer experimental opportunities which will contribute to progress in the analysis and practical use of plant genes. Amongst the areas where one would expect substantial progress within the near future are a) gene transfer into cereals and other recalcitrant crops, b) gene transfer into chloroplasts and mitochondria, c) gene isolation via complementation of deficiencies, via insertion mutagenesis, and via selection for dominant characters, d) gene inactivation via insertion mutagenesis or gene replacement, e) isolation of replication origins, construction of selfreplicating vectors, and analysis of replication, f) homologous recombination, and study of recombination and genome stability, g) analysis of regulatory DNA sequences and introns, h) analysis of transacting elements and celltype-specific gene expression, i) analysis of the effect of chemical modifications of genes, j) chromosome tagging with cytologically visible and biochemically selectable marker genes, k) analysis of position effects on gene expression, l) gene transfer across cell walls, m) identification and transfer of genes of agronomic or biochemical importance, n) identification and analysis of genes important in the regulation of plant development, etc. It is obvious that all this is not restricted to the method of direct gene transfer, but that gene transfer via biological vector systems and future technical developments will be at least as important.

REFERENCES

Brisson N, Paszkowski J, Penswick JR, Gronenborn B, Potrykus I., Hohn T (1984) Expression of a bacterial gene in plants using a viral vector. Nature 310:511-514.

Crossway A, Oakes, JV, Irvine JM, Ward B, Knauf VC, Shewmaker CK (1986) Integration of foreign DNA following microinjection of tobacco mesophyll protoplasts. Mol Gen Genet 202:179-185.

Davey MR, Cocking EC, Freeman J, Pearce N, Tudor I (1980) Transformation of Petunia protoplasts by isolated _Agrobacterium_ plasmids. Plant Sci Lett 18:307-313.

Deshayes A, Herrera-Estrella L, Caboche M (1985) Liposome-mediated transformation of tobacco mesophyll protoplasts by an _Escherichia coli_ plasmid. EMBO J 4:2731-2739.

Draper J, Davey MR, Freeman JP, Cocking EC, Cox BG (1982) Ti plasmid homologous sequences present in tissues from _Agrobacterium tumefaciens_ transformed Petunia tissues. Plant & Cell Physiol 23:451-458.

Fromm M, Taylor LP, Walbot V (1985) Expression of genes transferred into monocot and dicot plant cells by electroporation. Proc Natl Acad Sci USA 82:5824-5828.

Gheysen G, Dhaese P, Van Montague M, Schell J (1985) DNA flux across genetic barriers: The crown gall phenomenon. In: Hohn B, Dennis ES (eds) Genetic Flux in Plants, Plant Gene Research. Springer, Vienna New York, pp. 12-47.

Hain R, Steinbiss HH, Schell J (1984) Fusion of Agrobacterium and E. coli spheroplasts with Nicotiana protoplasts-direct gene transfer from microorganism to higher plant. Plant Cell Rep 3:60-64.

Hain R, Stabel P, Czernilofsky AP, Steinbiss HH, Herrera-Estrella L, Schell J (1985) Uptake, integration, expression and genetic transmission of a selectable chimeric gene by plant protoplasts. Mol Gen Genet 199:161-168.

Krens FA, Molendijk L, Wullems GJ, Schilperoort RA (1982) In vitro transformation of plant protoplasts with Ti-plasmid DNA. Nature 296:72-74.

Langridge WHR, Li BJ, Szalay AA (1985) Electric field-mediated stable transformation of carrot protoplasts with naked DNA. Plant Cell Reports 4:355-359.

Lörz H, Baker B, Schell J (1985) Gene transfer to cereal cells mediated by protoplast transformation. Mol Gen Genet 199:178-182.

Lurquin PF, Kado CI (1977) Escherichia coli plasmid pBR313 insertion into plant protoplasts and into their nuclei. Mol Gen Genet 154:113-121.

Miki B (1985) The development of microinjection technology. In: Galau GA (ed) Abstracts First International Congress Plant Molecular Biology. University of Georgia Center for Continuing Education, Athens, USA, p 28.

Müller T, Saul MW, Paszkowski J, Shillito RD, Potrykus I (1986) Direct gene transfer from a total N. tabacum genome, in preparation.

Mouras A, Saul MW, Potrykus I (1986) In situ hybridisation of a transformed gene in Nicotiana tabacum, in preparation.

Paszkowski J, Shillito RD, Saul MW, Mandak V, Hohn T, Hohn B, Potrykus I (1984) Direct gene transfer to plants. EMBO J 3:2717-2722.

Potrykus I, Shillito RD (1985) Protoplasts: isolation, culture and plant regeneration. In: Weissbach A and Weissbach H (eds) Methods in Enzymology, 118 (Plant Molecular Biology). Academic Press, Orlando, FL, USA.

Potrykus I, Saul M, Petruska J, Paszkowski J, Shillito RD (1985a) Direct gene transfer to cells of a graminaceous monocot. Mol Gen Genet 199:183-188.

Potrykus I, Paszkowski J, Saul M, Petruska J. Shillito RD (1985b) Molecular and general genetics of a hybrid foreign gene introduced into tobacco by direct gene transfer.Mol Gen Genet 199:169-177.

Potrykus I, Shillito RD, Saul M, Paszkowski J (1985c) Direct gene transfer: State of the art and future potential. Plant Mol Biol Rep 3:117-128.

Reich T, Iyer VN, Miki B (1985) Transformation of alfalfa protoplasts by intranuclear microinjection of Ti-plasmids. In: GAlau GA (ed) Abstracts First International Congress Plant Molecular Biology. University of Georgia, Center for Continuing Education, Athens, GA, USA, p 28.

Schocher RJ, Shillito RD, Saul MW, Paszkowski J, POtrykus I (1986) Co-transformation of unlinked foreign genes into plants by direct gene transfer. Biotechnology, submitted.

Shillito RD, Saul MW, Paszkowski J, Müller M, Potrykus I (1985) High Efficiency direct gene transfer to plants. Biotechnology 3:1099-1103.

Wienand U, Langridge P, Feix G (1981) Isolation and characterisation of a genomic sequence of maize coding for a Zein gene. Mol Gen Genet 182:440-444.

Wigler M, Sweet R, Sim GK, Wold B, Pellicer A, Lacy E, Maniatis T, Silverstein S, Axel R (1979) Transformation of mammalian cells with genes from prokaryotes and eukaryotes. Cell 16:777-785.

Yamada Y, Zhi-Qi Y, Ding-Tai T (1986) Plant regeneration from protoplast-derived callus of rice (*Oryza sativa* L.) Plant Cell Rep 5:85-88.

Plant Tissue and Cell Culture, pages 303–316

GENE TRANSFER IN CEREALS

Horst Lörz, Barbara Junker, Jeff Schell and Alicia de la Pena

Max-Planck-Institut für Züchtungsforschung
5000 Köln 30, Fed. Republic of Germany

Summary

Cell and tissue culture techniques and methods for the genetic manipulation of higher plants including the major cereal crops have progressed considerably in recent years. Plant regeneration from in vitro cultures, especially from multicellular explants, has been achieved with all the major cereal species. Somaclonal variation holds promise to isolate new variants and genotypes with improved characteristics. However, success in cereal protoplast culture is still limited and successful plant regeneration is reported so far for rice only. DNA-mediated stable transformation of cultured cells of wheat, maize and rice has been obtained by transforming protoplasts with isolated DNA. In these experiments plasmids have been used which contained a selectable chimeric gene coding for kanamycin resistance. A serious limitation in the wider application of DNA-mediated direct gene transfer to cereal crops remains the difficulty of regenerating plants from isolated protoplasts. Alternative transformation methods independent of protoplasts are therefore being sought by introducing foreign genes either via microinjection into multicellular embryogenic or morphogenic structures, or by delivering the isolated DNA directly into plants. High efficiency transformation of maize by a mixture of pollen and exogenous DNA has been reported, but so far without conclusive biochemical and molecular analysis. In contrast, transformation of rye has been achieved by DNA-injection into plants at a defined stage prior to meiosis and evidence for successful gene transfer is provided by specific enzyme activity of the introduced gene and positive DNA-hybridization.

INTRODUCTION

Most, if not all of the new techniques of plant biotechnology were developed first with Solanaceae species and in the past Gramineae and the cereals have been characterized as being recalcitrant in terms of in vitro culture and genetic manipulation (Lörz, 1986). Cereals and grasses, however, are by far the most important group of plants for human nutrition and animal feed. Thus it is not only an academic challenge but also of great economic relevance to apply the new biotechnological methods to these important species. The methods being discussed under the topic of in vitro manipulation of higher plants include in vitro culture techniques e.g. in vitro multiplication, pathogen elimination, in vitro selection, haploid production, plant regeneration from protoplasts, and, what will be discussed primarily in this report, genetic manipulation techniques. Variation and change in cereals is normally achieved by sexual means, namely crossing and selection. Alternatively, variability and genetic modification can now be created also by in vitro techniques and genetic engineering. The new techniques should however not be considered to replace or to compete with conventional plant breeding, but rather to complement plant breeders work (Austin et al., 1986). Several possibilities are listed (Table 1) on how to achieve new combinations of the genetic information of plant cells (by protoplast fusion) and how to introduce defined genes into a recipient cell.

Until now only a small number of isolated genes, which are of interest in a cereal breeding program, are available. Much more collaborative efforts between plant breeders, physiologists, pathologists, molecular and cell biologists will be necessary to define the genes of interest, to isolate them, to modify them, to reintroduce them into the recipient plants, and finally to breed a genetically engineered cereal crop.

Table 1. Genetic Manipulation of Higher Plants

PROTOPLAST FUSION
- somatic hybridization
- somatic cybridization
- limited gene transfer
VECTOR DEPENDENT GENE TRANSFER
- Agrobacterium
- DNA viruses
- RNA viruses
- fungal systems
- spheroplast fusion
DIRECT GENE TRANSFER
- DNA uptake into protoplasts
- microinjection of DNA into zygotic/somatic cells
- injection of DNA into plants
- pollen transforamtion

GENETIC MANIPULATION OF CEREAL PROTOPLASTS

Isolated protoplasts are a basic and most suitable tool in somatic cell genetics and genetic transformation of higher plants. In principle there should be no difference for their suitability in genetic manipulation experiments whether the protoplasts are derived from monocotyledonous or dicotyledonous species and from which tissue they have been isolated. Mesophyll protoplasts are most commonly isolated and cultured in the case of dicotyledonous species. However, so far culture and regeneration experiments with mesophyll protoplasts isolated from *Gramineae* (*Monocotyledoneae*) have failed. To overcome the culture and regeneration problem with cereal mesophyll protoplasts, in vitro cultures of numerous *Gramineae* have been used instead as a source for the preparation of protoplasts (Ozias-Akins and Lörz, 1984). The rationale behind this is the assumption that protoplasts express the same competence in respect to regeneration as the cells from which they have been isolated. In contrast to mesophyll tissue from fully developed leaves of cereals, embryogenic callus cultures and embryogenic suspensions derived from e.g. immature embryos, inflorescences or meristematic leaf segments, are competent for plant regeneration and thus provide a suitable material for the isolation of totipotent protoplasts. A major problem remains the frequent observation that during the process of establishing a cell culture suitable for the isolation of protoplasts, the originally embryogenic cultures loose this embryogenic and regenerative capacity. Only very recently has plant regeneration from protoplasts been described for rice by several authors (Fujimura et al., 1985; Coulibaly and Demarly, 1986; Yamada et al., 1986). In these reports and earlier publications describing "plantlet" regeneration of several grass species (Vasil, 1983), efficiency of protoplast plating and plant regeneration is low and reproducibility of these experiments rather difficult. More reproducible and efficient are protoplast culture systems limited to regeneration to the callus stage, which have been established for e.g. maize (Chourey and Zurawski, 1981), rice (Göbel et al., 1986), barley (Lührs and Lörz, 1986), wheat (Dudits, 1986) and triticale (Stolarz and Lörz, 1986).

Despite the present limitations in plant regeneration, cereal protoplasts have been used for somatic hybridization as well as transformation experiments. Direct gene transfer to plant cells mediated by DNA uptake into protoplasts was achieved first with tobacco protoplasts, and it has been shown that the transferred DNA is stably integrated and transmitted to the progeny in a Mendelian inheritance pattern (Paszkowski et al., 1984; Hain et al., 1985). This transformation method, applying chemical treatment (Krens et al., 1982) or electric pulses (Shillito et al., 1985) to stimulate DNA uptake, is a very general one and has also been used to successfully transform *Triticum monococcum* (Lörz et al., 1985), *Lolium multiflorum* (Potrykus et al., 1985), *Zea mays* (Fromm et al., 1986) and *Oryza sativa* (Uchimiya et al., 1986) protoplasts. In these experiments plasmids have been used which contained a selectable chimeric gene with the coding region of the aminoglycoside phosphotransferase type II (NPT II) gene. Transformed cells were selected in medium containing kanamycin and identified by detection of specific enzyme

(NPT II) activity. The original transformation efficiency with graminaceous protoplasts was rather low, but modifications in the DNA uptake procedure (Shillito et al., 1985) and the use of synchronized protoplast cultures (Meyer et al., 1986) increased the transformation efficiency significantly and 1% or more transformed colonies (based on the number of surviving protoplasts after DNA incubation) can now be achieved in tobacco.

Another direct way of transferring cloned genes into protoplasts is the microinjection of DNA. The procedure is most efficient when DNA is injected directly into the nucleus of protoplasts (Crossway et al., 1986; Miki et al., 1986). As this method is very labour intensive and only a limited number of protoplasts can be treated this way, the application of the procedure so far is limited only to protoplast systems with high plating efficiency (Lawrence and Davies, 1985).

Protoplasts isolated from different cereals and various tissues can be used also to study transient expression of foreign genes after transformation (Werr and Lörz, 1986; Junker et al., 1986). Transformation has been performed by incubating freshly isolated protoplasts with plasmid DNA in the presence of PEG and calcium ions. Different chimeric genes with either NPT II or CAT (chloramphenicol acetyltransferase) as enzyme coding component and promoter regions from either the nopaline synthetase gene (A.t.), the sucrose synthase gene (maize) or the 35S gene (CaMV) were used, and 2-10 days after DNA treatment the expression of the chimeric genes was shown by detection of either NPT II or CAT enzymic activity. The study of transient expression of introduced genes is not limited to protoplasts which can be cultured and start dividing after the treatment, but can be extended also to protoplast systems which normally do not develop in culture, such as maize or rice mesophyll protoplasts (Table 2; B. Junker, Cologne, in preparation). Transient expression studies provide suitable means to characterize promoter regions (promoter mutagenesis), and to analyze tissue and organ specific expression of foreign genes after transformation.

Table 2: Transient expression of foreign genes in cereal protoplasts isolated from leaves, roots or suspension cells.
Ca. 10^6 protoplasts are used for each transformation experiment with chimeric genes pNOS-NPT (pLGV1103, Hain et al., 1985) or pSuSy-NPT (pSKANI, Werr and Lörz, 1986) and exzyme activity analyzed after 48 hours.
(+ : dividing cells, positive enzyme activity; - :non-dividing protoplasts, no enzyme activity; N.D. : enzyme activity not determined)

Protoplasts	Division Capacity	pNOS-NPT	pSuSy-NPT
maize suspension	+	+	+
maize mesophyll	-	+	-
maize roots	-	+	N.D.
rice suspension	+	+	+
rice mesophyll	-	+	N.D.
barley suspension	+	+	N.D.
rye leaves	-	+	N.D.

IN VITRO MANIPULATION OF MULTICELLULAR STRUCTURES

The limitation concerning the genetic transformation of cereal protoplasts remains the difficulty of regenerating plants from protoplasts. It has been mentioned already that large numbers of plants from more or less all cereal crops can be regenerated if multicellular explants (in contrast to isolated single protoplasts) are used for initiating the in vitro cultures (Lörz, 1986). Consequently alternative transformation methods, independent of protoplasts, are being extensively studied. The approaches include the microinjection of DNA into morphogenetic cell clusters, into somatic embryos, and into unfertilized or fertilized egg cells and young zygotic embryos. Although the microinjection technique with glass capillaries is highly sophisticated and additional electric pulses might stimulate the uptake of DNA into intact cells (Morikawa et al., 1986), the limitation of this approach remains the difficulty due to low efficiency and low reproducibility of regenerating fertile plants from cellular structures small enough to be suitable for microinjection (Töpfer and Steinbiss, 1985).

VECTOR-DEPENDENT GENE TRANSFER

Foreign genes are most efficiently transferred to numerous dicotyledonous plant species with the natural vector system of the Ti-plasmid from _Agrobacterium_. The natural host range of _Agrobacterium_ is limited and so far only a few monocotyledonous species can be transformed with _A. tumefaciens_ (Hooykaas-Van Slogteren et al., 1984; Hernalsteens et al., 1984). Although intensive efforts have been made in several laboratories, none of the cereals as yet can be transformed with this vector system (A. Caplan, Gent, pers. comm.; Lörz et al., unpublished). Recently factors which are produced by plant cells (e.g. tobacco) and which induce the activity of the vir-genes on the Ti-plasmid have been isolated and chemically defined (Stachel et al., 1986). As these compounds could not be detected in cell suspensions of rice, maize or wheat (E. Messens, Gent, unpublished) the factor from tobacco is being included in treatments of cereal cells with Agrobacterium (L. Cabrera-Ponce, unpublished). A preliminary report has described the transformation of rice cells with _A. tumefaciens_ after fusion of rice protoplasts with A.t. spheroplasts (Baba et al., 1986). However, no molecular analysis is provided in this report and it has been found previously that the selection of rice transformants either based on hormone-independent growth or kanamycin resistance is not absolutely reliable and possible "escapes" from the selection scheme have to be taken into account (E. Göbel, Cologne, unpublished).

In addition to bacterial vector systems viral vectors based on DNA- as well as RNA-viruses have been developed for plant transformation, too. Cauliflower mosaic virus has been used successfully to replicate and to express a foreign gene in a dicotyledonous plant and this virus can be considered useful as a plant gene vector providing a self replicating system (Brisson and Hohn, 1986). Similarly, Wheat Dwarf Virus, a member of the gemini viruses which infects several _Gramineae_ species, has been cloned and reintroduced into wheat protoplasts. It has been shown that the cloned virus DNA replicates in wheat as well as maize protoplasts, and the replacement of the viral coat protein by the NPT II gene has led to the expression of a newly introduced, foreign gene after transfection of wheat and maize cells (Matzeit et al., in prep.). The process of plant transfection by WDV is strictly dependent on leaf hopper transmission; or the genetically engineered virus DNA has to be introduced directly into protoplasts. Both procedures, however, have severe constraints in respect to their wider applicability and further possibilities to introduce the modified viral genome directly into cereal plants are highly desirable.

DNA-MEDIATED DIRECT GENE TRANSFER INTO CEREALS

Considering the difficulties to regenerate plants from cereal protoplasts and the efforts necessary to handle totipotent in vitro cultures, all possible genetic transformation systems independent of in vitro cultures seem to be very attractive. For many years attempts have been made to introduce DNA into seeds, germinating seedlings, plants or pollen (Ledoux, 1975), but only since the use of cloned DNA and defined marker genes it has been possible to follow and to analyze the fate of the introduced genes.

Pollen transformation has been studied intensively with _Petunia_ and _Nicotiana_ (Hess and Dressler, 1984) and more recently also in maize (Ohta, 1986). Although formal genetic evidence has been provided for successful transformation, other researchers have failed to reproduce such experiments (Sanford et al., 1985). We have used for pollen transformation studies chimeric genes with the promoter region from nopaline synthetase and the NPT II gene coding for kanamycin resistance. This construct has been successfully used previously in protoplast transformation experiments. Within the population of ca. 300.000 tobacco seedlings derived from seeds, obtained after DNA treatment of pollen and pollination with these, pregerminated pollen grains, 5 kanamycin-resistant plants have been found. However, in none of these putative transformants it was possible to detect the introduced foreign gene in subsequent molecular analysis (Lörz and Schreier, unpublished). It was concluded therefore, that spontaneous variants or mutants have been isolated by the kanamycin screening, resembling a phenotype similar to the one being expected from the transformation experiment. As no phenotypic variants have been observed in the controls (ca. 200.000 seedlings) a non-specific mutagenic effect of the DNA-treatment can not be excluded.

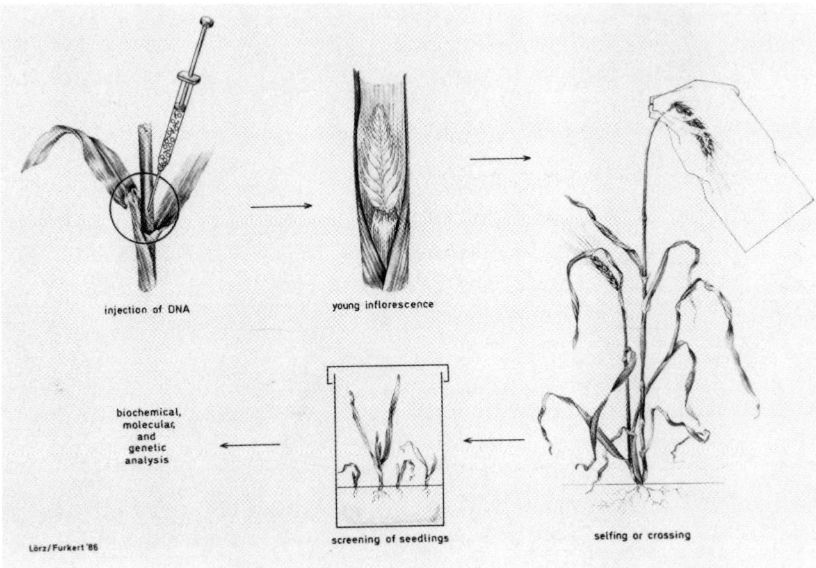

Fig. 1: Schematic summary of the transformation procedure of rye plants mediated by injection of isolated DNA into developing inflorescences.

In order to develop further alternative transformation methods avoiding the use of protoplasts, we attempted to incorporate exogenous DNA into cereal cells by means of injecting DNA directly into developing inflorescences. Previous work analyzing the effect of various drugs injected at different developmental stages of the rye germ cells had shown that the archesporial cells of rye are especially sensitive to exogenously applied toxin about two weeks prior to metaphase I (De la Pena et al., 1981; Puertas et al., 1984). This developmental stage might therefore represent a maximum for permeability and uptake capacity of foreign, exogenously applied molecules. Plasmid DNA (pLGV1103; pNOS-NPT II) coding for kanamycin resistance has been injected into developing inflorescences of *Secale cereale* cv. JNK, in field and glasshouse grown plants (Fig. 1). Flowering plants were crossed with each other and seeds from DNA-treated and control plants were germinated in the presence of kanamycin (10 μg/ml). Putative transformants developed normally and stayed green, whereas sensitive plants showed effects of "bleaching" and finally did not develop further. From the seven "green" seedlings selected so far (Table 3) two of them showed the presence of the active gene product (NPT II activity) and in two plants the physical presence of the foreign gene was demonstrated by Southern blot analysis of DNA (De la Pena et al., in preparation). The direct injection of DNA into developing rye plants provides an alternative transformation method totally independent from any in vitro culture system or specific transformation vectors. Further studies will therefore be extended to other cereal crops, and it will be of interest especially for species, from which up to now, plants can not be regenerated from isolated protoplasts.

Table 3: Transformation of rye (*Secale cereale* L.) mediated by injection of DNA into developing plants (preliminary data obtained until July 1986).

DNA: pNOS-NPT
ca. 200 plants injected
ca. 3000 seedlings screened
7 kan-resistant plantlets
2 positive for NPT II
2 positive in DNA analysis

ACKNOWLEDGEMENTS

Part of this work was supported by a research grant

from the Bundesministerium für Forschung und Technologie in Bonn (grant BCT-0362-2). A.d.l.P. gratefully acknowledges the fellowship provided by the Alexander von Humboldt-Stiftung in Bad Godesberg. We are thankful to Elke Göbel, Peter Brown and Janusz Zimny for valuable discussions and critical comments.

REFERENCES

Austin RB, Flavell, RB, Henson IE, Lowe HJB (1986). "Molecular Biology and Crop Improvement". Cambridge University Press.

Baba A, Hasezawa S, Syono K (1986). Cultivation of rice protoplasts and their transformation mediated by Agrobacterium spheroplasts. Plant Cell Physiol 27:463-471.

Brisson N, Hohn T (1986) Plant virus vectors: Cauliflower mosaic virus. Methods Enzymol 118:659-668.

Chourey PS, Zurawski DB (1981). Callus formation from protoplasts of a maize cell culture. Theor Appl Genet 59:341-344.

Coulibaly MY, Demarly Y (1986). Regeneration of plantlets from protoplasts of rice, Oryza sativa L. Z Pflanzenzüchtg 96:74-81.

Crossway A, Oakes JV, Irvine JM, Ward B, Knauf VC, Shewmaker CK (1986) Integration of foreign DNA following microinjection of tobacco mesophyll protoplasts. Mol Gen Genet 202:179-185.

De la Pena A, Puertas MJ, Merino F (1981) Bimeiosis induced by caffeine. Chromosoma 83:241-248.

De la Pena A, Schell J, Lörz H (1986). Transgenic cereal plants obtained by injection of DNA into inflorescences. Submitted.

Dudits D (1976). The effect of selective conditions on the products of plant protoplast fusion. In Dudits D, Farkas GL, Maliga P (eds): "Cell Genetics in Higher Plants", Budapest, Academiai Kiado, pp 153-162.

Fromm ME, Taylor LP, Walbot V (1986). Stable transformation of maize after gene-transfer by electroporation. Nature 319:791-793.

Fujimura T, Sakurai M, Akagi H, Negishi T, Hirose A (1985). Regeneration of rice plants from protoplasts. Plant Tissue Culture Letters 2:74-75.

Göbel E, Ozias-Akins P, Lörz H (1986). Cell and

protoplast culture in rice. In Aldersen P.G., Withers LA (eds): "Plant Tissue Culture and its Agricultural Applications", Kent: Butterworth, pp 359-365.

Hain R, Stabel P, Czernilofsky AP, Steinbiss HH, Herrera-Estrella L, Schell J (1985). Uptake, integration, expression and genetic transmission of a selectable chimeric gene by plant protoplasts. Mol Gen Genet 199:161-168.

Hernalsteens JP, Thia-Toong L, Schell J, Van Montagu M (1984) An Agrobacterium-transformed cell culture from the monocot Asparagus officinalis. EMBO J 3:3039-3041.

Hess D, Dressler K (1984) Bacterial transferase activity expressed in petunia progenies. J Plant Physiol 116:261-272.

Hooykaas-Van Slogteren GMS, Hooykaas PJJ, Schilperoort RA (1984). Expression of Ti plasmid genes in monocotyledonous plants infected with Agrobacterium tumefaciens. Nature 311:763-764.

Junker B, Baker B, Lörz H (1986). Culture and transformation studies with maize protoplasts. In Horn W, Jensen CJ, Odenbach W, Schieder O (eds): "Genetic Manipulation in Plant Breeding" Berlin: Walter de Gruyter, in press.

Krens FH, Molendijk L, Wullems GJ, Schilperoort RA (1982). In vitro transformation of plant protoplasts with Ti-plasmid DNA. Nature 296:72-74.

Lawrence WA, Davies DR (1985). A method for the microinjection and culture of protoplasts at very low densities. Plant Cell Reports 4:33-35.

Ledoux L (1975). "Genetic Manipulation with Plant Material." New York: Plenum Press.

Lörz H, Baker B, Schell J (1985). Gene transfer to cereal cells mediated by protoplast transformation. Mol Gen Genet 199:178-182.

Lörz H (1986). In vitro manipulation of cereals - limitations and prospects. In: "Cereals in a European Context", Proceedings 1st European Congress of Food Science and Technology, Chichester, U.K.: Ellis Horwood Lim., in press.

Lührs R, Lörz H (1986). Somatic embryogenesis, cell and protoplast culture of Hordeum vulgare L. (barley). In Horn W, Jensen CJ, Odenbach W, Schieder O (eds): "Genetic Manipulation in Plant Breeding" Berlin: Walter de Gruyter Publ., in press.

Matzeit V, Schäfer S, Van der Forst R, Gronenborn B

(1986). Regulation and expression of foreign genes by Wheat Dwarf Virus. In preparation.

Meyer P, Walgenbach E, Bussmann K, Hombrecher G, Saedler H (1985). Synchronized tobacco protoplasts are efficiently transformed by DNA. Mol Gen Genet 201: 513-518.

Miki BLA, Reich T, Dijak M, Brown DCW, Iyer VN (1986). In "Genetic Engineering of Plants and Microorganisms Important for Agriculture." Proceedings of EC Meeting, Wageningen, January 1986, p. 101.

Morikawa H, Iida A, Matsui C, Ikegami M, Yamada Y (1986). Gene transfer into intact plant cells by eletroinjection through cell walls and membranes. Gene 41:121-124.

Ohta Y (1986) High efficiency genetic transformation of maize by a mixture of pollen and exogenous DNA. Proc Natl Acad Sci USA 83:715-719.

Ozias-Akins P, Lörz H (1984) Progress and limitations in the culture of cereal protoplasts. Trends Biotechn. 2:119-123.

Paszkowski J, Shillito RD, Saul M, Mandak V, Hohn T, Hohn B, Potrykus I (1984). Direct gene transfer to plants. EMBO J 3:2717-2722.

Potrykus I, Saul MW, Petruska J, Paszkowski J, Shillito R (1985). Direct gene transfer to cells of a graminaceous monocot. Mol Gen Genet 199:183-188.

Puertas MJ, De la Pena A, Estades B, Merino F (1984). Early sensitivity to colchicine in developing anthers of rye. Chromosoma 89:121-126.

Sanford JC, Skubik KA, Reisch BI (1985). Attempted pollen-mediated plant transformation employing genomic donor DNA. Theor Appl Genet 69:571-574.

Shillito RD, Saul MW, Paszkowski J, Müller M, Potrykus I (1985). High efficiency direct gene transfer to plants. BioTechnol 3:1099-1103.

Stachel SE, Nester EW, Zambryski P (1986). A plant cell factor induces Agrobacterium tumefaciens vir gene expression. Proc Natl Acad Sci USA 83:379-383.

Stolarz A, Lörz H (1986). Somatic embryogenesis, cell and protoplast culture of Triticale (x Triticosecale Wittmack). In: Horn W, jensen CJ, Odenbach W, Schieder O (eds.) Berlin: Walter de Gruyter, in press.

Töpfer R, Steinbiss HH (1985). Plant regeneration from cultured fertilized barley ovules. Plant Science 41:49-54.

Uchimiya H, Fushimi T, Hashimoto H, Harada H, Syono K, Sugawara (1986). Expression of foreign gene in callus derived from DNA treated protoplasts of rice (*Oryza sativa* L.). Mol Gen Genet, in press.

Vasil IK (1983). Isolation and culture of protoplasts of grasses. Int Rev Cytol 16:79-88.

Werr W, Lörz H (1986). Transient gene expression in a *Gramineae* cell line. A rapid procedure to study plant promoters. Mol Gen Genet 202, 368-373.

Yamada Y, Yang ZQ, Tang DT (1986). Plant regeneration from protoplast-derived callus of rice (*Oryza sativa* L.). Plant Cell Reports 5:85-88.

Plant Tissue and Cell Culture, pages 317–329

AGROBACTERIUM-MEDIATED TRANSFORMATION OF PLANTS

Robert Horsch, Robert Fraley, Stephen Rogers, Joyce Fry, Harry Klee, Dilip Shah, Sheila McCormick, Jeanne Niedermeyer, Nancy Hoffmann

Plant Molecular Biology, Corporate Research Laboratories, Monsanto Company, St. Louis, MO, 63198

INTRODUCTION

In the 4 years since the last IAPTC Congress in Japan, two technologies have been developed for transferring isolated genes into plant cells and regenerating transgenic plants from those cells. One method is based on the capacity of plant protoplasts to incorporate and express exogenous DNA when treated with physical or electrical treatments to permeabilize their membranes. The other method is based on the natural gene transfer capacity of *Agrobacterium tumefaciens*, which can transform cells as well as protoplasts.

In theory, the delivery of exogenous DNA to protoplasts is without host limitation (Potrykus et al., 1985), while *A. tumefaciens* has been observed to tranform only dicotyledonous species and a few monocots in the lilly family (De Cleene and De Lay, 1976). In practice to date, a much broader range of transgenic plants have been produced with *A. tumefaciens*-mediated transformation (see table 1) since it can transform regenerable tissues with relative ease, while protoplasts of most species are still difficult or impossible to regenerate. The difficulties of DNA transfer and selection for transformed colonies during recovery and growth have proven to make low frequency regeneration from protoplasts a major problem. Only transgenic tobacco and carrot plants have been produced via protoplasts to date. Free DNA delivery to regenerable protoplasts may be the eventual key to transformation of important crops such as corn, wheat, or rice, but because of the practical ease of *A. tumefaciens*-mediated transformation, it is likely to remain the method of choice for routine production of transgenic plants for species within it host-range.

AGROBACTERIUM

A. tumefaciens causes crown gall disease by transferring a defined segment of DNA (T-DNA) from its tumor-inducing (Ti) plasmid into the nuclear genome of cells in an infected wound on many dicotyledonous plants (for recent reviews, see Bevan and Chilton, 1982; Depicker et al., 1983; Nester et al., 1984). The T-DNA contains genes that encode enzymes that catalyze the synthesis of the phytohormones IAA (Schroder et al., 1984; Thomashow et al., 1984; Akiyoshi et al., 1984) and IPA (Barry et al., 1984), an auxin and a cytokinin. This overproduction of phytohormones is responsible for the proliferation of wound callus into a gall which can harbor a population of the bacteria. Other T-DNA genes cause synthesis of unique metabolites called opines which *Agrobacterium* can catabolize, but which can not be utilized by many other microrganisms or plants. One strain of *A. tumefaciens* (A208) that we have used produces nopaline synthase, an enzyme that produces nopaline from arginine and alpha-ketoglutarate.

The movement of the T-DNA is mediated by genes in another region of the Ti plasmid. These virulence genes (Klee et al., 1983; Hille et al., 1984) are not themselves transferred with the T-DNA, but act in trans to cause the transfer (de Framond et al., 1983; Hoekema et al., 1983). The virulence genes are induced by phenolic compounds present in wounded plant cells (see Stachel et al., 1986). The T-DNA is defined by a 25 base sequence, the border sequence, which is present as a direct repeat outside the ends of the T-DNA (Barker et al., 1984). None of the genes or DNA sequences within the T-DNA are required for the transfer process (Garfinkel et al., 1981; Leemans et al., 1982).

These properties of the *Agrobacterium* DNA transfer system are invaluable for developing a powerful vector system for plant transformation (Fraley et al., 1985b). It is possible to remove all of the original T-DNA which was responsible for the disease and replace it with any DNA, as long as the border sequence is maintained (Zambryski et al., 1983). The DNA is inherently stable once in the plant genome since neither the border nor the virulence genes are transferred. Finally, since the borders define a discrete T-DNA segment, the frequency of co-transfer of the entire segment is very high. This means that your favorite genes are usually transferred along with the selectable marker used to identify the transformed cells or plants.

VECTORS

To harness this mechanism, we designed a shuttle vector (Fraley et al., 1983) that can replicate in *E. coli* where recombinant manipulations are easily handled, and then transferred into *A. tumefaciens* in preparation for transfer into plants. There are two basic modes of maintainance of the shuttle vectors in *Agrobacterium*: either by integration into the Ti plasmid by recombination at a region of DNA homology (Comai et al., 1983; Fraley et al., 1983; Zambryski et al., 1983) or by autonomous replication in trans to the Ti plasmid (Bevan, 1984; An et al, 1984; Klee et al., 1985). The former type of vector is referred to as a cis or integrating vector while the latter is called a trans or binary vector. In most cases they accomplish the same goals of shuttling genes from *E. coli* to *A. tumefaciens* in a T-DNA package that can then be transferred to plants.

The essential components of our vectors are shown in figure 1. The binary vector pMON505 (Horsch and Klee, 1986) is shown as an example. They include plasmid functions for replication (pBR322, RK2 Replicon) and/or integration (LIH) in the bacteria, as well as a bacterial selectable marker for spectinomycin and streptomycin resistance (Spc/Str^R) for genetic manipulations with the bacteria. They also include a chimeric selectable marker for plant antibiotic resistance such as neomycin phosphotransferase (NOS-NPTII-NOS) which makes transformed plant cells resistant to kanamycin, and a border sequence (RB) to identify the end of the new T-DNA. There is also a scoreable marker, the gene for nopaline synthase (NOS), that provides a powerful genetic tool for monitoring the presence of the T-DNA in plants and their progeny. The final feature is the most important of all: a site to add your favorite gene (YFG).

SELECTABLE MARKERS

Since only a small proportion of cells are transformed by *A. tumefaciens*, it is important to have a selectable marker to recognize the transformed cells and suppress the growth of wild-type cells. The first marker to be used was the neomycin phosphotransferase gene from *E. coli*. This gene has been used successfully in many systems, including mammalian cells, fungi, and *Dictyostellium*. In solanacious plants such as petunia, tobacco and tomato, it provides an excellent, unambiguous selectable marker.

The features of a good selectable marker are i) high level resistance of the transformed cells to concentrations of the drug that completely

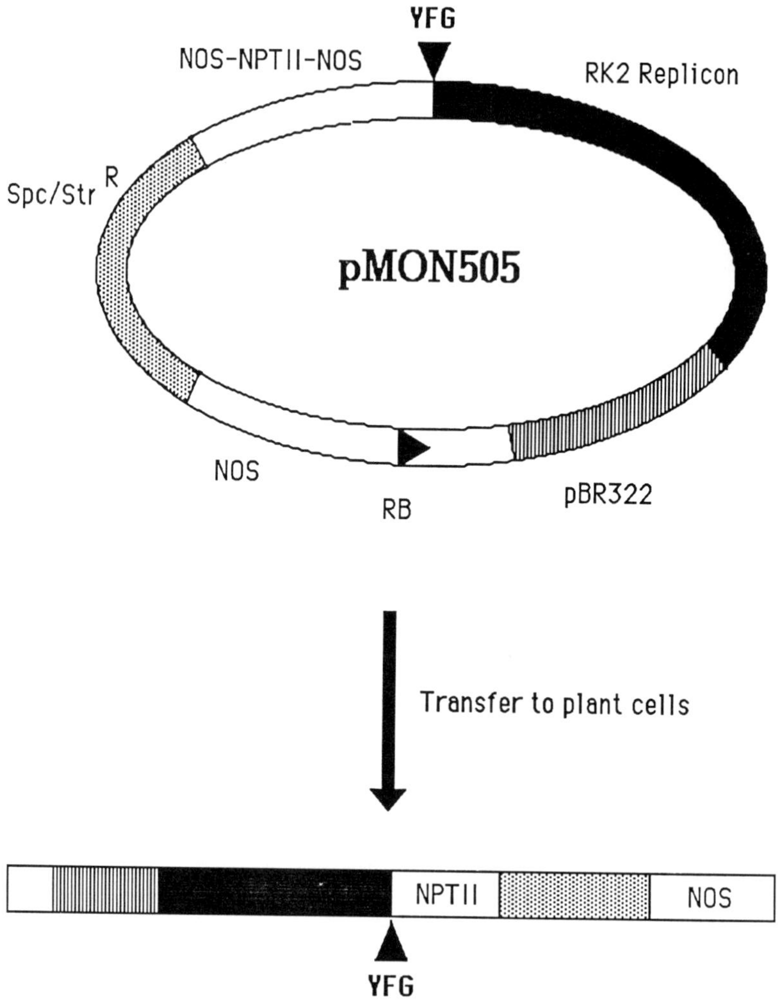

Figure 1. Binary vector pMON505.

inhibit wild-type growth, and ii) efficient selectability of rare transformants from a large excess of inhibited wild-type cells that surround them. Kanamycin-inhibited petunia cells do not appear to interfere with the growth of resistant cells in any way, and may actually facilitate their growth by acting as a nurse tissue. Single resistant clones can be obtained from petunia leaf discs transformed with attenuated mutants of *A. tumefaciens* (Horsch et al., 1986), demonstrating the efficacy of selection for this marker.

A second selectable marker that we have used is the mouse gene for dihydrofolate reductase. It encodes an altered protein that does not bind to methotrexate, a potent inhibitor of this essential enzyme. Petunia cells that contain and express this gene are able to grow on concentrations of methotrexate that completely inhibit the control cells (Eichholtz et al., in preparation). Selectability also appears to be good for methotrexate resistance.

LEAF DISC TRANSFORMATION

The simple and powerful interface between *A. tumefaciens* and plant cells is illustrated by the leaf disc transformation system (Horsch et al., 1985) where gene transfer, selection and regeneration are coupled together in an efficient process. Tobacco is an excellent host for *A. tumefaciens*, and also responds exceedingly well in culture. While the technique is most easily practiced with tobacco, it has been applied to number of other species (Table 1). This example will be described for tobacco, using a vector that confers kanamycin resistance such as pMON200 (Fraley et al., 1985a) or pMON505.

TABLE 1. Transgenic plants produced with *A. tumefaciens*-mediated transformation

Species	Reference
Nicotiana plumbaginifolia	Horsch et al., 1984
petunia	Horsch et al., 1985
tobacco	DeBlock et al., 1984
tomato	McCormick et al., 1986
lettuce	Michelmore, pers. comm.
celery	Michelmore, pers. comm.
poplar	Fillatti et al., this meeting
Arabidopsis thaliana	Lloyd et al., in press
Medicago varia	Deak et al., 1986
flax	McHughen, this meeting
Brassica napus	Fry et al., in preparation
sunflower	Everett, this meeting
cotton	Firoozabady, pers. comm.

Surface sterilized leaf discs or other axenic explants are infected with the appropriate strain of *A. tumefaciens* carrying the vector of choice, and cocultured on regeneration medium for 2 or 3 days. During this time the virulence genes in the bacteria are induced, the bacteria bind to the plant cells around the wounded edge of the explant, and the gene transfer process occurs. One empirical observation we have made is that for species other than tobacco, a nurse culture of tobacco cells increases the transformation frequency when used during the co-culture period. This may be due to more efficient induction of the virulence genes, or some other factor not yet understood.

After the transformation has occurred, the explants are transferred to regeneration/selection medium. This contains 500 ug/ml carbenicillin to kill the bacteria and the appropriate antibiotic to inhibit untransformed plant cells, usually kanamycin. During the next 3 weeks, the transformed cells grow into callus or differentiate into shoots via organogenesis. Between 3 and 6 weeks, the shoots develop enough to remove them from the explant and induce rooting in preparation for transfer to soil, figure 2.

Figure 2. Regeneration of Transgenic Petunia Shoots after Leaf Disc Transformation.

There are a number of technical issues that require careful attention, espescially when attempting to apply this technique on species other than petunia or tobacco. First, some of the shoots that grow in the presence of kanamycin do express and/or contain the foreign DNA. The reason(s) for these apparent escapes are not clear but may include loss of expression or loss of DNA during plant development, or incomplete selection due to cross-protection of wild-type cells by transformed cells nearby. What ever the reason(s), the problem can be dealt with by a second selction for ability to root in the presence of kanamycin (Horsch et al., 1985), and subsequent callus induction assays or opine assays for continued expression of the new genes.

A second problem arises in the interaction between *A. tumefaciens* and explants of some plants. While it is not certain what is happening, it is apparent that some plant species react more strongly to infection by the bacteria. For example, in *Lycopersicon esculetum* the titer of bacteria in the inoculum greatly influences the survival and vigor of explants and overall transformation efficiency (McCormick et al., 1986), while it makes little or no difference in tobacco or petunia (unpublished observations).

A third problem that may arise is the efficacy of the selectable marker used with a particular plant species or genotype. For example, the chimeric genes for expressing neomycin phosphotransferase II in plants work in all species tested, but do not confer kanamycin resistance at a selectable level in all species. In alfalfa, the enzyme can be detected in tumor tissues transformed with pMON200, but the cells do not grow any better in the presence of kanamycin than do wild-type cells, or transformed cells that lack the nptII gene altogether (J. Fry, unpublished data).

Position effects, that is, effects of the surrounding DNA or chromatin structure, result in differences in expression of the T-DNA genes in different transformants. The extreme of this is complete loss of expression during differentiation of the plantlet. This occurs in about a quarter to a third of the shoots that grew initially in the presence of 300 ug/ml kanamycin. One possible reason for failure to select against these escapes is that the kanamycin is not a good herbicide, and shoots can continue growth in its presence once they are large enough. It is usually necessary to screen several independent trangenic plants to identify the best expression of your favorite gene which sometimes, but not always, is correlated with the expression of the selectable marker.

INHERITANCE

Progeny of transgenic plants produced using this process always have inherited the T-DNAs found in the parent plant. This means that the L2 cell layer which produces the gametes in flowers has always been transformed, and that the L1 and L3 cell layers are not transformed with a different T-DNA(s). The only remaining possibility is that one or both of these tissues are not transformed at all. While we have not exhaustively searched for examples of this phenomenon, we have never observed it to occur, and would not anticipate any problem after-the-fact since progeny will be clonal anyway. This might be troublesome, if it occurs, during the intial selection for transgenic shoots, where non-transformed tissues might cause problems with morphogenesis.

Progeny usually inherit the T-DNA according to Mendelian rules for dominant genes. In cases where there is only one T-DNA, selfed progeny show a 3:1 ratio for nopaline or kanamycin resistance, while backcrossed progeny show a 1:1 ratio. This is also true for multiple T-DNAs that are present in a tandem array, all genetically linked. In a few cases, the there are multiple T-DNAs, some of which are unlinked. In these cases, selfed ratios are usually 15:1 (two unlinked loci) or higher (three or more unliked loci). In nearly all cases, the selected and unselected markers on a single T-DNA are coinherited.

We have genetically mapped the chromosomal position of the T-DNA in a number of independent transgenic petunias by using a special hybrid between two multiply-marked genetic testor strains (Wallroth et al., 1986). This F1 hybrid (VR) was produced at the Genetics Department, University of Amsterdam, by crossing strain V23 with R51. VR has at least one genetic marker that is heterozygous on each of its 7 chromosome pairs. In this way each chromosome is uniquely tagged and can be observed in a single generation by backcrossing to each of the original parents. In a sample of 9 transformants, the T-DNA was found in a different place in each, in four of the chromosomes. This is consistent with a random distribution, but does not eliminate bias or repeated dispersed sites for integration.

The T-DNA thus becomes a permanent, stable part of the genome of transgenic plants and behaves just as the other endogenous genes. The availability of plants with T-DNA inserts mapped in chromosomal locations brings a valuable new tool to plant genetics. These plants have an easily scoreable marker which can be used to map new mutants, and a selectable chromosome tag that might facilitate breeding or selection of interesting genes nearby.

ENGINEERING HERBICIDE TOLERANCE

The herbicide glyphosate is a potent inhibitor of an enzyme in the shikimate pathway in higher plants. The herbicide competitively blocks EPSPS (5-enolpyruvylshikimate-3-phosphate synthase) and kills plants as well as tissue cultures. We have engineered plants to be tolerant to otherwise lethal doses of glyphosate by a strategy of overexpression of EPSPS in transgenic plants. The method for acheiving this goal started in tissue culture to obtain the EPSPS protein, proceeded through molecular biology to obtain and modify the gene, and back into tissue culture to put the gene back into plants.

A glyphosate tolerant suspension culture of *Petunia hybrida* was selected by sequencial step-up from 0.1 mM glyphosate to 20 mM glyphosate over a period of several months (Steinrucken et al., 1986). The resistant cell line (MP4G) produces about 20 times more EPSPS than the sensitive parent culture (MP4). The EPSPS protein was purified form the MP4G culture and this pure protein was partially sequenced to obtain an amino acid sequence from its NH_2-terminus. A series of oligonucleotide probes was synthesized (in three families) that would cover all possible DNA sequences for the amino acid sequence. These families were then used to probe a northern blot of poly(A)+ mRNA from the two cell lines. One of the families of probes hybridized to a specific mRNA from the MP4G culture about 20 times more strongly than to the same size mRNA from the MP4 culture. This strongly suggested that this probe was specific for EPSPS and that the MP4G culture was overproducing the mRNA for this enzyme. The gene was then isolated from a cDNA library prepared form the MP4G culture. The cause of the overproduction of the EPSPS mRNA was an amplification of the gene during the sequencial adaptation of the culture to glyphosate (Shah et al., 1986).

Thus the suspension culture had provided an enriched source for purification of the protein, and a means to identify which probe (from a family of redundant probes) was the proper one to use to screen a cDNA library for the EPSPS gene. The culture was also used to prepare a cDNA library which was enriched for the EPSPS sequence.

A derivative of the pMON505 vector, pMON530, was used for *Agrobacterium*-mediated transfer of a chimeric EPSPS gene back into plant cells. It contains the NOS-NPTII-NOS gene for kanamycin resistance and an expression cassette consisting of the cauliflower mosaic virus (CaMV) 35S promoter and a nopaline synthase (NOS) polyadenylation signal. The complete open reading frame of the petunia EPSPS was cloned into pMON530 to create a chimeric EPSPS gene capabible of high level expression in plant cells (pMON546). The vector was then introduced into *A. tumefaciens* for transfer to plant cells.

Leaf discs transformed with vectors pMON505 and pMON546 both produced large amounts of kanamycin resistant callus. Small pieces of this callus were transferred to medium containing 0.1, 0.25, or 0.5 mM glyphosate. The callus transformed with pMON546 continued to grow at all concentrations of glyphosate while the callus transformed with pMON505 was inhibited. Glyphosate tolerance could also be demonstrated by direct selection for growth from infected leaf discs on the same concentrations of glyphosate. Four independent transgenic plants containing pMON546 were selected on kanamycin and grown in soil before spraying with a lethal dose of glphosate. These plants were sprayed with formulated glyphosate (containing a surfactant) at a rate equal to 0.8 pounds per acre, twice the dose needed to kill all control plants. All four transformed plants survived and continued to grow with slight chlorosis at the growing points, from which they later recovered. These experiments demonstrate that EPSPS is the prime target of glyphosate in plant cells and whole plants, since over expression of that one gene conferrs tolerance to the herbicide. These results have been extended to tobacco and tomato using the petunia gene construction in pMON546.

CONCLUSION

Molecular biology and gene transfer technology have opened the door to unprecented experimental power for understanding and manipulating plants (for example see: Broglie et al., 1984; Beachy et al., 1985a,b; Fraley et al., 1985b; Rogers et al., 1986). This power will build upon itself as we use it to discover the mechanisms of gene expression, protein structure and function, and biochemistry and physiology of plants. This knowledge and technology will rapidly increase our ability to intelligently predict beneficial changes to introduce into crops, as well as the means to do it.

REFERENCES

Akiyoshi D, Morris R, Hinz R, Mischke B, Kosuge T, Garfinkel D, Gordon M, Nester E (1983). Cytokinin/auxin balance in crown gall tumors is regulated by specific loci in the T-DNA. Proc. Natl. Acad. Sci. USA 80:407-411.

Akiyoshi D, Klee H, Amasino R, Nester E, Gordon M (1984). T-DNA of *Agrobacterium tumefaciens* encodes an enzyme of cytokinin biosynthesis. Proc. Natl. Acad Sci. USA 81:5994-5998.

An G, Watson B, Stachel S, Gordon M, Nester E (1984). New cloning vehicles for transformation of higher plants. EMBO J. 4:277-284.

Barker R, Idler K, Thompson D, Kemp J (1983). Nucleotide sequence of the T-DNA region from the *Agrobacterium tumefaciens* octopine Ti plasmid pTi15955. Plant Mol. Biol. 2:335-350.

Barry G, Rogers S, Fraley R, Brand L (1984). Identification of a cloned cytokinin biosynthetic gene. Proc. Natl. Acad. Sci. USA 81:4776-4780.

Beachy RN, Chen ZL, Horsch RB, Rogers SG, Hoffmann NL, Fraley RT, (1985a). Accumulation and assembly of soybean B-conglycinin in seeds of transformed petunia plants. EMBO J. 4:3047-3053.

Beachy RN, Abel P, Oliver M, De B, Fraley RT, Rogers SG, Horsch RB (1985b). Potential for applying genetic transformation to studies of viral pathogenesis and cross protection. In: Biotechnology in Plant Science: Relevance to Agriculture in the Eighties. (M. Zaitlin & A. Hollander, eds.) Academic Press, NY (in Press).

Bevan M, Chilton MD (1982). T-DNA of the *Agrobacterium* Ti and Ri plasmids. Ann. Rev. Genet. 16:357-384.

Bevan M (1984). *Agrobacterium* vectors for plant transformation. Nucleic Acids Res. 12:8711

Broglie R, Coruzzi G, Fraley RT, Rogers SG, Horsch RB, Niedermeyer JG, Fink CL, Flick JS, Chua NH (1984). Light-regulated expression of a pea ribulose-1,5-biphosphate carboxylase small subunit gene in transformed plant cells. Science 224: 838-843.

Comai L, Schilling-Cordaro C, Mergia A, Houck C (1983). A new technique for genetic engineering of *Agrobacterium* Ti plasmid. Plasmid 10:21- .

Deak M, Kiss GB, Koncz C, Dudits D (1986). Transformation of *Medicago* by *Agrobacterium* mediated gene transfer. Plant Cell Reports 5:97-100.

DeBlock M, Herrera-Estrella L, Van Montagu M, Schell J, Zambryski P (1984). Expression of foreign genes in regenerated plants and their progeny. EMBO J. 3:1681.

De Cleene M, De Lay J (1976). The host range of crown gall. Botanical Rev. 42:389.

deFramond A, Barton K, Chilton MD (1983). Mini-Ti: a new vector strategy for plant genetic engineering. Bio/Technology 1:262- .

Depicker, A., M. Van Montagu and J. Schell (1983) In: T. Kosuge, C. Meredith and A. Hollaender, eds. Genetic Engineering of Plants: an Agricultural Perspective, Plenum Press, New York, pp.143-176.

Fraley RT, Rogers SG, Horsch RB, Sanders PR, Flick JS, Adams SP, Bittner ML, Brand LA, Fink CL, Fry JS, Galluppi GR, Goldberg SB, Hoffmann NL, Woo SC (1983). Expression of bacterial genes in plant cells. Proc. Natl. Acad. Sci. USA 80: 4803-4807.

Fraley RT, Rogers SG, Horsch RB, Eichholtz DA, Flick JS, Fink CL, Hoffmann NL, Sanders PR (1985a). The SEV system: a new disarmed Ti plasmid vector for plant transformation. Bio/Technology 3:629-635.

Fraley RT, Rogers SG, Horsch RB, (1985b). Genetic Transformation in Higher plants. CRC Critical Reviews in Plant Sciences 4:1-46.

Garfinkel D, Simpson R, Ream L, White F, Gordon M, Nester E (1981). Genetic analysis of crown gall: fine structure map of the T-DNA by site-directed mutagenesis. Cell 27:143-153.

Hille J, van Kan J, Schilperoort R (1984). *trans*-Acting virulence functions of the octopine Ti plasmid from *Agrobacterium tumefaciens.* J. Bact. 158:754-756.

Hoekema A, Hirsch P, Hooykaas P, Schilperoort R (1983). A binary plant vector strategy based on separation of vir- and T- region of the *Agrobacterium tumefaciens* Ti-plasmid. Nature 303:179.

Horsch RB, Fraley RT, Rogers SG, Sanders PR, Lloyd A, Hoffmann N (1984). Inheritance of functional foreign genes in plants. Science 223: 496-498.

Horsch RB, Fry J, Hoffmann NL, Wallroth M, Eichholtz D, Rogers SG, Fraley RT (1985). A simple and general method for transferring genes into plants. Science 227:1229-1231.

Horsch RB, Klee HJ, Stachel S, Winans SC, Nester E, Rogers SG, Fraley RT (1986). Analysis of *Agrobacterium tumefaciens* virulence mutants in leaf discs. Proc. Nat'l. Acad. Sci. 83:2571-2575.

Horsch RB, Klee HJ (1986). Rapid assay of foreign gene expression in leaf discs transformed by *Agrobacterium tumefaciens*: role of T-DNA borders in the transfer process. Proc. Nat'l. Acad. Sci. 83: 4428-4432.

Klee H, Montoya A, Horodyski F, Lichtenstein C, Garfinkel D, Fuller S, Flores C, Peschon J, Nester E, Gordon M (1984). Nucleotide sequence of the tmr genes of the pTiA6NC octopine Ti plasmid: two gene products involved in plant tumorigenesis. Proc. Natl. Acad. Sci.USA 81:1728-1732.

Klee H, White F, Iyer V, Gordon M, Nester E (1983) Mutational analysis of the virulence region of an *Agrobacterium tumefaciens* Ti plasmid. J. Bacteriol. 153:878-

Klee HJ, Yanofsky MF, and Nester EW, (1985). Vectors for transformation of higher plants. Bio/Technology 3,637-642.

Leemans J, Deblaere R, Willmitzer L, De Greve H, Hernalsteens JP, Van Montagu M, Schell J (1982). Genetic identification of functions of TL-DNA trancripts in octopine crown galls. EMBO J. 1:147-152.

Lloyd AM, Barnason AR, Rogers SG, Byrne MC, Fraley RT, Horsch RB (1986). Transformation of *Arabidopsis thaliana* with *Agrobacterium tumefaciens* using a gene conferring hygromycin resistance. Science (in press).

McCormick S, Niedermeyer J, Fry J, Barnason A, Horsch R, Fraley R (1986). Leaf disc transformation of cultivated tomato (*L. esculentum*) using Agrobacterium tumefaciens. Plant Cell Reports 5:81-84.

Nester E, Gordone M, Amasino R, Yanofsky M (1984) Ann. Rev. Plant Physiol. 35:387-413.

Potrykus I, Shillito RD, Saul MW, Paszkowski J, (1985). Direct gene transfer state of the art and future potential. Plant Mol. Biol. Reporter 3:117-128.

Rogers SG, Horsch RB, Fraley RT (1985). Gene transfer in plants: production of transformed plants using Ti plasmid vectors. Methods in Enzymology, 118:627-640.

Rogers SG, Bisaro DM, Horsch RB, Fraley RT, Hoffmann NL, Brand L, Elmer JS, Lloyd A (1986). Tomato Golden Mosiac Virus A Component DNA Replicates Autonomously in Transgenic Plants. Cell 45: 593-600.

Shah DM, Horsch RB, Klee HJ, Kishore GM, Winter JA, Tumer NE, Hironaka CM, Sanders PR, Gasser CS, Aykent S, Siegel NR, Rogers SG, Fraley RT (1986). Engineering herbicide tolerance in transgenic plants. Science 233: 478-481.

Schroder G, Waffenschmidt S, Weiler E, Schroder J (1984). The T-region of Ti plasmids codes for an enzyme synthesizing indole-3-acetic acid. Eur. J. Biochem. 138:387-391.

Stachel SE, Timmerman B, Zambryski P (1986). Generation of single-stranded T-DNA molecules during the initial stages of T-DNA transfer from *Agrobacterium tumefaciens* to plant cells. Nature 322:706-712.

Steinrucken H, Schulz A, Amrhein N, Porter C, Fraley R (1986). Overproduction of 5-enolpyruvylshikimate-3-phosphate synthase in a glyphosate-tolerant *Petunia hybrida* cell line. Arch. Biochem. Biophys. 244:169.

Thomashow L, Reeves S, Thomashow M (1984). Crown gall oncogenesis: evidence that a T-DNA gene from the *Agrobacterium* Ti plasmid pTiA6 encodes an enzyme that catayzes synthesis of indoleacetic acid. Proc. Natl. Acad. Sci. USA 81:5071-5075.

Wallroth M, Gerats AGM, Rogers SG, Fraley RT, Horsch RB (1986). Chromosomal localization of foreign genes in *Petunia hybrida*. Mol. Gen. Genet. 202:6-16.

Zambryski P, Joos H, Genetello C, Leemans J, Van Montagu M, Schell J (1983). Ti plasmid vector for the introduction of DNA into plant cells without alteration of their normal regeneration capacity. EMBO J. 2:2143.

Plant Tissue and Cell Culture, pages 331–339

USING MOLECULAR MARKERS TO ANALYZE GENOME ORGANIZATION AND EVOLUTION

Robert Bernatzky and Steven D. Tanksley

Department of Plant Breeding and Biometry,
Cornell University, Ithaca, New York 14853

INTRODUCTION

Molecular markers have tremendous potential for analyzing problems in plant genetics and breeding (Tanksley 1983, Burr et al. 1983). Molecular markers are already used in animal systems as diagnostic tools for diseases (Newmark 1984). Compared with morphological markers, these markers have the advantage of being generally codominant, naturally-occurring and free of epistatic and environmental effects. A few of the potential applications are in facilitating interspecific gene transfer, identifying sexual and parasexual hybrids, measuring genetic variability in plant populations and analyzing the genetics of quantitative characters. We describe here the development of molecular markers and a genetic linkage map in tomato.

MATERIALS AND METHODS

The description of plant materials, isozyme and genetic analysis has been previously reported (Bernatzky and Tanksley 1986a). cDNA cloning, restriction digestion, electrophoresis, Southern blotting and hybridization has been reported elsewhere (Bernatzky and Tanksley 1986b, Bernatzky and Tanksley 1986c).

RESULTS AND DISCUSSION

Genome Organization

The data base for genetic mapping of DNA fragments was

developed from isozymes which had been previously mapped (Tanksley and Rick 1980). In order to obtain segregation data on DNA fragments, polymorphism for length of restriction fragments had to be established for the two parents used in the cross for mapping (in this case *L. esculentum* and *L. pennellii*). This was accomplished by digestion of the two parents with various restriction enzymes, electrophoresis and Southern blotting of the DNA to nylon membranes and hybridization of the membranes with a nick-translated DNA probe (Figure 1a). After identifying an enzyme which produces a polymorphism, DNA of the backcross or F2 progeny were similarly digested, blotted and probed (Figure 1b). Tests of independent assortment of these fragments with the isozymes or other mapped DNA markers was performed using chi-square analysis and linkage values were determined by the maximum likelihood equations of Allard (1954).

A linkage map of more than 150 molecular markers is shown in Figure 2. The sources of the markers are as follows: CD - random tomato cDNA sequences (Bernatzky and Tanksley 1986a); TG - random tomato genomic sequences (Tanksley and Miller, unpublished); Act - Actin-related sequences (Bernatzky and Tanksley 1968d); R45S, Rbcs and Cab - ribosomal RNA, small subunit of ribulose biphosphate carboxylase and the major chlorophyll a/b binding protein coding regions (Vallejos et al. 1986); PC - random pepper cDNA sequences (Tanksley and Bernatzky, unpublished); CR - repeat sequences, one of which was isolated from a tomato genomic clone of the Cab-1 gene cluster (Bernatzky, Tanksley and Pichersky, unpublished). The remaining loci are isozymes (Tanksley and Rick 1980). Chromosomes 9 and 11 do not as yet have markers assigned to them since there were no mapped isozymes for these chromosomes segregating in the crosses used. However, there are unassigned linkage groups indicated at the bottom of Figure 2, some of which probably belong to these chromosomes. This is currently besing tested using a set of primary trisomics. If a marker is situated on a chromosome (ie. 9 or 11) then there will be a 50% increase in DNA dosage for its primary trisomic as compared to the other trisomics.

Most of the DNA loci on the map are derived from random cDNA clones produced from tomato leaves (Bernatzky and Tanksley 1986a and b). Two important features of this set of markers should be noted. First, the majority (52%) of the cDNA clones hybridize to single loci and of these many

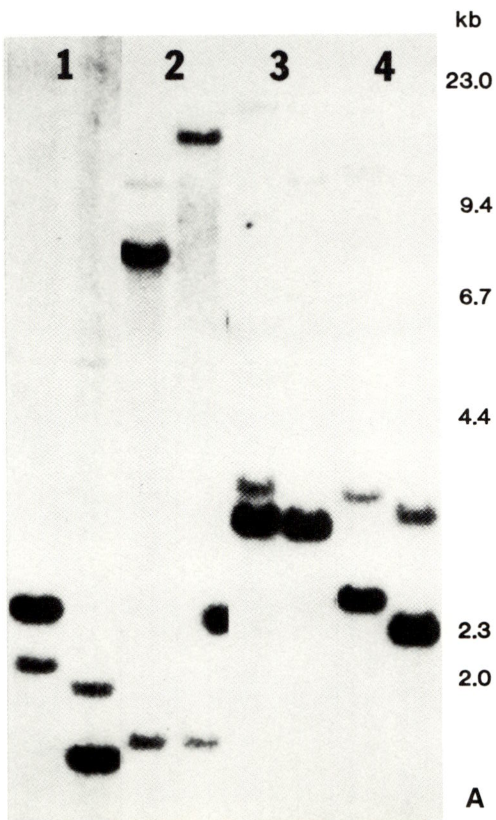

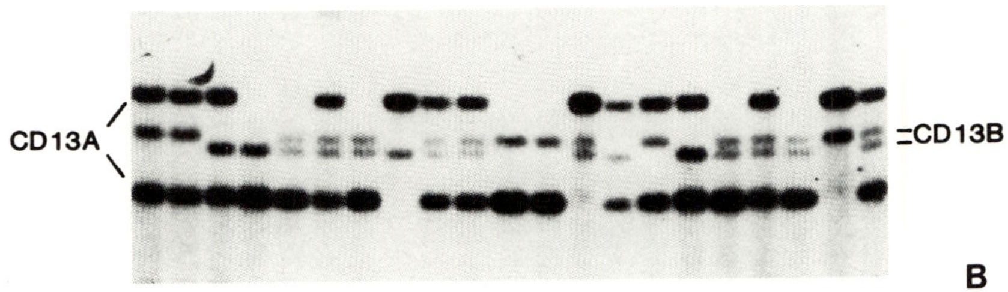

Figure 1 A. Restriction enzyme survey of *L. esculentum* and *L. pennellii* genomic DNA digested with various restriction endonucleases and probed with radiolabelled plasmid 3-87 (CD13). 1 *DraI* digest; left lane, *L. esculentum*, right lane, *L. pennellii*. 2 *EcoRI*. 3 *EcoRV*. 4 *HindIII*. Molecular weights derived from lambda DNA digested with *HindIII*.

B. Autoradiograph of F2 progeny probed with plasmid 3-87 (CD13). DNA was digested with *DraI*. Reprinted in part from Bernatzky and Tanksley 1986b.

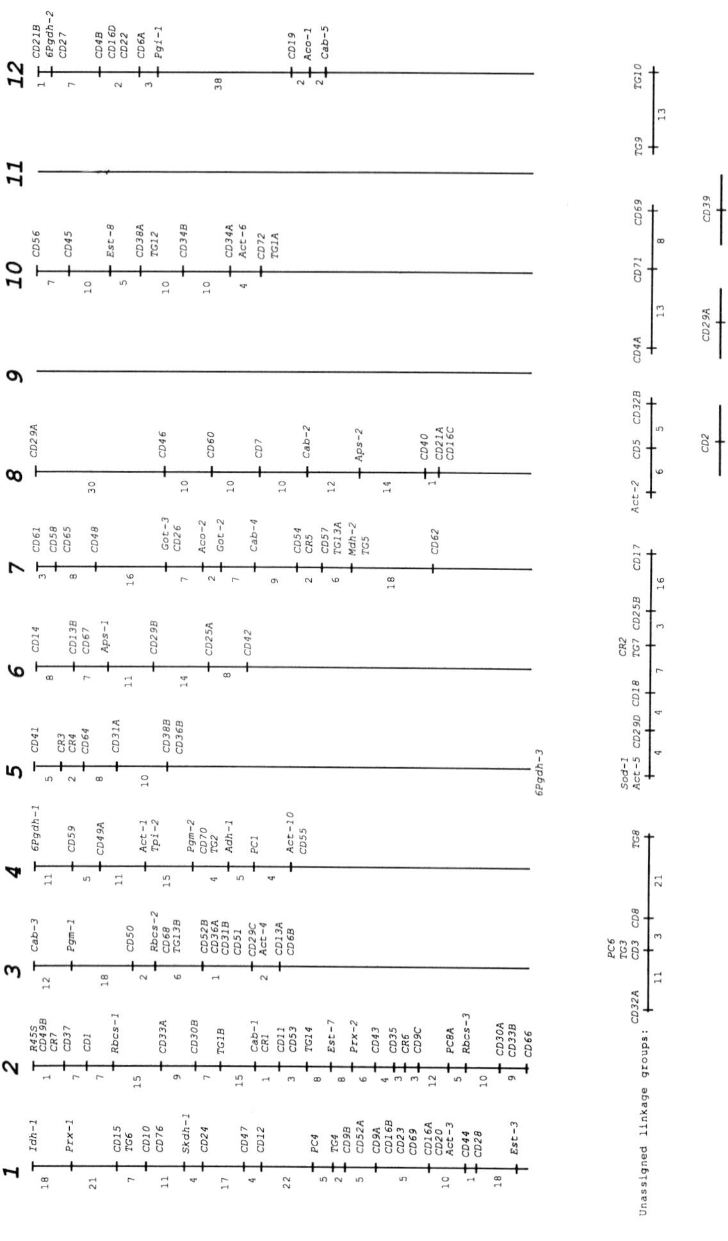

Figure 2. Linkage map of DNA and isozyme loci. Bold numbers at top refer to chromosome number and numbers along chromosomes refer to map distance in cM (not to scale). See text for description of gene symbols.

contain single copies of the gene per locus (Bernatzky and Tanksley 1986b). This indicates that most of the genes in tomato are reprsented once in the genome, rather than as multigene families, and as such are a good source of unique molecular markers. This also argues against polyploidy in the history of this genus although the basic chromosome number is high (n=12).

The second feature of these cDNA loci is that, of the clones that hybridize to more than one locus, most of the duplicate loci are unlinked (Figure 2)(Bernatzky and Tanksley 1986a). This suggests that there is a mechanism for dispersal of DNA fragments throughout the genome which may or may not be causally related to the duplication process itself.

There are some indications of non-random distribution of DNA markers among the chromosomes. Chromosomes 1 and 2 contain a higher proportion of markers than the other chromosomes. This may be the result of the fact that these are the largest chromosomes cytologically and may contain a greater proportion of DNA. There is also a large number of markers clustered within a few map units on chromosome 3 (from Rbcs-2 to CD6B). It is possible that this reflects a cytogenetic difference between *L. esculentum* and *L. pennellii* in this region which suppresses recombination. However, the distance between some of the markers has been checked in the progeny of a cross between *L. esculentum* and *L. cheesmanii*, which are much more closely related species, and similar linkage relationships have been found (Bernatzky and Tanksley, unpublished). The clustering in this region will continue to be investigated.

Changes in Locus Number

Approximately 5% of the cDNA clones identify loci in one tomato species which do not appear to exist in the other. In each case, the clone hybridized to more than one locus but one of the loci was mapped on the basis of a single allele donated by only one species. The locus is considered null for the other species and the data was analyzed in a dominant (presence of fragment) and recessive (absence of fragment) manner. That the duplicated locus is considered species-specific is a consequence of not being able to find an allelic fragment in one of the species. This, however, is a function of the restriction enzymes used and confidence in its absence from one species is dependent on an exhaustive search with many enzymes. Such an attempt is in progress for two multilocus clones and allelic counterparts for one locus of each has not been found. Of these clones, one locus is duplicated in L. esculentum and the other in L. pennellii. The absence of a locus in one species may actually represent a deletion which could be tolerated since these clones do hybridize to other, possibly functional loci. Phylogenetic analysis of the Lycopersicon genus could suggest which situation (duplication or null) is ancestral and which is derived. If these loci are true duplications then they may represent valuable sequences for determining the mode of duplication and the mechanism of gene dispersion throughout the genome since they are unlinked to the duplicate loci.

The locus number per clone also differs between tomato and a closely related genus, pepper (Capsicum 2n=24), approximately 10% of the time (Tanksley and Bernatzky, unpublished). One of the most notable differences is for the locus that codes ribosomal RNA. In tomato there is one locus (R45S) on chromosome 2 (Figure 2)(Vallejos et al. 1986). There are two, unlinked major loci in pepper and possibly a third minor locus that shows loose linkage with one of the other loci (Tanksley and Bernatzky, unpublished). Other changes in locus number occur in either pepper or tomato. A significant observation in this respect is that changes in locus number do not necessarily reflect differences in genome size. Although pepper has nearly four times the DNA content per haplod nucleus (2.76 pg vs 0.74 pg for tomato)(Galbraith et al. 1983) the number of loci detected per clone is approximately the same as in tomato and, when a difference in number does occur, tomato is as likely as pepper to show the higher number.

Molecular Markers in Plant Genetics and Breeding

The molecular markers developed so far cover more than 90% of the tomato chromosomes based on an estimate of 1200 cM for the entire genome. This collection of markers provides a powerful tool for locating, through linkage, genes or chromosome segments of interest. At present this molecular map is being used to analyze the introgression of segments from a wild relative, L. chimiliewskii, into cultivated tomato that may have positive effects on the levels of soluble sugars in mature fruit (Tanksley and Hewitt, unpublished).

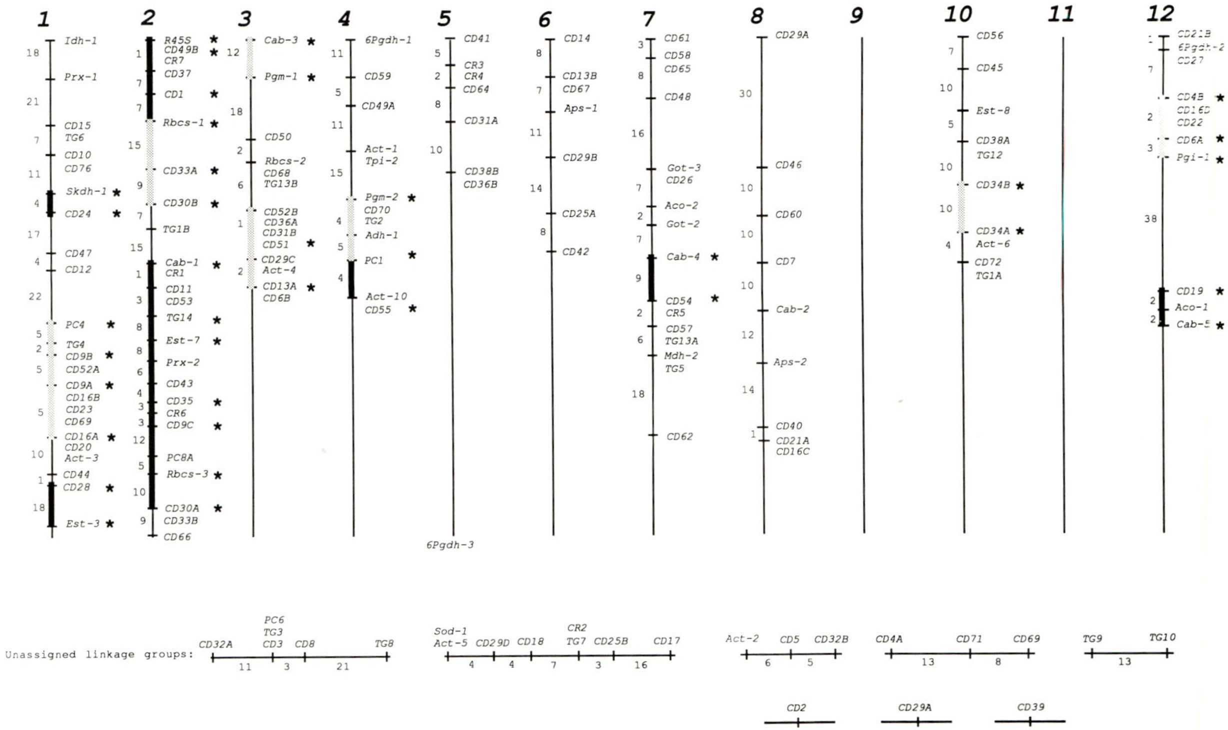

Figure 3. Linkage map of tomato contrasting regions of homology with pepper. Bold lines indicate conserved linkage in pepper and broken lines refer to loci that are linked in tomato but are not linked in pepper.

It is also being employed in a project to locate chromosome deletions produced from pollen irradiation (Zamir and Tanksley, unpublished). The map has been successfully used to locate and evaluate the stability of Ti plasmid insertions in transgenic tomato plants (Chyi et al. 1986).

We are currently using the cDNA clones to develop a comparative linkage map in pepper (Tanksley and Bernatzky, unpublished). Tomato and pepper have the same basic chromosome number (n=12). One of the most striking features of this comparative work is that the linkage relationships of these loci is more scrambled than conserved between these two genera (Figure 3). Besides the non-conserved regions illustrated in Figure 3, there are very tightly linked markers in the pepper genome most of which are on different chromosomes in tomato (data not shown). These data, along with the dispersion of duplicated loci in tomato indicates that the tomato genome is very fluid and that changes in locus number or position can occur without changes in the basic chromosome number.

REFERENCES

Allard RW (1954). Formulas and tables to facilitate the calculation of recombination values in heredity. Hilgardia 24:235-278.

Bernatzky R, Tanksley SD (1986a). Toward a saturated linkage map in tomato based on isozymes and random cDNA sequences. Genetics 112:887-898.

Bernatzky R, Tanksley SD (1986b). Majority of random cDNA clones correspond to single loci in the tomato genome. Mol Gen Genet 203:8-14.

Bernatzky R, Tanksley SD (1986c). Methods for detection of single or low copy sequences in tomato on southern blots. Pl Mol Biol Rep 4:37-41.

Bernatzky R, Tanksley SD (1986d). Genetics of actin-related sequences in tomato. Theor Appl Genet 72:314-324.

Burr B, Evola SV, Burr FA (19833). The application of restriction fragment length polymorphism to plant breeding. In Setlow JK, Hollaender A (eds): "Genetic Engineering: Principles and Methods", Plenum Publishing, pp 45-59.

Chyi YS, Jorgensen RA, Goldstein D, Tanksley SD, Loaiza-Figueroa F (1986). Locations and stability of *Agrobacterium*-mediated T-DNA insertions in the *Lycopersicon* genome. Mol Gen Genet, in press.

Galbraith DW, Harkins KR, Maddox JM, Ayres NM, Sharma DP, Firoozabady E (1983). Rapid flow cytometric analysis of the cell cycle in intact plant tissues. Science 220:1049-1051.

Newmark P (1984). Molecular diagnostic medicine. Nature 307:11-12.

Tanksley SD (1983). Molecular markers in plant breeding. Pl Mol Biol 1:3-8.

Tanksley SD, Rick CM (1980). Isozyme gene linkage map of the tomato: applications in genetics and breeding. Theor Appl Genet 57:161-170.

Vallejos CE, Tanksley SD, Bernatzky R (1986). Localization in the tomato genome of DNA restriction fragments containing sequences homologous to the rRNA(45s), the major chlorophyll a/b binding polypeptide and the ribulose bisphosphate carboxylase genes. Genetics 112:93-105.

Section VI.

Applications to Crops

Plant Tissue and Cell Culture, pages 343–358

BREEDING FOR DISEASE RESISTANT CROP PLANTS BY CELL CULTURE TECHNIQUES

G. Wenzel, M. Bolik, S. Deimling, S.C. Debnath, B. Foroughi-Wehr and R. Schuchmann

Institute for Resistance Genetics, Federal Biological Research Centre for Agriculture and Forestry, D-8059 Grünbach, Federal Republic of Germany

INTRODUCTION

Genetically fixed resistance is the cheapest and most prophylactic procedure of plant protection against a number of diseases. A rather high level of resistance has already been reached by classical procedures like selection and combination breeding. It is discussed here, how far cell and tissue culture can increase the efficiency of central steps in this process. This means in particular, whether genetic variation, selection and genome combination can be improved by in vitro approaches. On the examples of barley and potato techniques for the incorporation of resistances against viral and fungal diseases by in vitro culture methods are discussed. This includes monogenically inherited resistance to barley yellow mosaic virus (BaYMV) and polygenically inherited characters like field resistances to potato viruses X, Y, and leaf roll (PVX, PVY, PLRV) or fungi like _Phytophthora_ and _Fusarium._ The most important advantages of cell and tissue culture steps in the breeding process are: (1) Freedom from the effects of the natural environment; (2) the ability to handle large numbers of individuals in a very small space; and 3) the ability to work with haploids, the simpler genomes of which allow uncovering recessive traits and additive characters.

In this review two areas of research will be distinguished: Techniques making use of somatic cells and techniques using gametophytic stages. Both processes create new variability either via somaclonal variation (Larkin and Scowcroft,

1981), or via the normal meiotic segregation, and thus fulfill the first demand of plant breeding. For selection of resistant genotypes it is of no principle importance whether the new character has been caused by an induced mutation or a segregation event.

Breeding strategies aiming at disease resistance into used preferrably single genes that were simply inherited. However, it became obvious that such simple resistance favored the rapid coevolution of pathotypes with virulence on these cultivars. Consequently recent programs focus on quantitatively inherited traits, which are normally based on several or many genes. It would be of utmost importance if here, where classical procedures reach limitations, tissue culture could contribute.

STRATEGIES USING SOMATIC CELLS

Rapid Propagation and Meristem Culture

The oldest practical application of tissue culture techniques for disease control is rapid propagation of pathogene-free plants, normally starting from meristem cultures. The use of rapid propagation techniques for potato has opened new practical ways for a cheaper maintenance breeding and is used increasingly by the commercial companies (Wriedt, 1984). In this context techniques for curing vegetatively propagated crops, like the potato, from viruses has gained new emphasis, particularly under economic aspect to reduce the cost. In a comparative study for cleaning clones from PVS, PVX and PLRV, it could be demonstrated that only in PVX the size of the meristem is very critical, while for the purification from PVS even 2mm tips could give success (Meyer and Foroughi-Wehr, in prep.). These larger tips regenerated much easier and never produced callus, while the ones smaller than 0.2mm tend to proliferate unorganized. Since a given genotype must remain stable, it is important to prohibit this callus formation.

Selection in Somaclones at the Plant Level

Starting in the 1950s, the natural variability of plants was increased by mutation techniques. This resulted for some decades in mutation breeding programs. It was an obvious next step to determine, whether the variability expressed in plants of major crops, when regenerated from tissue culture could serve the same aim. This somaclonal variation was observed in plants regenerated from organ cultures, calli, and protoplasts (Larkin and Scowcroft, 1981). As this paper mainly concentrates on barley and potato, other plants, where new sources of resistance have been detected after in vitro culture of somatic cells are listed in Table 1.

TABLE 1. Examples for successful selections of disease-resistant plants from somatic cells in vitro

Plant	Disease	Reference
Without in vitro selection		
Potato	Alternaria	Shepard 1980
	Phytophthora	Shepard 1980
	PLRV	Thomson 1986
	PVX	this paper
	PVY	Thomson 1986
	Streptomyces	Thomson 1986
Sugarcane	Fijivirus	Coleman and Nickell 1964
	Helminthosporium	Krishnamurthi and Thaskal 1974
With in vitro selection		
Dianthus	Fusarium	Buiatti et al. 1985
Glycine	Phytophthora	Ebel et al. 1976
Maize	Helminthosporium	Gengenbach et al. 1977
Medicago	Fusarium	Hartmann et al. 1984
Oats	Helminthosporium	Briggs et al. 1984
Potato	Fusarium	Schuchmann 1985a
	Phytophthora	Behnke 1979
Rape seed	Phoma	Sacristan 1985
Rice	Helminthosporium	Ling et al. 1985
Tobacco	Alternaria	Thanutong et al. 1983
	Phytophthora	Helgeson et al. 1972
	Pseudomonas	Thanutong et al. 1983

A system that starts with single cells can help to circumvent problems of cross feeding and chimerism, and thereby increases the chance finding valuable mutants. Furthermore, the number of individuals can be increased to 10^6, which is within the range of the natural mutation rate of 10^{-6}. Once the 4x potato protoplast system was highly reproducible, Shepard (1980) examined the usefulness of protoplasts for creating new genotypes and found a tremendous variability in potato clones from the tetraploid 'Russet Burbank'. Although many of the variants were grossly aberrant, some seemed to be better than the starting material in several characters, but were genetically unstable even during vegetative propagation. Thomson et al. (1986) regenerated isolated protoplasts of three potato varieties ('Maris Piper', 'Foxton', and 'Feltwell') with the objective to find specific improvements in resistances to common scab (Streptomyces scabies), PVY, and PLRV. They produced about 10,000 calli and found 'Feltwell' to be the most responsive variety. From 'Feltwell' 875 calli regenerated, 235 of which were tested in the field. Although several clones performed better in single traits than 'Feltwell', no protoclone was found which had a better performance than 'Feltwell' for a combination of the three characters examined.

In own experiments protoclones from in vitro shoot cultures of four microspore-derived dihaploid (2n = 2x = 24) clones were screened for PVX resistance. The doubled monohaploid clones AH 78.6147, AH 82.4521/6, AH 82.4551/6 and AH 84.4566/6 were used as homogeneous source for protoplast isolation (Debnath et al., 1986). Protoplasts were cultured until several hundred calli were formed. 200 calli were randomly selected from each clone and cultured for shoot regeneration. One random shoot from each callus was established in MS-medium (Murashige and Skoog, 1962) and transferred to green house. Besides chromosome counts, plant height, number of nodes per plant, leaf fusion and length and width of leaflets, tuber number and weight, the reaction to PVX was determined. Mechanical inoculation was carried out on the leaves of young plantlets with mixtures of three different X-virus strains and evaluated by ELISA (Casper and Meyer, 1981). Whereever applicable, statistical analysis of data was performed following randomized block design. The ploidy levels and the reaction to PVX are given in Table 2. Different clones produced very small proportion of resistant plants through protoplast isolation and culture. All resistant protoclones, except three of AH 82.4551/6 were very poor in all other characters studied. The performance of the three best ones was statistically identical with the donor clone AH 82.4551/6 for all other characters studied. These three protoclones were diploid.

All these experiments take benefit of the large number of individual genotypes which can be handled in a Petri dish, but as the screening for the resistance has to be performed in the

TABLE 2. Percentages of plants with different ploidy level and reaction to PVX disease in the protoclones of four dihaploid potato clones.

Protoplast donor clone	Ploidy level			Reaction to PVX	
	4X	2X	Aneuploids	Resistant	Susceptible
AH 78.6147	33.5	24.5	42.0	2.5	97.5
AH 82.4521/6	55.0	30.0	15.0	4.5	95.5
AH 82.4551/6	27.5	50.0	22.5	5.5	94.5
AH 84.4566/6	24.0	28.0	48.0	3.0	97.0

field, this advantage is lost finally. Somaclonal variation without an in vitro selection system has nearly the same limitations as classical mutation breeding.

Selection at Cell or Callus Level

Somaclonal variation may be superior to simple mutation breeding, when a selection pressure is applied during the in vitro phase. The common advantage to the pathogen over the host, namely its ability to shift to a virulent population rapidly, is offset somewhat by the fact that numbers of the host cells can be about as large and flexibility for changes in the host cell population is about as great as for the pathogen. However, to achieve this requires a powerful in vitro screening system and a very large number of individual host cells. Typical of many older reports (e.g. Keen and Horsch, 1972) is that there was no relationship between the reaction of a plant against pathogens in tissue culture and the reaction of the same genotype as a plant or vice versa, i.e. the calli "resistant" in vitro did not give rise to resistant plants. Keen and Horsch (1972), based on their work with soybean and *Phytophthora megasperma*, warned against the use of unnatural host-pathogen systems for screening. The same did Vardi et al.(1986) which used nucellar calli from *Citrus* cultivars with known resistance to *Phytophthora citrophthora* for a test with cultures filtrates. Sensitivity of the four calli to the culture filtrate of the fungus was in reverse order to what is known on the susceptibility of the cultivars in vivo. The results indicate that the culture filtrate of *P. citrophthora* can not be used as a selection tool in vitro.

In contrast to these negative reports of no relationship between reactions of protoplasts, calli, and plants, Helgeson et al. (1972) prooved that tobacco calli resistant to *Phytophthora parasitica* could be obtained from resistant plants, and that susceptible plants yielded only susceptible calli. Deaton et al. (1982) showed further that the resistance to *P. parasitica* was expressed quantitatively in the calli. The disease expression in vitro appeared closely related to expression at the whole plant level. In her early experiments on the selection of *Phytophthora infestans* resistant potato clones Behnke (1979) used unfractionated liquid fungal growth medium, which contained exotoxins being responsible for the virulent reaction. Survivors were transferred up to five times to identical toxic media before regeneration was induced. From more than 40,000 calli 36 plants were regenerated and showed resistance in the greenhouse test. The selected clones had significant smaller local lesions following mechanical inoculation with *Phytophthora* spores, than controls, but the number of sporangia formed did not differ. The resistance was unspecific, and is probably quantitatively inherited. The final field tests clearly demonstrated that the selected clones were inferior to the starting material in many characters including resistance to other diseases making the screening for *Phytophthora* difficult (Wenzel, 1985). Probably during the callus phase many mutations took place - most of them being deleterious.

In the present work we use purified toxins, which were extracted from the fungal growth medium by ultrafiltration. The active compound has a molecular weight of 20,000 daltons, is heat insensitive and insoluable in organic solvents (Stolle and Schöber, 1984). For estimating the toxicity of a specific preparation of that toxin, the difference in fresh weight of rapidly in vitro propagated shoots on MS medium was measured in the presence and absence of the toxin after three weeks. A distinct difference in the growth pattern was observed among calli derived from dihaploid potato clones with different levels of field resistance: On MS medium at a concentration of 2mg/L 2,4-D, a medium that results in a rather loose callus growth, callus from the resistant source grew rather normal with the presence of high toxin concentrations but turned black, in contrast to the callus from the susceptible genotype, which grew poorly or not at all, depending on the toxin concentration (Foroughi-Wehr and Stolle, 1985). Tuber tissue from resistant clones also turned black in the presence of toxin. From black and dying callus, pieces may survive and can perhaps be regenerated. Among the types that grew slowly, as found by Behnke (1979), some mutated, grew normally, and developed into plants with a resistance to the first phase of infection. It is probable that specific in vitro screening methods will be needed for each phase of the infection process.

Considerable information has accumulated in the chemical nature of the toxins produced in culture by _Fusarium_ species. At the same time the reproducibility in 2x potato protoplast regeneration has been increased (Debnath et al., 1986) and experiments could be started using the crude exotoxin to screen potato protoplast suspensions for _F. coeruleum_ and _F. sulphureum_ (Schuchmann, 1985a). To insure reproducible concentrations of toxin in screening experiments, it is necessary to monitor the activity of the phytotoxin fraction. Schuchmann (1985b) estimated the activity of the _Fusarium_ toxin fraction by measuring the inhibition of the respiratory activity in potato cell suspension cultures; a toxin concentration 100-500 times higher than the basic value for selection stops completely respiration within 30 min. The relative concentration of _F. coeruleum_ and _F. sulphureum_ toxins, extracted with ethylacetate and resuspended in methanol, was directly correlated with the inhibition of the respiration activity. When screening for resistance to _Fusarium_, the standardized concentration of toxin allows 1-5% of protoplasts to form walls, to divide, and to continue their regeneration up to the plantlet stage. In this experimental series mainly two different dihaploid clones were used, namely H 75.1207/7, H 78.551/1. From each clone about 1,500 protoclones were regenerated. According to the rapid passage through the callus level all clones looked after regeneration phenotypically rather uniform being either still dihaploid or having doubled to the tetraploid level. Retesting of leaves or of secondary callus revealed a significantly increased stability against the _Fusarium_ toxins. The effect was either measured as fresh weight increase of the callus in the presence of the toxin (Fig. 1) or as the size of lesions after toxin application to the leaf surface. In all experiments the selected protoclones were less susceptible to the toxin treatment than the controles. The resistance was always stronger in those cases, where the first selection and the test were performed with toxins from the same _Fusarium_ species. When, however, the first selection was made using the toxin from _F. sulphureum_ and in the test _F. coeruleum_ extracts were used, the reaction was not significant, although a trend was clearly visible, indicative of an unspecific reaction also besides a specific one. The selected clones are propagated this year in the field, to deliver a uniform tuber generation, which can be tested for Fusarium resistance in the routine way.

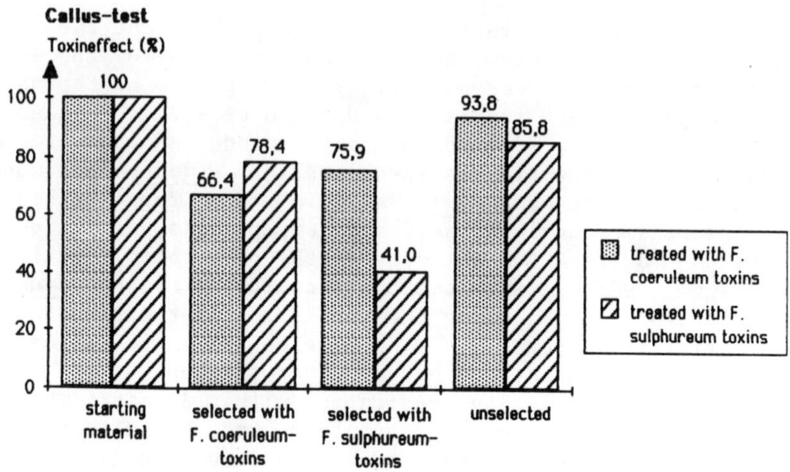

Figure 1. Reaction of secondary callus from selected protoclones of the dihaploid potato clone H 78.551/1 on medium containig toxins of Fusarium species. The reduction in fresh weight increase is shown relative to the starting material.

The early Phytophthora selection and this Fusarium system differ beside the starting material, callus or protoplasts, strongly in the exposure time to the selective agent. The callus was transferred several times to media with selection pressure, while the protoplasts were treated only once, shortly after their cell wall formation. The long treatment guarantees only a few escapes, but many mutations; the short one will result in many escapes, but fewer mutations. We think that the short treatment is preferrable, as it results in more normal clones, and although a second screening is necessary, the probability to maintain the genotype of the donor material and just add a few new characters is higher.

SELECTION IN CELL POPULATIONS OF GAMETOPHYTIC ORIGIN

The most suitable source of heterogeneous cell populations are gametophytic cells from hybrids. Both, egg cells and microspores, can be regenerated into haploid plants, thus facilitating selection.

Use of Egg Cells and Embryos

In most crops, the regeneration of egg cells is only possible after fertilization, but a report is available for regeneration of wheat plants from egg cells (Wang and Kuang, 1981) and we got a few haploids from unfertilized egg cells of winter or summer barley hybrids, together with one diploid (Wright-Turner and Foroughi-Wehr, unpublished). Screening in such unfertilized egg cell populations has not been successful yet. After fertilization screening of young embryos resembles closely procedures of seedling selection but has all advantages of in vitro culture. Each embryo, although being diploid, results from a sexual process and represents a different genotype - a striking difference to the selection somatic cells. In barley and wheat we screened immature embryos on fusaric acid containing B5 medium (Gamborg et al., 1968). Two persent of 2mm large wheat embryos (10 days after pollination) could survive on media with 0.4mM fusaric acid, while in barley 0.2mM was the critical concentration. As in potato it remains to be tested whether the regenerated plants tolerate a Fusarium attack or are just less susceptible to the toxin. Such early screening systems, although not being a typical tissue culture approach, are of utmost importance under applied breeding aspects.

Use of Microspores

Normally there are many more microspores than megaspores or egg cells produced by a plant, which means that heigher numbers of individuals can be screened on the male side. Their regeneration is possible either within the anther (anther culture) or in isolated culture (isolated microspore culture or shed pollen culture). Cereals represent a major group of crops with a sexual reproduction cycle, where tissue culture techniques have a direct practical application. Barley varieties obtained via doubling haploids are already available in Canada (Ho and Jones, 1980). Commercial barley breeders in Germany have now started programs using the androgenetic procedure for the rapid incorporation of resistances into winter barley lines. Of particular interest is the incorporation of resistance to the soil borne BaYMV. Barley in central and northern Europe and also in Japan is frequently damaged by this virus and the barley growing-areas affected with BaYMV is increasing. Grain yield from susceptible varieties may be 30 to 70% lower than the yield of resistant ones. Since the virus cannot be controlled by chemicals rentably, this restriction on yield can be elimated only by growing resistant cultivars. The use of a haploid step in the breeding program offers a rapid approach to incorporate BaYMV resistance into commercial cultivars (Foroughi-Wehr and Friedt, 1984). Nearly 5,000 androgenetic lines were produced from microspores from F_1 anther donor hybrids of crosses between susceptible lines and resistant cvs. such as `Franka'. Of 2,668 A-lines tested, 54% were resistant to BaYMV. The resistance to BaYMV was combined with other important characters within three to five years, demonstrating the time gained by use of haploids. Since 1984 increasing numbers of homozygous doubled haploid lines carrying the BaYMV resistance were released to practical breeding companies (Tab. 3).

TABLE 3. Protocol of androgenetic doubled haploid production in winter barley for the rapid incorporation of BaYMV resistance.

Year	Material	Numbers	Type
1981-84	P_1 (S) x P_2 (r)	59	parents
1982-85	F_1	53	crosses (anther donor hybrids)
1982-84		737,465	anthers plated
1982-85	A_1	4,847	doubled haploid lines
		2,668	lines BaYMV tested
		1,221	lines (S) susceptible
		1,447	lines (r) resistant

A_3 lines given to breeders

1984	196 (122 r)	**1985**	236 (163 r)	**1986**	2,050

It is generally accepted today that a stepwise reduction of the ploidy level of potato offers the advantage of a simpler inheritance and also a better chance to combine qualitatively inherited characters in this plant (Wenzel et al., 1984). Data now available demonstrate that field resistance to PVY and PLRV is maintained and expressed on all ploidy levels over a five years' period, which means that both, monogenically and polygenically inherited resistance, are maintained during the successive haploidization process. When a genome has passed the monohaploid level, an additive character being homozygous can be transferred efficiently to the next generation via hybridization in the same way that a monogenic character can be transferred. Similar results have been obtained for nematode resistance, particularly resistance to Globodera pallida (Uhrig and Wenzel, 1986). In a vegetatively propagated crop, such as potato, the F_1 hybrid can be propagated as a new cultivar.

The haploid technique has been critisized for starting from an F_1, thereby allowing just one recombination event and drastically reducing the variability and the chances of new gene combinations. This criticism is valid if the parents are genetically very different so that several recombinations are required. In case where the breeder intends to keep most of the genome together, the haploid procedure works. If, on the other hand, frequent recombinations are necessary in order to break repulsion linkages, then the starting material should be in the F_3 generation, or if a good early screening system is used, then possibly in the F_2.

In crops where anther culture has been successful it has become possible to regenerate isolated microspores, e.g. barley (Sunderland and Xu, 1982); wheat (Wei, 1982; Datta and Wenzel, 1986) and potato (Uhrig, 1985). In most cases, the microspores are not isolated mechanically from the anther but they rather are shed initially into liquid culture media. It appears that isolated microspore cultures offer a more effective system to regenerate a random sample from the microspore population than does anther culture. This is particularly important when resistance is linked with low plasticity. Isolated microspores open up the possibility to combine haploid techniques with single cell methods. Cereals offer the best system for isolated microspore culture, as they have a very thin exine that allows the observation of early regeneration processes. From a total of about 50,000 uninucleate barley microspores suspended in an ovary-conditioned medium, 0.2% divided and about 30% of these developed into callus, and some into plantlets (Köhler, 1984; Köhler and Wenzel, 1985). As a screening procedure 0.1mM fusaric acid was applied as the selective agent when the cultures were in the microcallus stage. Four of 1,000 calli were found to be resistant. The conductivity of the culture medium after addition of 1mM fusaric acid was used to test for rate of survival; in unselected controls, the conductivity increased up to 30%, but in preselected calli, by only 8% (Köhler, 1984). This system has not yet been tested with segregating microspore populations.

COMBINATION OF RESISTANCES BY PROTOPLAST FUSION

Since protoplast fusion has become a tool in in vitro systems, there is considerable interest in how useful this technique might be in resistance breeding programs. Somatic fusion allows the asexual addition of two heterozygous genomes without any meiotic recombination. This tool is very important where the need is to combine resistances that are polygenic and inherited quantitatively. Since amphidiploid i.e. polyploid hybrids are of no immediate practical interest it will usually be necessary to reduce the ploidy level before or after protoplast fusion to the level desired in the crop. Furthermore, propagation procedures must be developed that will allow the continous growth of such a heterozygous genotype without immediate segregation. Among the field crops these prerequisites are adequately fulfilled in potatoes only. The fusion techniques are of particular interest with this plant, since breeding at the dihaploid level is routine. The first tetraploid hybrid from fusion of two 2x protoplasts was produced by Austin et al. (1985); by this method, these workers transferred the resistance to PLRV from _Solanum brevidens_ into _S. tuberosum_ (x) _S. brevidens_ hybrids. From the 11 hybrids produced nine showed the resistance, while two were susceptible.

Under applied aspects it would be desirable to combine quantitatively inherited resistances. The _Phytophthora_ resistance is quite probably such a quantitatively inherited trait (Behnke, 1979) and for _Fusarium_ resistance quantitative factors are involved as well. For both resistant in vitro cultures are available. After fusion and cell wall formation we add now the _Fusarium_ toxins and when small colonies are formed the _Phytophthora_ toxins are added. It is expected that by this procedure not necessarily combined resistances are selected but that the double treatment enriches presumptive fusion hybrids.

CONCLUSION

The high genetic variability observed for cultures in vitro as well as correlations observed sometimes between in vitro and in vivo response of a host to pathogen support the use of in vitro approaches in resistance breeding. However, as correctly stated by Ingram in 1977, although substantial progress in dual growth of host and pathogen has been made, it is still not certain that future progress in tissue culture provides a simplified experimental system. More basic research is required for a wider application and for progress in unconventional breeding techniques, to overcome empirical and pragmatic strategies. A constructive understanding of regulatory processes acting early in the higher plant when changed by a pathogen is needed, and until such results are available, the trial and error approach must be followed. Quantitative tests are needed for those examples of resistance that are quantitatively expressed. Here, in vitro procedures with plants grown in artificial environments offer the highest reproducibility and consequently will become increasing more popular. The use of preselected microspores can facilitate the rapid incorporation and early identification of resistances based on additive gene effects. Generally it is not variability but the selection efficiency that limits progress of breeders. Until now this is also true for most unconventional programs. It is easy to create variability classically or in vitro, e.g. via somaclones; the problem is, which individuals to select as the more resistant ones, and which one to ultimately grow in the field.

Acknowledgements

The support of parts of this research by grants of the BMFT to M.B., the DAAD to S.C.D and the GFP to B.F.-W. is gratefully acknowledged.

REFERENCES

Austin S, Baer MA, Helgeson JP (1985). Transfer of resistance to potato leaf roll virus from Solanum brevidens into Solanum tuberosum by somatic fusion. Pl Sci Let 39: 75-82.

Behnke M (1979). Selection of potato callus for resistance to culture filtrates of Phytophthora infestans and regeneration of resistant plants. Theor Appl Genet 55: 69-71.

Briggs SP, Scheffer RP, Haug AR (1984). Selective toxin of Helminthosporium victoriae: Thermal relationships in effects on oat tissues and protoplasts. Phytopathology 74: 768-73.

Buiatti M, Scala A, Beltini P, Nascari G, Morpurgo R, Bogani P, Pellegrini G, Gimelli F, Venturo R (1985). Correlations between in vivo resistance to Fusarium and in vitro response to fungal elicitors and toxic substances in carnation. Theor Appl Genet 70: 42-47.

Casper R, Meyer S (1981). Anwendung des ELISA Verfahrens zum Nachweis pflanzenpathogener Viren. Nachrichtenbl Deut Pflanzenschutzd 33: 49-54.

Coleman RE, Nickell LG (1964). cit. Heinz DJ, Krishnamurthi M, Nickell LG, Maretzki A (1977). Cell tissue and organ culture in sugarcane improvement. In Reinert J, Bajaj YPS (eds): "Plant Cell, Tissue, and Organ Culture," Berlin, Heidelberg, New York: Springer Verlag, pp 3-17.

Datta SK, Wenzel G (1986). Isolated microspore derived plant formation via embryogenesis in Triticum aestivum L. Plant Sci (in press).

Deaton WR, Keyes GJ, Collins GB (1982). Expressed resistance to black shank among tobacco callus cultures. Theor Appl Genet 63: 65-70.

Debnath SC, Schuchmann R, Wenzel G (1986). Regeneration capacity of potato protoplasts isolated from singel cell derived donor plants. Acta Bot Neerl (in press).

Ebel J, Ayers AR, Albersheim P (1976). Host-pathogen interactions. XII. Response of suspension-cultured soybean cells to the elicitor isolated from Phytophthora megasperma var. sojae, a fungal pathogen of soybean. Plant Physiol. 57: 775-79.

Foroughi-Wehr B, Friedt W (1984). Rapid production of recombinant barley yellow mosaic virus resistant Hordeum vulgare lines by anther culture. Theor Appl Genet 67: 377-82.

Foroughi-Wehr B, Stolle K (1985). Resistenzselektion in vitro am Beispiel des Systems Kartoffel/Phytophthora infestans (Mont.)de Bary, Nachrichtenbl Deut Pflanzenschutzd 37: 170-73.

Gamborg OL, Miller RA, Ojima K (1968). Nutrient requirements of suspension cultures of soybean root cells. Exp Cell Res 50: 151-58.

Gengenbach BG, Green GE, Donovan CM (1977). Inheritance of selected pathotoxin resistance in maize plants regenerated from cell cultures. Proc Natl Acad Sci USA 74: 5113-17.

Hartmann CL, McCoy TJ, Knons TR (1984). Selection of alfalfa (Medicago sativa) cell lines and regeenration of plants resistant to the toxin(s) produced by Fusarium oxysporum f.sp. medicaginis . Pl Sci Lett 34: 183-94.

Helgeson JP, Kemp JD, Haberlach GT, Maxwell BP (1972). A tissue culture system for studying disease resistance:the black shank disease in tobacco callus cultures.Phytopathology. 62:1439-43.

Ho KM, Jones GE (1980). Mingo barley. Can J Plant Sci 60: 279-80.

Ingram DS (1977). Applications in plant pathology. In Street HE (ed): "Plant Tissue and Cell Culture," Oxford: Blackwell, pp 463-500.

Keen NT, Horsch R (1972). Hydroxyphaseolin production by various soybean tissues: a warning against the use of unnatural host-parasite systems. Phytopathology 62: 439-442.

Köhler F (1984). Regeneration und Selektion isolierter Mikrosporen. Ph.D.Thesis, Köln.

Köhler F, Wenzel G (1985). Regeneration of isolated barley microspores in conditioned media and trials to characterize the responsible factor. J Plant Physiol 121: 181-91.

Krishnamurthi M, Thaskal J (1974). Fiji desease resistant *Saccharum officinarum* var. Pindar subclones from tissue cultures. Proc Int Soc Sugarcane Technol 15: 130-137.

Larkin PJ, Scowcroft WR (1981). Somaclonal variation - a novel source of variability from cell cultures for plant improvement. Theor Appl Genet 60: 197-214.

Ling DH, Vidhyaseharan P, Borromeo ES, Zapata FJ, Mew TW (1985). In vitro screening of rice germplasm for resistance to brown spot disease using phytotoxin. Theor Appl Genet 71: 133-135.

Murashige T, Skoog F (1962). A revised medium for rapid growth and bioassays with tobacco tissue culture. Physiol Plant 15: 473-97.

Sacristan MD (1985). Selection for disease resistance in *Brassica* cultures. Hereditas Suppl 3: 57-63.

Schuchmann R (1985a). In vitro Selektion auf *Fusarium* Resistenz bei der Kartoffel. PhD Thesis TU München.

Schuchmann R (1985b). Methode zur relativen Konzentrationsbestimmung von *Fusarium* Toxinen durch Messung der Respirationsrate. Nachrichtenbl Dtsch Pflanzenschutzd 37: 81-84.

Shepard JF (1980). Mutant selection and plant regeneration from potato mesophyll protoplasts. In Rubinstein I, Gengenbach B, Green CE (eds): "Emergent Techniques for the Genetic Improvement of Crops," Minneapolis: Univ Press, pp 185-219.

Stolle K, Schöber B (1984). Wirkung eines Toxins von *Phytophthora infestans* (Mont.)de Bary auf Kartoffelgewebe. Potato Res 27: 173-184.

Sunderland N, Xu ZH (1982). Shed pollen culture in *Hordeum vulgare.* J Exp Bot 33: 1086-1095.

Thanutong P, Furusawa I, Yamamoto M (1983). Resistant tobacco plants from protoplast-derived calluses selected for their resistance to Pseudomonas and Alternaria toxins. Theor Appl Genet 66: 209-215.

Thomson AJ, Gunn RE, Jellis GJ, Boulton RE, Lacey CND (1986). The evaluation of potato somaclones. In Semal J (ed): "Somaclonal Variations and Crop Improvement," Dordrecht: Martinus Nijhoff Publ, pp 236-243.

Uhrig H (1985). Genetic selection and liquid medium conditions improve yield of androgenetic plants from diploid potatoes. Theor Appl Genet 71: 455-460.

Uhrig H, Wenzel G (1986). Breeding for virus and nematode resistance in potato through microspore culture. In Bajaj YPS (ed): "Biotechnology in Agriculture & Forestry", Berlin:Springer Verlag, (in press).

Vardi A, Epstein E, Breiman A (1986). Is the Phytophthora citrophthora culture filtrate a reliable tool for the in vitro selection of resistant Citrus variants? Theor Appl Genet 72: 569-74.

Wang, CC, Kuang BJ (1981). Induction of haploid plants from the female gametophyte of Hordeum vulgare L. Acta Bot Sinica 23: 329-330.

Wenzel G (1985). Strategies in unconventional breeding for disease resistance. Ann Rev Phytopathol 23: 149-72.

Wenzel G, Uhrig H, Burgermeister W (1984). Potato - a first crop improved by the application of microbiological techniques? In Arber W, Illmensee K, Peacock WJ, Starlinger P (eds): "Genetic Manipulation: Impact on Man and Society," Cambridge: University Press, pp 127-137.

Wriedt G (1984). Die Gewebekultur als Möglichkeit der Erhaltungszucht in einem praktischen Zuchtbetrieb aus biologischer und ökonomischer Sicht. In Winiger FA, Stöckli A (eds) Abst Conf Pap Triennial Conf EAPR, Wageningen: EAPR p 384.

Plant Tissue and Cell Culture, pages 359–365

APPLICATION OF PLANT TISSUE AND CELL CULTURE TO FIELD CROPS IMPROVEMENT, THE UP-AND DOWN-SIDE

Wallace D. Beversdorf
Allelix Inc.
6850 Goreway Drive
Mississauga, Ontario
L4V 1P1

INTRODUCTION

Biotechnologies, a complex collection of modern genetic, cellular and molecular biology tools have potential value for understanding, manipulating, modifying and protecting agricultural plant and animal species. Although the recent history of biotechnology can be easily documented, the accelerating pace of technological achievement in biotechnology and most other areas of modern science makes it difficult to predict their future. It is an unlikely that 25 years ago anyone would have predicted the breakthroughs in cellular and molecular biology that have been achieved through 1986. It is similarly difficult to predict what breakthroughs will contribute to crop genetic improvement or crop management products by the turn of the century.

In an attempt to understand the potential impact of modern genetic cellular and molecular tools on crop improvement we must first recognize that each of these biotechniques have some obvious potentials and some obvious inherent limitations. In addition, potential applications of these technologies are clouded by our limited understanding of gene expression, promotion and regulation, by the rapid emergence of new concepts, by contradictory claims of application potential, by frequently broaden or poorly defined patent claims and licenses agreements and by social concerns over application of biotechnology. Many of these problems transend the laboratory but will be resolved with time.

RELATIONSHIP TO CLASSIC PLANT BREEDING

What is happening in biotechnology related to agriculture is too complex to cover in depth in a single presentation. Efforts in many areas, from acquiring very basic understanding of molecules related to gene expression; development and application of DNA cloning, labeling, sequencing and synthesis techniques and development and application of DNA delivery vectors to the development and application of in vitro techniques for fertilization, mutagenesis, selection, haploid induction, parasexual hybridization, and organism regeneration, all have potential for applica- tion to plant genetic improvement.

Plant genetic improvement or plant breeding has traditionally involved five general operations: formation of geneti- cally variable populations, genetic improvement of populations through the application of selection pressure (to change gene frequencies), stabilization of selected populations, evaluation of selected populations and multiplication of basic seed stocks. Technologies associated with each of these operations have evolved rapidly during the past century and have been responsible for the rapid and dramatic improvement in the major food, feed and fiber crops of modern agriculture. The potential impact of emerging technologies may be assessed by their potential to improve the efficiency or to overcome the limitations of any of the these basic breeding operations.

FORMATION OF GENETICALLY VARIABLE POPULATIONS

To early agrarian societies, nature provided a deverse array of genetic variability. Regardless of the claims modern technologists, we can thank early agricultural producers for transforming the deverse array of natural variability among plant species into the productive crop forms that make up the vast majority of modern agriculture. With improved understanding of genetics that emerged in the latter half of 1800's and early 1900's, breeding technologies (now considered traditional or classic), included hybridization or mutagenesis followed by recombination and segregation of genes to provide genetically variable populations. This variability of course is the fuel upon which all modern plant breeding programs have come to rely upon.

Hybridization has been limited by natural sexual compatibility barriers, allowing recombination of only those nuclear genes available within a species. Three emerging technologies have demonstrated potential to overcome these natural limitations: embryo rescue (to overcome abortion of interspecific hybrids between closely related species) parasexual hybridization (protoplast fusion), and introgression via gene transformation (insertion of foreign or synthesized genes into a species). Model examples of these applications, although infrequent, have adequately demonstrated their potential for overcoming natural sexual barriers (Fraley et al. 1984, Pelletier et al, 1983, Shillito et al, 1985). In addition, recombination of cytoplasmic organelles, not possible in species with uniparentally inherited cytoplasms, has been achieved through protoplast fusion technology (Pelletier et al, 1983, Fluhr et al. 1983).

Mutagenesis has been less popular than hybridization in forming genetically variable populations primarily because the frequency of desirable mutations is very low and therefore difficult to identify. Somaclonal variation associated with cell and protoplast cultures (Larkin and Scowcroft, 1981) has provided new variability in some species. Although the majority of such variants have not undergone genetic analysis (Reisch, 1983) some potentially useful spontaneous variants have emerged (Evans et al, 1984, Bhatia et al, 1985).

Molecular genetics techniques are also emerging to enhance the frequency of induced mutations, assist in isolation of mutant (disrupted) DNA and to potentially direct specific mutations _in situ_ (Carlson, pers. comm.). Emergence and refinement of these molecular tools covered significantly alter classic approaches to generating new variability for simply controlled plant characteristics, although a considerably greater understanding of the genetic basis of multigenic characteristics will be required before hybridization/recombination is abandoned by plant breeders as a means of generating genetic variability.

GENETIC IMPROVEMENT VIA SELECTION PRESSURE

Selection in classical plant breeding is intended to increase the frequency (or isolate) desirable genes or combinations of genes (genotypes). Improvement of genetically variable populations has been accomplished through phenotype/genotype selection at the whole plant, family or population level.

In vitro selection technology combined with spontaneous or active mutagenesis has been effective in altering or isolating genetic variability for characteristics expressed at the isolated-cell level (Chaleff, 1983). For such traits as herbicide, salt and disease toxin tolerance and amino acid over-producers, *in vitro* selection technology may be a powerful selection tool particularly in haploid or gametic cultures.

Many agronomic characteristics such as productivity, adaptation, harvestability are governed by the interaction of complex environmental, physiological and multigenic factors which are not well understood at a cellular or molecular level. As a result, traditional approaches to selection may be supplemented by *in vitro* and molecular manipulation but will not be replaced by these technologies in the foreseeable future.

STABILIZATION OF GENOTYPES IN GENETICALLY VARIABLE POPULATIONS

Stabilization of genetically variable populations is a prerequisite to faithful sexual reproduction of improved genotypes from generation to generation. Improved plant genotypes or populations of genotypes are stabilized by two general methods in traditional breeding: Inbreeding to approach homozygousity, as in development of pure lines (as in wheat and soybeans) or inbred parents of hybrids (as in maize); or by random mating within selected populations to approach Hardy-Weinburg Equilibrium in outcrossing species (as in synthetic varieties of alfalfa, polish rapeseed and rye).

Anther or microspore culture have been very useful in providing haploids or spontaneous doubled haploids as a source of homozygous lines directly with considerable time savings over traditional inbreeding in some species (Chuong and Beversdorf, 1985).

EVALUATION OF GENETICALLY IMPROVED POPULATIONS

Emerging biotechnologies may provide specialty chemicals useful in the evaluation of improved genotypes for specific constituent chemicals (eg. toxins) or for evaluation of pathogen levels in seed stocks, grain or vegetative products. Evaluation of agronomic characteristics such as yield, maturity, adaptation, winter survival, etc., will undoubtedly continue on a whole plant or population basis, as it has historically.

GENOTYPE MULTIPLICATION AND DISTRIBUTION

Desired gentoypes (varieties) are traditionally multiplied through sexual processes of controlled seed production. These procedures vary with the reproductive nature of the species and the type of variety under multiplication (homozyous line, controlled hybrid or heterozygous synthetic variety). *In vitro* embryogenesis may someday allow production of artificial or synthetic seed in some crops, although the complexity and cost of synthetic seed production will probably limit its utility to multiplication of only extremely valuable genotypes that cannot be multiplied by conventional technologies (eg. self incompatible inbreds, genetic male-sterile inbreds or extremely valuable hybrid genotypes).

CONCLUSION

Each of the biotechniques indicated above has serious inherent limitations that preclude general utility in plant genetic improvement. No combination of biotechniques available today can rival traditional plant breeding in terms of efficiency or potential contribution to economic plant improvement. Specific biotechniques however, have useful applications within the breeding operations of a growing number of economic species.

For biotechnology to make significant contributions to agricultural productivity in the foreseeable future, realization of potential applications must be understood by agriculture researchers. Similarly, biotechnologists seeking agricultural applications for emerging biotechnologies must attempt to understand the factors limiting agricultural productivity, or seek and accept the guidance of agricultural experts.

Cooperation will ultimately lead to increases in the productivity and quality of agricultural products in a manner similar to the benefits achieved by the cooperation of plant breeders, agronomists, pathologists, agricultural chemists, engineers, and statisticians.

REFERENCES

Bhatia BS, Joshua DC & Mathews H (1985). Somaclonal variation: A genetic interpretation based on the rates of spontaneous chromosomal observations and mutations. Trends Pl. Res. 317-326.

Chaleff RS (1983). Isolation of agronomically useful mutants from plant cell cultures. Science 219:676.

Chuong VP, Beversdorf WD (1985). High frequency embryogenesis through isolated microspore culture in *Brassica napus* and *B. carinata*. Plant Sci. 39:219-226.

Evans DA, Sharp WR, Medina-Filho HP (1984). Somaclonal and gametoclonal variation. Amer. J. Bot. 7:759-774.

Fluhr R, Aviv D, Edelman M, Galun E (1983). Cybrids containing mixed and sorted-out chloroplasts following interspecific somatic fusions in *Nicotiana*. Theoret. Appl. Genet. 65:289.

Fraley RR, Horsch RB, Matzke A, Chilton MD, Chilton WS, Sanders PR (1984). *In Vitro* transformation of petunia cells by an improved method of co-cultivation with *A. tumefaciens* strains. Plant Mol. Biol. 3:371.

Larkin PJ, Scowcroft WR (1981). Somaclonal variation - a novel source of variability from cell cultures for plant improvement. Theor. Appl. Genet. 60:197-214.

Pelletier GC, Primard F, Vedel F, Chetrit P, Remy R, Rosselle, Renard M (1983). Intergeneric cytoplasmic hybridization in Cruciferae by protoplast fusion. Mol. Gen. Genet. 191:244.

Reisch B (1983). Genetic variability in regenerated plants. pp. 748-769. In:Handbook of Plant Cell Culture Vol. 1 Collier Macmillan, London.

Shillito RD, Saul MW, Paszowski J, Muller M, Potrykus I (1985). High efficiency direct gene transfer to plants. Biotechnology 3:1099-1103.

Plant Tissue and Cell Culture, pages 367–382

IN VITRO PROPAGATION OF TREE SPECIES

M. Boulay

AFOCEL domaine de l'Etancon 77370 - Nangis, France
Department of Pomology, University of California, Davis, California 95616

ABSTRACT

In vitro cultivation of forest tree species has developed dramatically over the last ten years. At the level of micropropagation the two methods of (axillary budding and adventitious budding) are now commonly used. The first one mainly for hardwood species and the second for conifers. Nevertheless, commercial applications remain limited. Limitations are based on the need to rejuvenate elite trees selected at a mature stage and to develop the appropriate juvenile-mature correlations. In vitro models are useful to understand the rejuvenation process. *Sequoia sempervirens* seems well-suited for that purpose and useful biochemical criteria have been defined (e.g. K/Ca ratio, peroxydase content, effect of cytokinins, and levels of endogenous hormone).

Another limitation of *in vitro* methods is the production cost of the resulting plantlets. It is now believed by many that somatic embryogenesis in coniferous species and hardwood e.g.Eucalyptus can be useful for decreasing this cost.

INTRODUCTION

The clonal propagation of forest tree species has progressed significantly over the past ten years. Numerous species are now propagated vegetatively, including coniferous which are considered difficult to root. However, improvements in vegetative propagation have not solved problems linked with the maturation of selected trees. The potential for the clonal propagation of these trees and specially conifers are limited to selections done in the first five years during clonal or progeny testing. The most appropriate strategy involves the propagation of seedlings from seed orchards, without previous individual selection (bulk program) or to select a sufficient number of clones which are maintained juvenile by repropagation until the results of field testing are known.

The increasing rate of success for *in vitro* techniques in woody ornamentals and fruit trees species has encouraged scientists working with forest tree species. Nevertheless, a major reason for slow development arises because most forest nurseries are traditionally oriented more towards seedling production rather than towards rooted cuttings. The latter have significant advantages and work in this area is just beginning at a commercial level. Furthermore, the low production costs for reforested plants as compared to the market price of ornamental or fruit plants, the lack of selected clones and technical difficulties are also responsible. I will review several applications and develop future prospects with special emphasis on the main coniferous species in reforestation.

I. In vitro micropropagation

In most cases the methods used include adventitious budding and/or axillary budding. Callusing and neoformation of buds are rarely used, though these have been achieved on some forest tree species (*Ulmus*, liquidambar, aspen, *Betula*, *Sequoia sempervirens*, etc.). The list of species which are propagated by *in vitro* methods has been published over the years (Karnonsky, 1981; Brown and Sommer, 1981; David, 1982; Boulay, 1986) and there is now more.

1. *Adventitious budding*. This method has been often used in coniferous species, particularly for the multiplication of seeds or seedlings. After the work of Sommer *et al*, 1975 (pine) and Cheng, 1975 (Douglas fir),

Campbell and Durzan, 1976, (spruce), cotyledons, hypocotyls and excised embryos have been the main source of explants. In most cases regeneration of whole plants has been obtained. The number of successes is now still increasing and one can say that practically all coniferous species give adventitious budding on embryo explants. *Abies* sp. remain difficult. In angiosperms numerous species respond to such a method (eucalyptus, populus, aspen, etc.). Often there is a passage through a more pronounced callus stage than in conifers.

Using adult material most researchers choose explants from leaf primordia of dormant buds (Von Arnold and Erickson, 1976; Bonga, 1981; Misson *et al*., 1982) or needles (Borman and Jansson, 1980; Ball, 1978). Often the initiation of meristemoids or buds is possible but the elongation of such buds is difficult. In fact, because of this problem, whole plants have rarely been regenerated from adult mother trees. *S. sempervirens* is one exception to this generalization.

This technique has a great potential for multiplication but several problems remain. The first is the problem of obtaining true-to-type genetic copies which is the main purpose of clonal forestry. The work of Patel and Berlyn (1982) has shown a genetic instability among the buds obtained using this method. Nevertheless in their work the adventitious budding was not direct and passed through a callusing stage.

A lot of plantlets have now been regenerated (Aitken-Christie, T.A. Thorpe, 1984) and none has mentioned significant genetic abnormalities. Evidence supports the view that direct adventitious budding can be a useful tool for multiplication of most of the coniferous species starting from seed material. Second, the variation in the growth behaviour of resulting plants remains uncertain. It has been noted that some copies are not always juvenile. Some clones manifest plagiotropic growth and poor vigour. Others exhibits orthotropic growth and vigour typical of young seedling. These problems arise from rough handling of clonal stock during the propagation procedure. This heterogeneity between the plantlets coming from adventitious budding is also noted by Leach (1979) in *Pinus taeda*. He reports that in field tests, the growth of *in vitro* plantlets is more heterogenous than a population of seedlings. McKeand *et al.* (1984) following the growth of two kinds of plants (plantlets versus seedlings) found that plantlets have a greater mortality (up to 38%) and less shoot growth than conventional seedlings. By experimentation they concluded that differences in height were mainly due to differences in root morphology than root physiology. Root systems of *in vitro* plantlets are less branched and had no tap roots. These differences lead to variations in ion absorption capacity. It seems that similar results (i.e., less growing of plantlets) have been obtained on other species (Smith, 1985). In such experiments it is difficult to distinguish the contributions of growth conditions during acclimatization process, which can influence greatly the root system quality. It is also possible that a better selection of *in vitro* shoots before rhizogenesis or a better condition of acclimatization before field planting test can overcome these problems.

2. *Axillary budding*. This method is most used particularly in angiosperms and some commercial application has been done. The *in vitro* multiplication of wild cherry tree, aspen, poplar, eucalypt, coast redwood, giant sequoia, *Thuya*, *cunninghamia Lanceolata* have found some development.

In coniferous species, the method is not always efficient because it is sometimes impossible (specially in species like - Douglas-fir, Norway spruce for example) to stimulate an axillary bud at the base of each needle. As a result the multiplication rate is very low.

In *Pinus* species the use of needle fascicles as initial explant has been developed (David *et al.*, 1977). Mass propagation by this means is still limited to young, selected trees. Recently on *Pinus radiata* K. Horgan (1985) has used the entire dormant bud. After pulling off the scales, the preformed shoot is cut into cylinders of 1 cm height and preformed brachyblastes are allowed to develop on a multiplication medium. Similar results have been obtained on Sugar pine (Gupta and Durzan, 1985). Using this method even mature trees can be propagated, but at a much slower rate. The resulting plantlets show maturation which limits their use to seed-orchard and study of gene x environmental interactions and not to reforestation.

Nevertheless the multiplication of mature selected trees using axillary buds has succeeded with Angiosperm trees e.g. Wild cherry (Cornu *et al*. 1981), Aspen (Ahuja, 1984) Eucalyptus (Gupta *et al*, 1981; Bennett and McComb, 1982; Boulay, 1983) Teak (Gupta *et al*., 1980). For these species pretreatment of the selected mature tree is not always necessary and in most cases successive transfers on multiplication medium (i.e., cytokinin) improve the rooting ability of the *in vitro* produced shoots. Nevertheless, Cornu *et al* (1981) and Boulay (1983) note that taken explants from suckers, or sprouts on stumps, or grafting adult scion on to juvenile seedlings and even regrafting improve the initial reaction of explant and makes faster the multiplication.

For coniferous species the choice of explant seems more important. For Douglas-fir Gupta and Durzan, (1985) have cultivated and regenerated whole plantlets from pieces of mature Douglas-fir stem, but the rooting ability is quite low (5 to 10%). The experiment of de la Goublaye de Nantois (1980) using grafting and regrafting on to young seedlings and Reynoird (1983) using cytokinin sprays on the stock plant have shown promise for mature Douglas-fir trees (75, and 90 years old) but the axillary budding multiplication are still limited perhaps by the lack of appropriate medium.

This result must be considered with those of Bekkaoui *et al*. (1986) who found that continous transfer of young Douglas-fir shoots on medium without growth regulators but with activated charcoal added leads to a clonal degeneration. Degeneration is characterized by stem swelling, apical bud dormancy and a high endogenous rate of zeatine and zeatine riboside (see Table 1).

Table 1

Lots	AiA	%V	ABA	%V	2 + 2R	%V
A_1	1.07		2.77		0.17	
A_{15}	1.49	+39	4.82	+74	0.72	+323
$B_{15}N$	1.05		4.55		0.62	
$B_{15}G$	0.55	-48	2.92	-36	2.19	+253

Hormonal composition (µg/gramme dry weight) of Douglasfir microcuttings %V: Variation percentage of A_{15}/A_1 and $B_{15}G/B_{15}N$.
A_1: one subculture on basal medium - A_{15}: 15 transfers on basal medium. $B_{15}N$: 15 transfers on basal medium, microcuttings are normal without degeneration. $B_{15}G$: 15 transfers on basal medium, microcuttings present degeneration (see text) from Bekkaoui *et al*. 1986.

In this case increasing the number of transfer does not improve the multiplication ability. This contrasts to the work on angiosperm or Sequoia sempervirens (H de la Goublaye de Nantois 1981; Verschoore-Martouzet 1985; Fouret *et al* 1984). As in adventitious budding and depending on the starting material, the multiplying of mature coniferous species results in different behaviour of resulting plantlets. Ball *et al*. (1978) found that explants from the stem of plagiotropic shoots never developed into soil-grown plantlets, and shoot tips from upper bole plagiotropic shoots produce plagiotropic plantlets which can be transferred to soil. By contrast, explants from basal orthotropic shoots produced plantlets which can be transferred to soil and give orthotropic shoots and growth similar to seedlings.

With basal sprouts Boulay (1978) came to a similar conclusion. In vitro plants from such explants grow as well as seedlings. This view is supported by five years of field testing. For *Cunninghamia lanceolata* the results are similar (Engelmann, 1982). Working with two clones one juvenile (one-year-old) and one mature, he found that the juvenile type produced two kinds of plantlets. One looks like normal seedlings but the second is heavily branched and lacks a leader. All copies of the older clone are plagiotropic. For this clone the rooting percentage improved during successive *in vitro* transfers but none of the applied treatment to the mature tree are able to overcome plagiotropism. Nevertheless, some propagules of the mature tree displayed morphological typically juvenile characteristics. This means that this problem can be overcome.

2. Rejuvenation and Use of Sequoia Sempervirens as a Model System

The *in vitro* propagation of mature trees is not solved for most of the coniferous species, specially if one to make plantlets for reforestation (good vigour and orthotropic growth). For this reason the problem of rejuvenation (in the sense of obtaining copies from an adult tree which looks like seedlings) still remain.

The advantage of *Sequoia sempervirens* is based on the regeneration of whole plantlets from different parts of the tree. Propagules present different characteristics in maturation. Juvenile explants are obtained from: 1) sprouts at the base of the trunk; 2) by pretreatment of the mother tree; 3) after several transfers in vitro on to cytokinin medium.

Ball *et al*. (1978), Boulay (1978) and and H. de la Goublaye (1981) have shown that making transfer on to cytokinin medium improves the axillary budding ability of piece of stem and also orthotropic growth and vigour of resulting plantlets. Verschoore-Martouzet (1985) has attempted to define chemical and biochemical criteria to evaluate juvenility or rejuvenation. She finds that two can be used: K/Ca and peroxydase activity/total protein ratios of the bud content.

For the first parameter under in vivo conditions there is variation depending upon the season but always the youngest material has the highest K/Ca ratio (see Table II).

Table II.
In vivo K/Ca ratio depending upon age, location on tree and in vivo pretreatment. Legend : A : mature tree. 80 years old clone. From Verschoore-Martouzet (1985)

Characteristics of Explants	K/Ca
Seedlings:	4.54
A : grafted basal sprouts	3.29
A : grafted lower branch	3.35
A : lower branch	0.96
A : grafted 15m high branch	1.95
A : 15 m high branch	0.75

In front of this result one can see that the K/Ca ratio characterizes the physiological age of an explant on the tree and that grafting on to young seedling raises the K/Ca ratio to a level similar to seedling.

By transfers _in vitro_ on cytokinin medium the K/Ca ratio increases with the number of transfers (Table III) and after four transfers, differences are still limited between seedlings and two different locations on the mature tree.

Table III.
In vitro evolution of K/Ca ratio according to the number of transfers. Legend : A : mature tree. 80 years old clone. From Verschoore-Martouzet (1985)

		Characteristics of explants	
Number of Transfers on Cytokinin Medium	seedling	A Lower branch	B 15 m high branch
0	4.5	1.03	1.08
1	6.4	1.6	3.0
2	8.5	4.4	4.2
3	9.8	7.5	6.5
4	10.4	9.2	8.8

On this medium there is an increase in K/Ca ratio with transfers even for seedling material. The increasing ratio is also correlated with the rooting ability of the explant coming from mature tree. After 6 transfers the difficult-to-root old clone can be rooted at a percentage comparable to explants coming from 6-month-old seedling. (See Table IV).

Table IV
Best results of rooting percentage from differently aged trees and location on the same tree, and effect of grafting. A : mature tree 80 years old. From Verschoore-Martouzet (1985)

Characteristics of Explants on original tree	Rooting Percentage

Seedling	80
A : basal sprouts; 2 transfers	75
A : grafted 15m high branch; 4 transfers	40
A : 15 m high branch; 6 transfers	55
A : lower branch; 6 transfers	70

For peroxydase activity/total protein a similar pattern is found. During the first 28 days in vitro there is a peak in this value at 7 days for seedlings explants and 14 days for the other. The maximum value of the ratio is classified as following: seedlings in vitro explants > sprouts explants > basal branch explants > 15 m branch explants.

Increasing the number of transfers increases the maximum activity of explants from mature tree and curve resemble those obtained from juvenile seedlings explants.

These two parameters correlate well with what we called rejuvenation. Nevertheless some differences remain with plagiotropism, specially with explants coming from 15m high branch. This could mean that all physiological characteristics are not reversed or are partially reversed by such methods.

Fouret (1984) has done similar work but working with 3 clones (1, 50, and 600 years old) and using different periods between transfers on cytokinin medium, she finds that 100% of rooting is obtained for the difficult to root clone at the third transfer. In these experiments the resulting plantlets remain plagiotropic once transferred to soil (in comparison plantlets from 1 and 50 year old clones are othotropic).

Taking endogenous hormone analysis (Fouret _et_ _al_., personal communication) they find different concentrations of IAA and ABA between the three different clones and that AiA/ABA is smaller in the old clone. After transferring clones to a cytokinin medium this ratio increase as the rooting percentage increase. Rooting is improved by a decreasing of ABA content and in a similar way the endogenous content of IAA increases in the old clone. Studying the peroxydase content the same authors find that increasing the number of transfer increase the number of peroxydase isozymes (mainly cationic). This coincides with an increase in the total peroxydase activity in the stem explants.

An interesting study using this model system was done by Walker (1985). Using shoot apices (meristematic dome + 2 or 3 needles primordia) as source explant and as repropagating organ for successive transfers, she found increased juvenility in front of parameters as: rhizogenesis, adventitious budding capacity on in vitro leaves, length of shoot, monthly rate of multiplication, number of formed leaves on shoots and decreasing in the meristem size. But even using this method some resulting plantlets of the old clone remained plagiotropic. She concluded that all physiological parameters have not been reversed and that plagiotropic growth can arrive as a secondary phenomenon which is established under acclimatization conditions (less vigourous root system which leads to weakness of stem and poor growth).

Although the problem of plagiotropic habit arises in plantlets of old clones, most of the plantlets coming from one-hundred-year-old selected trees can be grown in the field. This system can interest plant molecular biologists to work with it and to find criteria which are responsible of such maturation and revision of this stage.

3. Somatic Embryogenesis

Somatic embryogenesis has been obtained now for some species on which Corylus avellana (Radojevic et al., 1975), liquidambar styraciflua (Sommer and Brown, 1980), Santalum album (Lakshmi-Sita et al., 1979), Pawlonia-Tomentosa (Radojevic, 1979), Vitis, and in Prunus root-stock (Hansen and Durzan, 1986). More recently, Ou Quyang et al., 1981 on Eucalyptus and Gonzalez et al., (1985) on Castanea sativa have also obtained somatic embryogenesis. This method can be very useful in propagating tree species for two main reasons. First, if one is able to obtain from a mature tree somatic embryogenesis we can assume that true rejuvenation has been achieved. Also, most problems linked with maturation and behavior of plantlets in the field will be overcome. The second point concerns the possibility of maintaining in small volume, large quantities of somatic embryos in liquid culture. The improved maintenance of such culture and the recovery of bipolar structure which can be encapsulated, and which can develop root and shoot at the same time could decrease the cost of produced plantlets. This process compete with improved seeds. Moreover with pseudo seeds we can give to the nurseryman a product perhaps easier to handle than plantlets coming from micropropagation.

Working with some frost resistant species of eucalyptus (Boulay, 1986 unpublished data) we have obtained somatic embryogenesis using internodes of sterile germination. The technique includes callusing on MS medium supplemented with glutamine and NAA and K or NAA and BAP. After 2 or 3 monthly transfers on this medium for increasing the white, friable callus a transfer on MS half strength without hormones and with activated charcoal leads to bipolar structures which evolve rapidly to root shoots and finally to small plantlets. All this is done on solid media. Trials with liquid culture have failed to regenerate whole plants. Some problems have been encountered. The most important one is the rapid conversion to plantlets and also the obtaining of twin, and malformed embryos. More research must be done to try to synchronize the different stages of embryos and to slow down their maturation process.

For coniferous species somatic embryogenesis has known a rapid development in the last two years. Subsequently on Norway spruce (Hakman and Von Arnold, 1985; Gupta and Durzan, 1986) Larix, (Nagmani and Bonga, 1985) *Pinus lambertiana* (Gupta and Durzan, 1986) loblolly pine (Gupta and Durzan, 1986) embryogenic callus cell lines have been obtained. Some of them have regenerated whole plants. On Norway spruce, these cell lines have been maintained for one year without losing their embryogenic capacity (Gupta and Durzan, 1986). It seems that this phenomenon will be general on coniferous species which are able to give polyembryony by cleavage (Gupta and Durzan, personal communication). These authors have also demonstrated that cells in the embryonal-suspensor mass produce an embryogenic cell line, and that embryogenic capacity is still kept if the time between two transfers is not too long (10 to 12 days). While globular somatic polyembryos have been obtained in suspension culture, improvement must be made mainly for liquid culture to be able to recover large numbers of mature embryos (i.e., cotyledons stage). Encapsulation of embryos has been reported (Gupta and Durzan, 1986).

An important problem for this method to be used at a commercial level is to ensure that the variability of genetic copies is acceptable, and there is a need to plant these plantlets in "clonal" test fields.

Moreover this method is also still limited to immature or mature seeds, this means that it is a useful tool to multiply hand crosses or select progeny, but not selected trees. Nevertheless the knowledge of somatic embryogenesis, the definition of all stages and of different parameters may someday be applied to explants taken from mature trees.

One possible application is to freeze preserve embryogenic cell lines. Dividing each cell line in two lots, one in gene bank and the other for regeneration, clonal field tests can be conducted with regenerated embryogenic plantlets. Once the results of field tests are known, after 6 to 10 years the multiplication of cell lines and production of plantlets from the best clone can be attempted.

CONCLUSION

We have not indicated in this paper experiment on protoplast, somatic hybridization, or gene transfers which are still in the research lab area. Progress has also been achieved in these different fields.

In vitro methods for forest tree species do not yet have the commercial applications evident for fruit trees. Nevertheless there are still some angiosperm species from which commercial production is done, for example species of high value (i.e. walnut, wild cherry). In forestry coniferous species in vitro produced plantlets are mainly used as liners, stock plants. In most cases propagated conifers are very efficient cutting donor plants which permits a rapid scale up of selected clones after selection.

ACKNOWLEDGEMENT

The author wants to thank Dr. D.J. Durzan and Dr. P. Gupta in reviewing the manuscript. Thanks is also given to Mrs. Verschoore-Martouzet and Fouret for giving their results.

REFERENCES

Ahuja MR (1984). A commercially feasible micropropagation method for aspen silvae genetica. 33:174-176.

Aitken-Christie J, Thorpe TA (198)6. Clonal propagation Gymnosperms in cell culture and somatic cell genetics of plants. Cell Culture and Somatic Cell Genetics of Plant Vol. 1 Academic Press pp. 82-95.

Ball EA, Morris DM, Rydelius JA (1978). Cloning of sequoia sempervirens from mature trees through tissue culture. Round Table Conf. "in vitro" Multiplication of Woody Species. Gembloux (Belgium) June 6-8, 1978. p. 181-226.

Bekkaoui F, Maldiney R, Pilate G, Boulay M, Franclet A (1986). Degenesescence clonale du douglas (Pseudotsuga menziesii) cultive in vitro; relation avec la teneur en elements mineraux en auxine, acide abscissique, zeatine and zeatine riboside. C.R. Ac. Sciences, Paris, in press.

Bennet IJ, McComb JA (1982). Propagation of Jarah (Eucalyptus marginata) by organ and tissue culture. Australian Forest Research 12(2):121-128.

Bonga JM (1981). Organogenesis in vitro of tissue culture from mature conifers. In Vitro 17:511-518.

Borman CH, Jansson E (1980). Organogenesis in cultured Pinus sylvestris tissue. Pflanzen. Physiol. 96:1-6.

Boulay M (1978). Multiplication rapide du Sequoia sempervirens en culture in vitro. AFOCEL 1977:38-65.

Boulay M (1983). Micropropagation of frost resistant eucalyptus. Proceeding of a Workshop on Eucalyptus in California, June 14-16, Sacramento, USDA For. Serv. General Technical Report. P.S.W. 69:102-107.

Boulay M (1986). Conifer micropropagation: Applied Research and Commercial aspects. To be published in Tissue Culture in Forestry. Bonga, J. and Durzan, D.J. (Eds.).

Campbell RA, Durzan DJ (1976). Vegetative propagation

of Picea glauca by tissue culture. Can. J. For. Res. 6:240-243.

Cheng TY (1975). Adventitious bud formation in culture of douglas-fir (pseudotsuga menziesii Mirb. Franco) Plant Sci. Lett. 5:97-103.

Cornu D, Riffaud JL Capelli P (1981). In vitro propagation of wild cherry tree (Prunus avium L.). In Colloque International Sur la Culture in vitro des Essences Forestieres AFOCEL ED.

David A (1982). In vitro propagation of gymnosperms: In Tissue Culture in Forestry. J.M. Bonga and D. J. Durzan (Eds.); Martinus Nijogg/Dr. Junk W. Publishers pp. 72-108.

David A, David H. 1977. Manifestation de diverses potentialite organogenes d'organes ou de fragments d'organes de pin maritime (pinus pinaster Sol.) en culture in vitro. C.R. Acad. Sci. 284:627-630.

Engelmann F (1982). Multiplication vegetative in vitro de clones juveniles et matures de Cunninghamia lanceolata. D.E.A. Univ. Pierre et Marie Curie Paris, p. 43.

Fouret Y (1984). Micropropagation in vitro du Sequoia sempervirens Endl.: Effet de la frequence des repiquages sur le rajeunissement de trois clones issus de materiel d'age different et recherche de critere precoce de juvenilite. D.E.A. Univ. Pierre Et Marie Curie Paris VI, p. 43.

Gonzalez ML, Vieitez AM, Vieitez E (1985). Somatic embryogenesis from chesnut cotyledon tissue cultured in vitro. Scientia Horticulturae, 27:97-103.

Gupta PK, Nadgir AL, Mascarenhas AF, Jagannathan V (1980). Tissue culture of forest trees: clonal multiplication of tectona grandis (teak) by tissue culture. Plant Science Letters 17:259-268.

Gupta PK, Mascarenhas AF, Jagannathan V (1981). Tissue culture of forest trees clonal propagation of mature trees of eucalyptus citriodora Hook by tissue culture. Plant Science Letters 20:195-201.

Gupta PK, Durzan DJ (1985). Shoot multiplication from mature trees of Douglas-fir (Pseudotsuga menziesii) and sugar pine (pinus lambertiana). Plant Cell Reports 4:177-179.

Gupta PK, Durzan DJ (1985). In vitro establishment and multiplication of juvenile and mature Douglas-fir and Sugar pine. To be published in Acta Horticulturae.

Gupta PK, Durzan DJ (1986). Somatic polyembryogenesis

from callus of mature Sugar pine embryos. Biotechnologie 4:643-645.

Gupta PK, Durzan DJ (1986). Plantlet regeneration via somatic polyembryogenesis from subcultured callus of mature embryos of Picea abies. In press In Vitro.

Gupta PK, Durzan DJ (1986). Somatic polyembryogenesis in callus and suspension cultures from callus of immature embryos of pinus taeda (in preparation).

Hansen K, Durzan DJ (1986) unpublished.

Hakman I, Von Arnold S (1985). Plantlet regeneration through somatic embryogenesis in Picea abies (Norway spruce). J. Plant Physiol. V. 121, p. 149-158.

Horgan KJ (1985). In vitro multiplication of mature pinus radiata in ICTCWG Meeting, August 1985 IFR, New Zealand. To be published in New Zealand Journal of Forestry.

Karnosky DF (1981). Potential for forest tree improvement via tissue culture. BioScience 31:111-120.

La Goublaye de Nantois H de (1981). Vieillissement et rajeunissement chez le Sequoia sempervirens Endl. en relation avee la propagation vegetative. These 3 IEME cycle, Universite de Paris VI, p. 170.

La Goublaye de Nantois T de (1980). Rajeunissement chez le Douglas (pseudotsuga menziesii) en vue de la propagation vegetative. Etude sur la plagiotropie des parties aeriennes et racinaires. D.E.A. Universite de Paris VI, 44 p.

Lakshmi-Sita G, Raghava Ram NV, Vaidyanathan CS (1979). Differentiation of embryoids and plantlets from shoot callus of sandal wood. Plant Sci. Lett. 15:265-270.

Leach GL (1979). Growth in soil of plantlets produced by tissue culture loblolly pine. Tappi 62:59-61.

MeKeand SE, Allen HL (1983). Nutritional and root development factors affecting growth of tissue culture plantlets of loblolly pine. Physiol. Plant, 61:523-528.

Misson JP, Coumans M, Giot-Wirgot P, Gaspar TH (1982). Induction de bourgeons adventifs de Picea pungens en culture in vitro. Z. Pflanzenphysiol. 107:161-167.

Patel KR, Berlyn GP (1982). Genetics instability of multiple buds of pinus coulteri regenerated from tissue culture. Can. J. F. Res. 12:93-109.

Quyang O, Peng HZ, Li QQ (1981). Studies on the development of embryoid from eucalyptus callus. Scientia Silvae Sinicia, 17:1-7.

Radojevic L (1979). Somatic embryos and plantlets from

callus cultures of Pawlonia Tomentosa Steud. Z. Pflanzenphysiol. 91:56-62.

Radojevic L, Vujicic R, Nesovic M (1975). Embryogenesis in tissue culture of Corylus avellana L. Z. Pflanzenphysiol. 77:33-41.

Reynoird JP (1983). Effet de pretraitement par greffages successifs ou par pulverisation de cytokinine sur la reactivation de douglas age (Pseudotsuga menziesii) en vue de la propagation vegetative. incidence de quelques facteurs du milieu de culture sur la croissance in vitro d'individus juveniles. D.E.A. Universite de Paris VI, 40p.

Sommer HE, Brown CL, Kormanik PP (1975). Differentiation of plantlets in long leaf pine (pinus palustris Mill) tissue cultured in vitro. Bot. Gaz. 135:196-200.

Sommer HE, Brown CL (1980). Embryogenesis in tissue cultures of sweetgum (Note). For. Sci. 26:257-260.

Smith D (1985). Field testing comparison between seedlings and plantlets of Pinus radiata. In ICTCWG Meeting, August, 1985 I.F.R. Rotorua, New Zealand. To be published in New Zealand Journal of Forestry.

Von Arnold S, Eriksson T (1979). Bud induction of adventitious buds on buds of Norway spruce (Picea abies) grown in vitro. Physiol. Plant 45:29-34.

Verschoore-Martouzet B (1985). Etude de la variation topophysique au cours du clonage de Sequoia sempervirens Endlicher. These de 3eme cycle. Universite Pierre et Marie Curie, Paris VI, 146 p.

Walker N (1985). Quelques aspects de la micropropagation de Sequoia sempervirens: culture de meristemes, rejuvenilisation, acclimatation. D.E.S. Universite Pierre et Marie Curie, Paris VI, 74 p.

Plant Tissue and Cell Culture, pages 383–393

RECENT TRENDS IN THE APPLICATION OF TISSUE CULTURE TO ORNAMENTALS

Pierre C.A. Debergh

Department of Horticulture, State University of Gent, Coupure links, 653, B-9000 Gent, Belgium

INTRODUCTION

The first plants masspropagated by tissue culture were ornamentals (Morel, 1965), and they also lie at the root of the first commercial tissue culture facilities.

Most of the initial research work as well as commercial experience was gained with cutflowers (*Cymbidium*, *Dianthus*, *Chrysanthemum*) and in the seventies it evolved mainly in the domain of pot plants. The eighties were mainly characterized by an expansion in the area of the woody ornamentals. A prognosis is that at the end of this decade more than 3000 species will be micropropagated and the majority will be ornamentals. In other words it is impossible to give an overview of the different micropropagated ornamental species and it is even not necessary as there are a lot of good review articles covering this field (Bhojwani & Razdan, 1983; George & Sherrington, 1984; Hughes, 1980).

As ornamentals are a trend setter in micropropagation, it is more interesting to cover in this review the major new orientations, problems and eventual solutions, since one thing is clear: micropropagation is evolving from a discipline for good chief cooks to a more and more scientifically based discipline.

PROPAGATION METHODS

The most often used technique to micropropagate ornamental plants is to start with more or less big shoot tips and to stimulate axillary branching in the successive sub-cultures. Although plants in the families *Gesneriaceae* and *Begoniaceae* are most frequently multiplied from buds regenerated directly on explants and cultured material (George & Sherrington, 1984).

There is also a regain of interest for sawing *in vitro*, especially for endangered species, e.g. terrestrial orchids (Van Waes & Debergh, 1986a & b), and ferns (Pierik et al., 1986).

Especially for woody ornamentals, e.g. roses, but also for potplants, e.g. *Dizygotheca*, micropropagated plants are grown on their own roots, and are evaluated and compared with the classical grafted material.

Martin et al. (1981) published their results on different types and species of roses after 4 years of field trials and concluded that the propagation can be speeded up with a factor 10.000, the product was equivalent to grafted roses and a salable product can be produced in 6 months instead of 30 months. Verdegaal (1984) reported on a visit to the Delbard nursery (France) and was told that *in vitro* propagated roses have a production which is 10 to 20% higher, they produce more shoots, bloom earlier, and have a better resistence. The results of Reist (1985), in a first year plantation of cv's 'Sonja' and 'Mercedes', are less positive: later productivity, too much branching and lowered production of first quality flowers.

Apparently embryogenesis doesn't withhold the attention of many people working with tissue culture of ornamentals. This lack of interest is probably motivated by the problems of genetic instability experimented in this field with other species (e.g. with *Carica*: Litz & Conover, 1983; *Apium graveolens*: Orton, 1985; *Citrus*: Navarro et al., 1985).

MEDIA

The almost generalized use of MS-medium (Murashige & Skoog, 1962) in the sixties and seventies, makes more and more place for modifications of this formulation. Most often the nitrogen content is lowered, but it is also very often used in dilutions.

The Woody Plant Medium (Lloyd & McCown, 1980)is popular for woody ornamentals, but also others gain in interest, e.g. the Olive medium (Rugini, 1984).

STAGE 0 : PREPARATION OF THE MOTHER PLANT

Two main types of reasons motivate the use of stage 0: (1) physiological and (2) pathological.

The physiological status of a plant can be influenced in different ways. Recently much attention has been paid to novel plant growth regulator delivery systems. One of the first techniques used was spraying the mother plants with cytokinins (up to 100, and even 800 mg/l BA), (Franclet, pers. comm.). It is also possible to inject the motherplants with BA (up to 500 mg/l)(Maene & Debergh, 1981 non publ. results), or to put the explant sources in a forcing solution which contains sucrose, 8-hydroxyquinoleine citrate, BA and GA3 (Read, 1985) or to incubate the primary explants in solutions containing different concentrations of BA (up to 440 µM)(Vogelmann et al., 1984). Such approaches allow to obtain more reactive explants in stage I.

In the development of standard procedures for micropropagation it is also important to standardize the developmental stage of the explant sources. A classical example, which is especially applicable to bulbous crops, is the fulfilment of the dormancy requirements (Van der Linde & Van Aartrijk, 1986), but for other types of plants a standardized growing environment (e.g. light intensity, photoperiod, nutritional status) and appropriated pruning can be helpful.

Growing mother plants under more hygienic cultural conditions can help to have a better control of the hygienic status of the explants (Debergh & Maene, 1981). Avoiding overhead watering, but providing water and nutrition with a trickle irrigation system or a sand bed can yield cleaner material from plants with an upright growth habit. When axillary buds are located near the base of the plant or even in the substrate it can be indicated to take away the substrate at the base of the plant and to allow the buds to develop for a few months in a cleaner and drier environment.

Apart from the beneficial effect on the hygienic and physiological status of the explants another major advantage is that a less severe sterilization procedure can be applied and this helps to minimize hypersensitivity reactions when starting the cultures.

BROWNING PROBLEMS

Hypersensitivity reactions are typical for plants which are submitted to stress situations. One of the best known hypersensitivity reactions under tissue culture conditions is browning of the tissue and/or the medium. Overcoming this problem isn't only limited to the use of anti-oxidants and/or adsorbents. Other types of interventions can also be helpful, a.o. control of the activity of definite enzymes (phenylalanine ammonialyase, peroxidases, phenolases) by appropriated temperatures and/or light conditions, or by lowering the content of definite mineral elements (Mn, Cu). Also the choice of the cytokinins or the salt concentration of the medium can be involved in this control (Debergh, 1986).

BACTERIAL CONTAMINATIONS

Most tissue cultured ornamentals are traditionally propagated in a vegetative way. This has tremendous repercussions on the sanitory status. Apart from virus diseases it are especially bacterial contaminations which mortgage successes in micropropagation systems. There is a lot of evidence to accept that endogeneous bacteria are responsable (Debergh & Maene, 1984a,b) but others do not agree (Boxus & Terzi, 1985).

A lot of reports presented at the symposia of Gembloux (Belgium, 1985) and Beltsville (U.S.A., 1985) proposed the use of antibiotics in the culture medium as a possibility to overcome infections, but in most cases no clear prove was provided that there is really a bactericidal effect. Most often the time or the number of sub-cultures between the treatment with antibiotics and the final evaluation of the cultures was too short to allow clearcut interpretations. In many cases and with different types of antibiotics, we have only been able to observe a bacteriostatic effect.

In a few reports Alcide (TM) has been mentioned to be an effective bactericidal product, a.o. for _Phoenix dactylifera_ (Maurice et al., 1985), for tropical hardwood stock (Mathias et al., 1985) and for _Gerbera_ (Podwyszynska &

Hempel, 1984).

In a lot of cases the presence of bacteria is only manifested late in culture, and this hinders the interpretation and evaluation even more.

As the addition of antibiotics in the medium doesn't yet give convincing results, an alternative is to use detection media which allow to clearly show the presence of bacteria. Different medium supplements were proposed in literature (a.o. Beauchesne et al., 1977; Boxus & Terzi, 1985). If no bacteria are revealed on such a supplemented medium, this doesn't allow to say that the culture is bacteria free, it only allows to say that no bacteria were detected on such a medium.

In conclusion, bacterial contamination remains problematic. More research efforts are needed to develop bactericidal products and/or to develop polyvalent test media.

PHYSIOLOGY OF TISSUE CULTURED PLANTS

Since the publication of our research work on vitrification in _Cynara scolymus_ (Debergh et al., 1981), many people identified their physiological problems in culture as vitrification. This resulted in many publications and in different approaches to explain and/or to overcome the problem (Debergh et al., 1981; Kevers & Gaspar, 1985; Paques, 1986; Wardle et al., 1983; Ziv et al., 1983).

During the last five years there were also many papers describing the anatomical and morphological characteristics of tissue cultured plant showing difficulties upon transfer from _in vitro_ to _in vivo_ (Donnelly & Vidaver, 1984; Vieitez & Ballester, 1985; Wetzstein & Sommer, 1982).

There is a striking parallelism between both phenomena: non-functional stomata, absence of cuticular wax and a disturbed vascular system are observed. That's why we think that it is reasonable to formulate the hypothesis that both phenomena are induced by the same causal agent, and that vitrification is only a more pathological expression of the difficulties that plants have to adapt to _in vivo_ conditions. In our opinion the causative agent is the absence of a gradient in water retention capacity in the gasphase of the container (Maene,1985; Maene & Debergh, 1985a). In a closed tissue culture container the relative humidity is almost 100%, so it doesn't make sense for the plant to develop functional systems to avoid dessication and to transport water. A shoot will only do so when a gradient in relative humidity is installed, that's to say when the leaves can develop in an atmosphere which is not saturated with water vapour. In almost every culture room light sources are installed beneath each shelf and this is a source of heath which stimulates the evaporation of water from the culture medium. The water condenses on the coldest surface, i.e. the lid and the wall of the container and this is an indication that the atmosphere is saturated with water vapour. By creating the reverse situation a more appropriated ecosystem, which allows the development of normal functioning leaves, is installed. We have been able to confirm this for _Calathea ornata_, _Iris germanica_, rhubarb, _Prunus avium_, _Oreopanax nymphaeifolia_, _Ficus benjamina_, _Dahlia_, and others.

Another kind of abnormal physiology results from the accumulation of gaseous substances, such as ethylene and carbon dioxide (De Proft et al., 1985) and ethanol (Thomas & Murashige ,1979). The use of a container which allows a certain amount of exchange with the surrounding can solve such problems.

MECHANIZATION VS ROBOTIZATION

Answers to the problem of labor saving are mechanization and/or robotization. The choice is largely dependent on the type of enterprise. Robotization can be indicated if one is working with a specified prototype of plant with a well controled growth habit. However most tissue culture laboratories for ornamental plants deal with many plant types, and moreover a laboratory manager can very often appreciate the skill of a bench operator. This is to say that in a lot of cases, if not most, it is necessary to interpret the way of cutting the explant. Can a robot do so for many plants?

Another positive question is: in which stage of the propagation cycle will robotization be introduced? For most plants the propagation stage is the manipulation of clusters of shoots or buds and the cost price of this manipulation can be divided by the number of off-springs from this explant. In fact the manipulations in the propagation stage are relatively cheap.

The impact of labor craft becomes much more important when single shoots are manipulated in the elongation and/or rooting stage. By developping an appropriated system of adding liquid media to induce elongation and/or rooting in final stage II-cultures, the manipulation of single shoots is avoided. The addition of liquid media does not necessitate robotization but can easily be solved by adapted mechanical devices and can be performed in the growing room. This again will save labor (Maene & Debergh, 1985b).

Another area where mechanization can be an issue is the filling of containers with media. The actual trend to use disposable plastic boxes as container creates problems at this level. The sterile containers need to be filled with already sterilized medium and most often this is actually done by manual dispension of the medium in the containers. Some prototypes apparatus, which allow mechanical dispension of agar media in plastic containers, are already used, but it still remains a research area to develop.

If the actual interest to favour the use of stagnant liquid media, instead of agarized media, is given concrete form, the problem of dispension will even become very easy.

HARDENING STAGE

The weaning stage starts in the laboratory. A lot of problems in the hardening off-stage can probably be nullified when good quality material has been produced in the latter tissue culture stages. Rooting in vivo, which also allows to save a lot of labor, will then also become less problematic for many plants (Debergh & Maene, 1981).

In many cases, if not most, micropropagation techniques do not allow the production of homogeneous plants. Two kind of approaches can help to solve this handicap. A first possibility is to use a substrate which allows flexibility, such as plug systems, artificial substrates (rockwool, cellulose, foam, ...), jiffy,...With such a system plants can be graded when they are being saled.

A second possibility is to use tissue cultured plants as micro-mother plants from which uniform micro-cuttings are harvested. Advantage is taken from the juvenility characteristics of the material.

REFERENCES

Beauchesne G, Albouy J, Morand JC, Daguenet J (1977). Multiplication de clones de Pelargonium hortorum et Pelargonium peltatum à partir de culture de méristème pour l'obtention de plants sains. Acta Hortic 78: 397-402.

Beltsville (1985). Proceedings of the Symposium "Tissue culture as a plant production system for horticultural crops," Beltsville (USA): in press.

Bhojwani SS, Razdan MK (1983)." Plant Tissue Culture: Theory and Practise." Amsterdam, Oxford, New York, Tokyo: Elsevier, pp 502.

Boxus Ph, Terzi FM (1985). Big losses due to bacterial contaminations can be avoid in mass propagation scheme. In Boxus Ph (ed): "Book of abstracts Symposium -In vitro problems related to mass propagation of horticultural crops", Gembloux Belgium: pp 4.

Debergh P (1986). Micropagation of herbaceous plants. In Anderson PG (ed): "Symposium Micropropagation in horticulture - Practice and commercial aspects," Univ Nottingham (UK): in press.

Debergh P, Harbaoui Y, Lemeur R (1981). Mass propagation of globe artichoke (Cynara scolymus): evaluation of different hypothesis to overcome vitrification with special reference to water potential. Physiol Plant 53: 181-187.

Debergh P, Maene L (1981). A scheme for commercial propagation of ornamental plants by tissue culture. Scientia Hortic 14: 335-345.

Debergh P, Maene L (1984a). Pathological and physiological problems related to the in vitro culture of plants. Parasitica 40: 69-75.

Debergh P, Maene L (1984b). Preparation of tissue cultured plants for rooting and establishment in vivo. In Novak FJ, Havel L, Dolezel J (Eds): "Plant tissue and cell culture. Application to crop improvement,"Prague, pp 487-496.

De Proft M, Maene J, Debergh P (1985). Carbon dioxide and ethylene evolution in the culture atmosphere of Magnolia cultured in vitro. Physiol Plant 65: 375-379.

Donnelly VA, Vidaver W (1984). Leaf anatomy of red raspberry transferred from culture to soil. J Amer Soc Hort Sci 109: 172-176.

George EF, Sherrington PD (1984). "Plant Propagation by tissue culture." Eversley: Exegetics Ltd, pp. 709.

Gembloux (1985). Proceedings of the Symposium "In vitro

problems related to mass propagation of horticultural crops," Gembloux Belgium, Acta Hortic: in press.

Hughes KW (1980). Ornamental Species. In Conger BV (ed): "Cloning agricultural plants via in vitro techniques," Boca Raton, Florida: CRC Press Inc, pp 5-50.

Kevers C, Gaspar T (1985). Soluble, membrane and wall peroxidases, phenylalanine ammonialyase and lignin changes in relation to vitrification of carnation tissues cultured in vitro. J Plant Physiol 118: 41-48.

Litz RE, Conover RA (1983). High-frequency somatic embryogenesis from Carica suspension cultures. Ann Bot 51: 683-686.

Lloyd G, Mc Cown B (1980). Commercially-feasible micropropagation of mountain laurel, Kalmia latifolia by use of shoot-tip culture. Proc Int Plant Propagtors Soc 30: 421-427.

Maene L (1985). "Optimalisering van de commerciële vermeerdering van planten in vitro." Brussels: I.W.O.N.L.-publication: pp 47.

Maene L, Debergh P (1985a). Optimalization of the transfer of tissue cultured shoots to in vivo conditions. In Boxus Ph (ed): "Symposium - In vitro problems related to mass propagation of horticultural crops," Gembloux Belgium, Acta Hortic: in press.

Maene L, Debergh P (1985b). Liquid medium additions to established tissue cultures to improve elongation and rooting in vivo. Plant Cell Tissue Organ Culture 5: 23-33.

Martin C, Carré M, Vernoy R (1981). La multiplication végétative in vitro des végétaux ligneux cultivés: Cas des rosiers. C R Acad Sc Paris 293: 175-177.

Mathias PJ, Alderson PG, Leaky RRB (1985). Bacterial contamination in cultures of tropical hardwood species. In Boxus Ph (ed): "Book of abstracts - Symposium In vitro problems related to mass propagation of horticultural crops," Gembloux Belgium: pp 1.

Maurice V, Vandercook CE, Tisserat B (1985). Automated plant surface sterilization system. Physiol Vég 23: 127-133.

Morel G (1965). Producing virus-free Cymbidiums. Am Orchid Soc Bull 29: 495-497.

Murashige T, Skoog F (1962). A revised medium for rapid growth and bioassays with tobacco tissue cultures. Physiol Plant 15: 473-497.

Navarro L, Ortiz JM, Juarez J (1985). Aberrant Citrus plants obtained by somatic embryogenesis of nucelli cultured in vitro. HortScience 20: 214-215.

Orton TJ (1985). Genetic instability during embryogenic

cloning of celery. Plant Cell Tissue Organ Culture 4: 159-169.

Paques M (1985). Culture model to study 'vitrification': present results. In Boxus Ph (ed): "Symposium In vitro problems related to mass propagation of horticultural plants." Gembloux Belgium, Acta Hortic: in press.

Pierik RLM, Oerlemans APC, Verhoeven MAP, Mentjox MA, Molenbergh H (1986). Zaai van varens in kweekbuizen. Vakblad Bloemisterij 10: 48-49.

Podwyszynska M, Hempel M (1984). Identification and elimination of "slowly growing" bacteria from a micropropagated Gerbera. In Boxus Ph (ed): "Book of abstracts - Symposium In vitro problems related to mass propagation of horticultural plants," Gembloux Belgium , pp 3.

Read P (1985). Novel Plant Growth Regulator Delivery Systems for in vitro culture of horticultural plants. In Boxus Ph (ed): "Symposium - In vitro problems related to mass propagation of horticultural crops," Gembloux Belgium, Acta Hortic: in press.

Reist A (1985). Culture in vitro (civ) en pépinière de rosiers: une alternative au bouturage ou au greffage des variétés. Rev Suisse Vitic Arboric Hortic 17: 361-364.

Rugini E (1984). In vitro propagation of olive (Olea europaea L.) cultivars with different rootability and medium development using analytical data from developing shoots and embryos. Scientia Hortic 24: 123-134.

Thomas SDD, Murashige T (1979). Volatile emissions of plant tissue cultures I. Identification of the major components. In Vitro 15: 654-658.

Van der Linde PCG, Van Aartrijk J (1986). Micropropagation of specific Crops: Bulbs. In Alderson PG (ed): "Symposium - Micropropagation in Horticulture - Practice and commercial Problems." Univ Nottingham (UK): in press.

Van Waes J, Debergh P (1986a). Adaptation of the tetrazolium method for testing the seed viability and scanning electron microscopy study of some Western European orchids. Physiol Plant 66: 435-442.

Van Waes J, Debergh P (1986b). In vitro germination of some Western European orchids. Physiol Plant: in press.

Verdegaal J (1984). Weefselkweekrozen op steenwol hebben toekomst. Vakblad bloemisterij 15: 58-59.

Vieitez AM, Ballester A (1985). Anatomical and chemical studies of vitrified shoots of chestnut regenerated in vitro. Physiol Plant 65: 177-184.

Vogelmann JC, Bornman CH, Nissen P (1984). Uptake of benzyladenine in explants of Picea abies and Pinus sylvestris. Physiol Plant 61: 513-517.

Wardle K, Dobbs E, Short K (1983). In vitro acclimatisation of aseptically cultured plantlets to humidity. J Amer Soc Hort Sci 108: 386-389.

Wetzstein H, Sommer H (1982). Leaf anatomy of tissue cultured Liquidambar styraciflua during acclimatisation. Amer J Bot 69: 1579-1586.

Ziv M, Meir G, Halevy A (1983). Factors influencing the production of hardened glaucous carnation plants in vitro. Plant Cell Tissue Organ Culture 2: 55-65.

Plant Tissue and Cell Culture, pages 395–406

PROPAGATION AND GENETIC IMPROVEMENT OF TEMPERATE FRUITS: THE ROLE OF TISSUE CULTURE

Michael G. Mullins

Department of Agronomy and Horticultural Science,
University of Sydney, N.S.W. 2006,
Australia

INTRODUCTION

The application of tissue culture to agriculture and horticulture has been the subject of scores of review articles and dozens of conferences. The main applications in temperate fruit crops are virus elimination and production of pathogen-free stock, rapid multiplication of plants, production of haploids and homozygous diploids, production of cybrids and novel hybrids by protoplast fusion and as a vehicle for genetic transformation of plants. Some of these procedures are already well established in horticulture, for example, micropropagation and the use of aseptic methods for phytosanitary purposes, but others are still at an early stage of development, especially those associated with plant improvement. It is worth commenting that many writings on the potential of tissue culture and plant biotechnology in primary industry have emanated from molecular geneticists or botanists with little understanding of crop plants, crop husbandry or crop breeding. The literature is pervaded by a feeling that only a few trivial technical problems remain and that we are on the threshold of a latter-day agrarian revolution. This is wishful thinking. Temperate fruits, in particular, are subject to special circumstances which affect or limit the use and usefulness of tissue culture methods for both plant propagation and plant improvement. Substantial benefits will accrue to the fruitgrowing industries from the further development of tissue culture but advocacy for *in vitro* techniques must be tempered by an appreciation of the biological, historical and economic realities of horticultural production.

THE SPECIAL CIRCUMSTANCES OF FRUITGROWING

Most temperate fruit plants are highly heterozygous outcrossers and do not breed true from seed, and selected clones (cultivars) are perpetuated by vegetative propagation. In many temperate fruit species these clonal cultivars are of great antiquity and have been propagated for many human generations. The grapevine, Cabernet Sauvignon, is reputed to be of Roman origin (Bouquet, 1982) and most of the world's apple crop is produced by genotypes which arose before 1900 (Knight and Alston, 1969). Most of the agronomic and gastronomic characters which constitute a good fruit are controlled by subtle combinations of genes of minor effect and genetic variance due to non-additive causes is significant. Seedlings of woody perennial plants, whether of zygotic or somatic origin, have a long juvenile phase during which they do not form flowers (5-8 years in apple) and the slow turnover of generations is a major limitation to progress in fruit breeding. Other technical problems in fruit improvement include self-incompatibility, parthenocarpy and the subjective nature of selection criteria. Fruit plants are bred for quality rather than for total fruit biomass and there are no quick, objective methods for determining quality in fresh fruits or in fruit products such as wine. The extreme sensitivity of the human palate to flavours in fruits and fruit beverages creates additional problems for fruit improvement and these problems are exacerbated by complex genotype x environment interactions (GxE). Some cultivars of fruit crops perform well in a wide range of environments but others perform best, or in particular ways, in particular environments. Favourable GxE's cannot be predicted but have to be found by experience. This leads to the final, if not major, difficulty faced by fruit improvers, that of conservatism. Conservatism at the level of consumer and producer finds its strongest expression in the wine industry where good quality has become equated with the winemaking characteristics of a narrow range of genotypes (eg Cabernet Sauvignon, Chardonnay, Pinot Noir), and where cultural methods have become enshrined by custom or by law (eg Appellation d'Origine Controllée). To be successful, a new cultivar of wine grape must be similar to or indistinguishable from the traditional cultivar that it is to replace. Fruit breeding programmes are thus very long term and costly and the perpetuation of existing genotypes by vegetative propagation has been reinforced by a combination of biological, economic and historical factors. In most fruit crops there has been useful progress in breeding and selection of rootstocks but improvement in scions has been based primarily upon production of pathogen-free propagating material and on clonal selection - the exploitation of genetic variation within cultivars.

From the foregoing it will be clear that tissue culture has two important roles in temperate fruit production (i) as a means of plant multiplication in which the genotype of the mother plant is reproduced with high fidelity in the propagules; (ii) as a vehicle for creating new genetic variation within existing cultivars and, thereby, providing new raw material for clonal selection.

MICROPROPAGATION OF POME AND STONE FRUITS

The technique of micropropagation has been applied successfully to the multiplication of clonal rootstock cultivars of *Malus*, *Pyrus* and *Prunus* and is now a routine operation in many commercial laboratories. Until recently, however, many of the leading scion cultivars of apple, pear and certain *Prunus* species were difficult, if not impossible to propagate as own-rooted trees. For example, the apples Jonathan, Richared and Granny Smith are exceedingly difficult to propagate from hardwood or softwood cuttings, as are the pears, Packham's Triumph, Beurré Bosc and Williams Bon Chrétien and the prune, D'Ente (syn. D'Agen). This apparent lack of ability to form adventitious root primordia is one of the reasons why the cultivars concerned are normally propagated by grafting onto rootstocks.

Work has been in progress in this laboratory for several years on the physiology of adventitious root formation in difficult-to-propagate apples, pears and prunes (Mullins, 1985).

Initially, it was thought that the procedures of micropropagation, in which the life of microcuttings is prolonged by culture *in vitro*, would lead, sooner or later, to formation of adventitious roots. This was not so and microcuttings, like conventional softwood or hardwood cuttings, were highly resistant to root formation.

Extensive experiments with basal media, phytohormones,

phenolic compounds, vitamins and other organic addenda failed to produce a simple recipe which induced roots in difficult-to-propagate apple cultivars. Eventually, it was found that mother cultures could be "conditioned" in vitro so that they produce axillary shoots with an enhanced capacity for adventitious root formation when made into microcuttings (Sriskandarajah et al., 1982).

In the apples, Jonathan, Richared and Granny Smith this conditioning process involved serial subculture on MS-medium with BA (10μM) under specific conditions of light (24h photoperiod, $90\mu Em^{2} sec^{-1}$) and temperature (26°C). In Jonathan, for example, only 8% of microcuttings from the initial culture produced roots but with increasing numbers of subcultures (each of 4 weeks' duration) there was a progressive increase in the rooting percentage. After 9 subcultures up to 95% of microcuttings formed roots. Similar effects of subculturing were found in the micropropagation of the prune cultivar, D'Ente, but the environmental conditions which led to enhanced rooting were markedly different from those which were effective for apple. For prune, enhancement of adventitious rooting occurred when shoots were subcultured at 23°C with 12h photoperiods. Root formation first occurred at the 5th subculture. At the 10th subculture and thereafter it was noted that roots were formed at the bases of axillary shoots whilst still growing in the proliferation medium containing BA (10μM), a compound which is normally inhibitory to adventitious root formation (Baleriola-Lucas and Mullins, 1984).

It appears that physiological changes can be induced in vitro in apple, prune and pear (Shen and Mullins, 1984), which predispose the tissues of microcuttings to formation of adventitious roots and the hormonal basis of these changes has been the subject of several studies (Takeno et al., 1982; Mullins, 1985). By use of the subculture procedures it is possible to obtain populations of microcuttings of the same genotype (eg Jonathan, Granny Smith) but with differing capacities of adventitious rooting and comparisons were made of endogenous levels of phytohormones in these "rooting" (R) and "non-rooting" (NR) microcuttings. In brief, subculture leads to a rapid decline in abscisic acid and gibberellin, compounds which inhibit rooting when applied to R-microcuttings, but the levels of free-IAA are unchanged. In addition, there are morphological and anatomical differences between NR- and R- microcuttings. Stems become less vascularized with increasing numbers of subcultures and the epidermis develops the ability to form anthocyanin pigments. The occurrence of a red coloration at the bases of R-microcuttings after they had been incubated for a few hours in auxin-rich media was closely associated with the commencement of root formation. A similar correlation between anthocyanin synthesis and adventitious rooting has been found in other woody species (Bachelard and Stowe, 1962).

Tissue culture-induced enhancement of adventitious root formation is strictly a transient phenomenon. Once microcuttings have become established as rooted plants the ability of cultured axillary shoots to form adventitious roots is either lost or greatly diminished. Rooted microcuttings of Jonathan and Richared were grown-on in large containers with standard potting media and they came into flower within 2.5 years. In these experiments the acquisition of rooting ability was thus not associated with reversion to the juvenile or non-flowering phase but the conditioning procedure may, perhaps, involve rejuvesescence or rejuvenation (see Schwabe, 1976).

GENETIC IMPROVEMENT OF TEMPERATE FRUIT CROPS

1. Tissue Culture and Random Genetic Variation

In many species, plants regenerated from callus or from cells exhibit considerable variation in morphological and physiological attributes (Scowcroft et al., 1983; Evans et al., 1984). This variability arises from gross changes in chromosome numbers or structure or from more subtle changes in the nuclear DNA which occur during the tissue culture process. The random spontaneous genetic variation which arises during plantlet formation *in vitro* is termed "somaclonal" variation. Somaclonal variants of sugar cane, potato, rice, wheat, barley, rape and some other crop species have been discovered which possess disease resistance and several other agronomically-interesting characters. These early reports led to much enthusiastic speculation on the potential of somaclonal variation in plant breeding but experience with many seed-propagated plants including cereals has been disappointing.

In temperate fruits, however, there is still great interest in somaclonal variation because it could provide a means of amplifying variability within the existing cultivar and, thereby, provide new opportunities for clonal selection. In addition to somaclonal variation, there are other sources of variation in long-established cultivars of vegetatively-propagated plants which are dependent upon tissue culture for their expression. First, most cultivars of fruit crops are subject to infection by viruses and virus-like organisms and one explanation of intraclonal variability is that plants contain differing loads of pathogens. It is well-established that tissue culture, in combination with heat-treatment, is effective in producing pathogen-free (or symptomless) plants (Galzy, 1964). Second, many fruit cultivars arose as somatic mutations and are chimeras. The rearrangements in chimeric structure which occur during plant regeneration in vitro are an additional source of clonal variation (Skene and Barlass, 1983). Third, ancient clones such as the traditional cultivars of grapevines are likely to have accumulated a considerable load of point mutations over the centuries and cell culture methods may provide the means by which normally covert variation can be expressed. Thus, tissue culture continues to hold great promise for fruit improvement in general and for viticulture in particular. In the case of grapevines these expectations are reinforced by the lack of realistic alternatives to clonal selection for cultivar improvement in most of the premium winegrowing regions of the world. Further discussion on the use of tissue culture for improvement of fruit crops will be restricted to grapevines.

Until recently, a major stumbling block to exploitation of tissue culture for grapevine improvement was the lack of efficient methods for regenerating plants from cultured cells. Callus cultures of the grapevine were first established by Morel (1944) but the first report of somatic embryogenesis, and of production of plantlets from cell suspensions of wine grape cultivars, was in 1976 (Mullins and Srinivasan, 1976). Organogenesis (Rajasekaran and Mullins, 1981) and a novel method of plant multiplication in vitro from fragmented apices (Barlass and Skene, 1979) have been reported subsequently, but the goal of high-frequency plantlet production from grapevine protoplasts has yet to be achieved. Similarly, the use of aseptic methods to produce haploid grapevines has yet to be successful. (Rajasekaran and Mullins, 1979, 1983). Meanwhile, procedures for producing somatic embryos and plantlets in vitro have undergone considerable refinement (Srinivasan and Mullins, 1980; Rajasekaran and Mullins, 1985), but progress in the exploitation of somaclonal variation is limited by technical difficulties in the regeneration of plantlets in sufficient numbers for large scale screening. Grapevines of most major cultivars have now been produced *in vitro* by somatic embryogenesis using nucellar tissues of unfertilized ovules (Mullins, 1978) or the vegetative tissues of anthers (Rajasekaran and Mullins, 1983). However, the yields of plantlets are generally low and the main problems to be overcome are dormancy and the poor "germination" of somatic embryos. The problem area is the gemmule, the incipient apical meristem. Elongation of roots occurs in most somatic embryos, but the production of shoots is highly irregular. There are marked differences among cultivars in the ease of plantlet production. Grenache and Shiraz have consistently produced greater numbers of vines *in vitro* than Cabernet Sauvignon and Chardonnay (Vilaplana and Mullins, unpublished). Information on hormonal changes in somatic embryos is being acquired (Rajasekaran et al., 1982; Takeno et al., 1983) and, hopefully, this will lead to a solution of the dormancy problem. Meanwhile, evidence of somaclonal variation in grapes has come primarily from research on genotypes which are highly regenerative *in vitro*, for example, Gloryvine, a *Vitis vinifera* x *Vitis rupestris* hybrid. Gloryvines raised from somatic embryos exhibit abnormalities such as dwarfism and albinism. Leaf shape is normally a highly stable character in grapes and it is the basis of ampelography (Galet, 1979), but plants produced *in vitro* show marked variations in leaf shape, including differences in petiolar sinuses and lobation. In addition, Gloryvines raised from somatic embryos show variation in sex-expression. Gloryvine is a male genotype but 3 vines among the 125 which were planted in a field trial have proved to be hermaphrodite and they have produced fruit for each of the last four years.

The occurrence of somaclonal or tissue-culture induced variation in *Vitis* has been established but the usefulness of this variation for genetic improvement of wine grapes remains to be proved. In 1977, twelve vines of "somatic Cabernet Sauvignon" were planted in the field with the collaboration of the New South Wales Department of Agriculture. This planting was a curiosity rather than an experiment but it has provided some interesting observations. Initially, the somatic vines were highly variable in growth and cropping, as are most newly-established grapevines, but they have become more uniform in recent years. A similar loss of variability with increasing age has been found in other grapevines regenerated from cells, for example, reversion to the ampelographically-accepted leaf-shape of the cultivar.

The original 12 somatic Cabernet Sauvignon are now vigorous vines and characteristic of their cultivar but in other respects they are unremarkable, at least to the naked eye. A more extensive field experiment was planted in 1983, to compare the growth and cropping of somatic- and normally-propagated grapevines. It will be two or three years before meaningful results are available.

The variation which arises in tissue culture is essentially random in nature and its successful exploitation is dependent upon the availability of rapid, accurate screening procedures. The development of these methods is relatively straightforward for characters such as disease resistance. Micropathogenicity tests are already available for selection _in vitro_ for resistance to downy mildew (_Plasmopara viticola_; Morel, 1948; Lee and Wicks, 1982) and powdery mildew (_Uncinula necator_; Klempka et al., 1984) and selection at the level of phytoalexin production is an interesting possibility for future research (Stein and Hoos, 1984). Experience in selecting at cellular level for NaCl-tolerant grapevines has been disappointing (Lebrun, Rajasekaran and Mullins, 1985). It must be emphasized that selection among somaclones for qualitative characters such as wine quality will remain as difficult and as time-consuming as conventional clonal selection with conventionally-propagated grapevines.

2. Tissue Culture and Directed Genetic Change

So far greatest progress in the practical application of recombinant DNA technology has been made with herbaceous annuals and the gulf between highly regenerative _Solanaceae_, for example, and recalcitrant woody perennial fruit plants is wide. Nevertheless, genetic engineering has many potential applications in fruit improvement. Provided that genes of interest can be identified they could be cloned and then inserted into the genomes of already well-established cultivars. This goal is still some way off but it seems to be achievable for pest and disease resistance, and for other single gene characters. In a dynamic field like molecular genetics it would be foolish to be dogmatic on what is and what is not possible, but it is difficult at present to envisage the direct manipulation of complex polygenic characters such as yield or quality by molecular methods. Another difficulty is that procedures for genetic transformation of temperate fruit plants have yet to be perfected.

Research is in progress in Sydney and elsewhere on the transfer of bacterial genes to grapevines. The genes concerned are of no viticultural significance but are experimental tools for the development of methods. The gene vectors in this research are specifically modified Ti-plasmids of the Crown Gall pathogen, *Agrobacterium tumefaciens*, It is probable that the occurrence of genetically-transformed grapevines will be reported in the near future but this is only a first step. The major challenge has yet to be faced, that of identifying and cloning a viticulturally-important gene and transferring it to a commercially-important cultivar. Even when this has been achieved there are still opportunities for disappointment because disruption of the genome by the introduction of foreign DNA could well impair the subtle qualities of traditional cultivars of wine grapes such as Cabernet Sauvignon or Chardonnay.

SUMMARY

Micropropagation of fruit tree rootstocks is a well-established commercial application of tissue culture and propagation *in vitro* can now be extended to most, if not all, of the leading "scion" cultivars of pome and stone fruits. The availability of own-rooted trees in large numbers, and at a unit cost which is considerably less than that of conventional grafted trees, will greatly facilitate the development of mechanised high-density planting systems for fruit production. Horticultural evaluation of own-rooted trees is in progress at many centres notably by R.H. Zimmerman, USDA, Beltsville, and at East Malling Research Station, England. This research is long term but early indications are that some cultivars are more successful than others when grown on their own roots (McVittie and Cook, 1984). So far, the uniformity of clonal cultivars produced by micropropagation appears to have been acceptable and there have been no reports of gross variation or loss of cultivar identify.

The use of rootstocks has many horticultural advantages,

including size-control of scions, pest resistance and effects on yield and quality, and it is probable that grafted trees will continue to be a mainstay of temperate fruit production. However, the availability of micropropagated own-rooted trees provides another management option for fruit-growers and new degrees of freedom in the design of highly automated systems of fruit growing. Improved understanding of the physiology of adventitious root formation in hitherto difficult-to-propagate genotypes of pome and stone fruit will enable the use of aseptic methods for production of pathogen-free stock, procedures which have proved to be highly successful in easily-propagated species such as grapes (Galzy, 1964). Finally, the use of micropropagation as a vehicle for mutation breeding of pome and stone fruits is worthy of further investigation.

The value of tissue culture for the genetic improvement of fruit plants has yet to be proven but the potential for these techniques in woody perennials, plants which are difficult to breed by conventional methods, is considerable. In grapevines, if not in other temperate fruits, systems for regeneration of plants in vitro are now sufficiently well developed for systematic investigations to be made of somaclonal variation and its contribution to clonal selection. Directed genetic change in woody perennial fruit plants is a promise of things to come.

REFERENCES

Bachelard EP, Stowe BB (1962). A possible link between root initiation and anthocyanin formation. Nature 194: 209-210.

Baleriola-Lucas C, Mullins MG (1984). Micropropagation of two French prune cultivars (Prunus domestica L.). Agronomie 4:473-478.

Barlass M, Skene KGM (1979). In vitro propagation of grapevine (Vitis vinifera L.) from fragmented shoot apices. Vitis 17:335-340.

Bouquet A (1982). Origine et évolution de l'encépagement français a travers les siècles. Prog Agric Viticole 99: 110-121.

Evans DA, Sharp WH, Medina-Filho HP (1984). Somaclonal and gametoclonal variation. Amer J Bot 71:759-774.

Galet P (1979). A practical ampelography: grapevine identification. Morton LT, translator. New York NY:Cornell University Press.

Galzy R (1964). Technique de la thérmotherapie des viroses de la vigne. Ann Epiphyt 15:245-256.

Klempka KC, Meredith CP, Sall MA (1984). Dual culture of grape powdery mildew (Uncinula necator Burr.) on its host (Vitis vinifera L.) Am J Enol Vitic 35:170-174.

Knight RL, Alston FH (1969). Developments in apple breeding. Rep E Malling Res Stn for 1968-1969. 125-132.

Lebrun L, Rajasekaran K, Mullins MG (1985). Selection in vitro for NaCl-tolerance in Vitis rupestric Scheele. Ann Bot 56:733-739.

Lee TC, Wicks T (1982). Dual culture of Plamopara viticola in grapevine and its application to systemic fungicide evaluation. Plant Dis 66:308-310.

McVittie A, Cook ME (1984). Progress in research in 1983. Rep E Malling Res Stn for 1983 (1984). 197-205.

Morel G (1944). Sur le développement de tissus de vigne cultivées in vitro. CR Seance Soc Biol 138: 62.

Morel G (1948). Cultures associées de tissus vegetaux et de parasites obligatoires. Ann Epiphyt 14:78-84.

Mullins MG (1978). Somatic embryos and plantlets of Cabernet Sauvignon grapes by apomixis in vitro. C R 2eme Symp Int Amelior Vigne (Bordeaux 1977) 11-15.

Mullins MG (1985). Regulation of adventitious root formation in microcuttings. Acta Hort 166:53-62.

Mullins MG, Srinivasan C. 1976. Somatic embryos and plantlets from an ancient clone of the grapevine (cv Cabernet Sauvignon) by apomixis in vitro J Exp Bot 27:1022-1030.

Rajasekaran K, Mullins MG (1979). Embryos and plantlets from cultured anthers of hybrid grapevines. J Exp Bot 30:399-407.

Rajasekaran K, Mullins MG (1981). Organogenesis in vitro in internode explants of grapevines. Vitis 20:218-227.

Rajasekaran K, Vine J, Mullins MG (1982). Dormancy in seeds and somatic embryos of Vitis: Effects of chilling on germination and ABA content. Planta 154:139-144.

Rajasekaran K, Mullins MG (1983). The origin of embryos and plantlets from cultured anthers of hybrid grapevines. Amer J Enol Vitic 34:108-113.

Rajasekaran K, Mullins MG (1985). Somatic embryo formation by cultured ovules of Cabernet Sauvignon grape: Effects of fertilization and of the male gameticide toluidine blue. Vitis 24:151-157.

Schwabe WW (1976). Applied aspects of juvenility and some theoretical considerations. Acta Hort 56:45-54.

Scowcroft WR, Larkin PJ, Brettell RIS (1983). Genetic variation from tissue culture. Use of Tissue Culture and

Protoplasts in Plant Pathology. Helgeson JP, Deveall BJ, eds. Academic Press Australia 139-162.

Shen XS, Mullins MG (1983). In vitro propagation of pear, Pyrus communis L. cvs Packham's Triumph, William's Bon Chretiền and Beurré Bosc. Scientia Hort 23:51-57.

Skene KGM, Barlass M (1983). Studies on the fragmented shoot apex of grapevine iv Separation of phenotypes in a periclinal chimera in vitro. J Exp Bot 34:1271-1280.

Srinivasan C, Mullins MG (1980). High-frequency somatic embryo production from unfertilized ovules of grapes. Scientia Hort 13:245-252.

Sriskandarajah S, Mullins MG, Nair YC (1982). Induction of adventitious rooting in difficult-to-propagate cultivars of apple. Plant Science Letters 24:1-9.

Stein U, Hoos G (1984). Induktions-und Nachweismethoden fur Stilbene bei Vitaceen. Vitis 23:179-194.

Takeno KT, Taylor JS, Sriskandarajah S, Pharis RP, Mullins MG (1982). Endogenous gibberellin and cytokinin-like substances in cultured shoot tissues of apple, Malus pumila cv Jonathan, in relation to adventitious root formation. Plant Growth Regulation 1:261-268.

Takeno K, Koshioka M, Pharis RP, Rajasekaran K, Mullins MG (1983). Endogenous gibberellin like substances in somatic embryos of grape (Vitis vinifera x V.rupestris) in relation to embryogenesis and the chilling requirement for subsequent development of mature embryos. Plant Physiol 73:803-808.

Plant Tissue and Cell Culture, pages 407–418

APPLICATION OF TISSUE CULTURE TO TROPICAL FRUITS

Richard E. Litz

University of Florida, Tropical Research and Education Center, Institute of Food and Agricultural Sciences, 18905 S.W. 280 Street, Homestead, FL 33031

Fruit-producing plant species are generally very heterogeneous, and have largely resulted from uncontrolled pollinations. Selections of elite plants have been propagated vegetatively in order to preserve the unique genetic complexes and heterogeneity that confer superior fruit and tree characteristics. Little is known about the genetics of these plants, particularly with respect to tropical fruit species. Disease, pest and environmental stresses supply constant pressure in the monoculture agroecosystems of the tropics, resulting in periodic, yet predictable crop-threatening situations. The application of tissue culture techniques for propagation and improvement of tropical fruits by conventional and somatic cell genetics approaches holds great promise which is slowly being realized (Krikorian and Cronauer, 1984; Litz *et al*., 1985).

The regeneration of woody plant species from cell and tissue cultures has been elusive despite 3 decades of research (Brown and Sommer, 1982). The nature of problems associated with differentiation of both angiosperms and gymnosperms from callus has recently been reviewed by Bonga (1982). Inherent in woody plant species is the relative absence of tissues with cells having morphogenetic potential in mature trees. Although *Coffea arabica* (Rubiaceae) and *Theobroma cacao* (Sterculiaceae) have been regenerated by somatic embryogenesis (Sondahl and Sharp, 1977; Litz, 1986) and *Morus indica* (Moraceae) by organogenesis (Mhatre *et al*., 1985) from leaf explants, these examples appear to be exceptions to the rule.

Nucellar adventive embryony (polyembryony) occurs in many woody plant species. In fact, this form of polyembryony has been identified in species representing 16 plant families

(Rangaswamy, 1982), many of which are largely comprised of tropical species (Litz, 1985). Although the *in vitro* regenerative potential of the nucellus, particularly of *Citrus* species, has been recognized for many years (Stevenson, 1956). The practical demonstration that other trees could be regenerated by somatic embryogenesis from the nucellus has occurred only recently (Table 1). Is the nucellus, particularly from tropical tree species having relatively large seeds, the explant of choice for regenerating woody plants by somatic embryogenesis from cell and tissue cultures?

TABLE 1. Regeneration of woody plant species by somatic embryogenesis from tissues of mature origin

Species	Family	Explant	Reference
Citrus	Rutaceae	nucellus (P)*	Stevenson (1966)
Citrus	Rutaceae	nucellus (M)	Rangan *et al*. (1968)
Coffea arabica	Rubiaceae	leaf	Sondahl and Sharp (1977)
Eriobotrya japonica	Rosaceae	nucellus (M)	Litz (1985)
Eugenia spp.	Myrtaceae	nucellus (P)	Litz (1984a)
Hevea brasiliensis	Euphorbiaceae	nucellus (M)	Carron and Enjalric (1985)
Malus domestica	Rosaceae	nucellus (M)	Eichholtz *et al*. (1979)
Mangifera indica	Anacardiaceae	nucellus (P)	Litz *et al*. (1982)
Mangifera indica	Anacardiaceae	nucellus (M)	Litz (1984b)
Myrciaria cauliflora	Myrtaceae	nucellus (P)	Litz (1984c)
Theobroma cacao	Sterculiaceae	leaf	Litz (1986)
Ribes rubrum	Saxifragaceae	nucellus (M)	Zatyko *et al*. (1975)
Vitis vinifera	Vitaceae	nucellus (M)	Mullins & Srinivasan (1976)

* M = monoembryonic, P = polyembryonic

Cells within the nucellus that are morphogenetically competent and which do not require a stimulus to become embryogenic *in vitro* have been described as pre-embryogenic determined cells (Evans *et al.*, 1981). The origin of these cells and the regulation of adventitious embryogenesis from them is not clearly understood. It is not known, for example, if nucellar cells are morphogenetically competent but quiescent until they are exposed to an external stimulus, or if they are cells that become competent in response to a stimulus. Frost and Soost (1968) have suggested that pollination and fertilization are the necessary triggering mechanisms for adventitious embryogenesis in *Citrus* ovules; however, van Pijl (1934) has clearly demonstrated that adventitious embryos in *Eugenia jambos* are formed prior to pollination and fertilization. The presence of probable inhibitors of nucellar adventitious embryony have been demonstrated in monoembryonic *Citrus* ovules (Esan, 1973), and extracts from the micropylar end of monoembryonic ovules can inhibit somatic embryogenesis from carrot callus (Esan and Murashige, 1972). Tisserat and Murashige (1977 a;b) attempted to correlate the effectiveness of certain naturally occurring regulators of somatic embryogenesis, e.g. ethylene, ethanol, etc. with the control of adventitious embryony in *Citrus* ovules.

In order to determine whether the regeneration of adventitious embryos from the nucellus *in ovulo* and of somatic embryos *in vitro* from nucellar callus are regulated in a similar manner, we have used mango as a model. Mango cultivars like those of *Citrus* can be either polyembryonic or monoembryonic, depending on their geographic origin. The mango ovule is large and has an easily excisable nucellus. Somatic embryogenesis from nucellar callus of both monoembryonic (Litz, 1984b) and polyembryonic (Litz *et al.*, 1982; 1984) mangos has been described. Embryogenic mango callus is globular and pseudobulbil-like in appearance (Litz *et al.*, 1984). Coconut water (Litz *et al.*, 1982) and 2,4-D (Litz *et al.*, 1984) have both been effective in stimulating growth of embryogenic callus from nucellar explants (Fig. 1). Litz and Schaffer (1986) have demonstrated that somatic embryogenesis from mango nucelli is not dependent upon 2,4-D as a stimulus. In the absence of coconut water or 2,4-D, somatic embryogenesis occurs at a relatively low frequency directly from the nucellar explant (Fig. 2) without an intermediate callus phase. The role of 2,4-D may therefore be to preferentially clone the pre-embryogenic determined cells of the nucellus until a sufficient cell density is present to permit effective cell interaction (Halperin, 1967) and organization. Two, 4-D also increases the efficiency of somatic embryogenesis by increasing the quantity of embryogenic cells. Subculture of globular somatic embryos onto medium without 2,4-D permits the development of cotyledons and the differentiation of a characteristic bipolar structure. The *in vitro* responses of mango nucelli are cultivar- or genotype-dependent, and are also dependent on the ovule type, i.e., monoembryonic or polyembryonic (Litz, 1985; Litz and Schaffer, 1986); nucelli from polyembryonic mango ovules are more embryogenic.

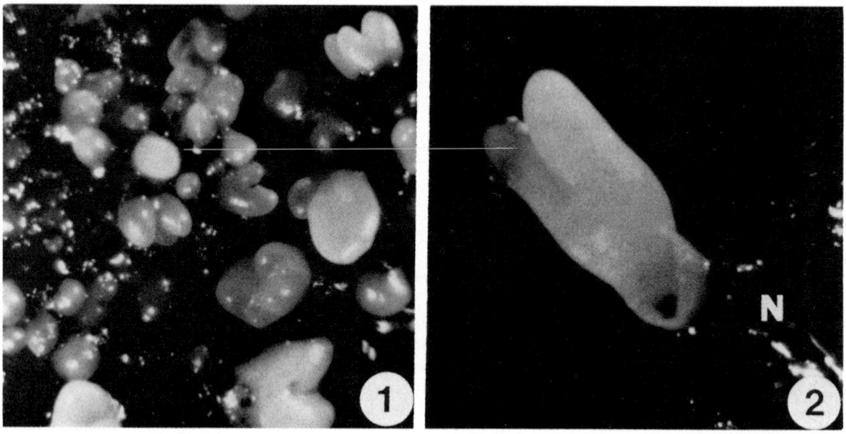

Figure 1. Embryogenic mango callus on medium with 20% (v/v) coconut water.

Figure 2. Somatic embryogenesis directly from cultured nucellus (N) of mango.

There is only a limited period annually during which somatic embryogenesis can be stimulated from mango nucelli *in vitro*. This period is defined by the growth and development of the zygotic embryo in monoembryonic ovules and of the adventitious embryos in polyembryonic ovules. It is equivalent to the period between the earliest detection of the embryo(s) and the filling of the embryo sac (Table 2). During this time the endogenous concentration of putrescine in the nucellus declines sharply in both polyembryonic and monoembryonic mango cultivars. Concentrations of both spermidine and spermine are much lower or are undetectable in monoembryonic mango nucellus. This decline in polyamine concentration corresponds with a decline in production of nucellar embryogenic callus and with the progressive deterioration of the nucellus as the embryo mass fills the embryo sac.

The possible mediation of somatic embryogenesis by polyamines has been the subject of several recent studies (Bradley *et al.*, 1984; Feirer *et al.*, 1984; Montague *et al.*, 1978; 1979). Using carrot as a model system, it has been possible to regulate somatic embryogenesis by the addition of polyamines to the growth medium (Bradley *et al.*, 1984) and to detect large increases in arginine decarboxylase activity during somatic embryogenesis (Feirer *et al.*, 1984). Therefore, studies were also initiated to compare the role of polyamines in adventitious and somatic embryogenesis, because of the apparent relationship that exists between endogenous polyamine concentrations in the nucellus at the time of culturing and somatic embryogenesis (Litz and Schaffer, 1986). Putrescine, spermidine and spermine have no positive effect on either callus induction or somatic embryogenesis from the cultured nucelli of polyembryonic and monoembryonic mangos (Litz and Schaffer, 1986). This is in apparent contradiction with results obtained in other studies in which carrot has been used as a model (Bradley *et al.*, 1984; Feirer *et al.*, 1984; Montague *et al.*, 1978; 1979). However, in all of these studies, somatic embryogenesis has been defined as the sequence of developmental events that follows the transfer of cultures from medium containing 2,4-D to medium without 2,4-D. Polyamine concentrations and associated enzymatic activity in these cultures have been observed to change dramatically (Feirer *et al.*, 1984).

Somatic embryogenesis from mango callus occurs while cultures are exposed to 2,4-D; the transition from globular to heart-shaped and more advanced stages of development is usually impeded by the presence of 2,4-D. Therefore, the changes in polyamine concentration that accompany the transfer of cultures from 2,4-D containing medium to 2,4-D-free medium probably do not reflect the metabolic events during somatic embryogenesis, but rather the altered metabolism during somatic embryo development from an early, globular stage to a later stage. Thus, it is improbable that polymines are stimuli for adventitious embryogenesis *in ovulo* or for somatic embryogenesis *in vitro* with mango.

TABLE 2. Effect of ovule development on induction of embryogenic nucellar callus

Cultivar	Type	Ovule development (length cm) Ovule	Embryo	Embryo:ovule	Polyamine ug/g fresh wt. ±SE Put	Spd	Sp*	% embryogenic nucellar callus
Lippens	M*	1.32	0.41	0.31	9.3±0.8	1.7±0.1	0.5±0.0	17.74
		1.59	1.17	0.74	1.6±0.1	0.1±0.1	0.0	7.94
		1.94	1.51	0.78	0.0	0.0	0.0	8.05
Cambodiana	P	1.65	0.71	0.43	11.9±1.0	0.1±0.1	0.0	29.35
		1.91	1.16	0.61	1.5±0.2	0.0	0.0	18.31
		2.25	1.60	0.71	0.1±0.1	0.0	0.0	6.32

* M = monoembryonic; P = polyembryonic; Put = putrescine; Spd = Spermidine; Sp = Spermine

Other tropical trees in addition to *Citrus* and mango have been regenerated by somatic embryogenesis from nucellar cultures *in vitro* (Table 1). These trees represent both monoembryonic and polyembryonic types, although in every case, the latter have been more responsive *in vitro*. Kaur *et al.* (1980) have suggested that apomixis, i.e., polyembryony, may be widespread in the tropical rain forest which has been the center of origin of not only *Citrus* and *Mangifera*, but also *Lansium domesticum*, *Garcinia mangostana*, several *Eugenia* species, etc. Characteristic of the tropical rain forest environment are high species diversity, low frequency of distribution of a particular species, barriers to self pollination and diseases, particularly of flowers and immature fruit, that adversely affect fruit set. Monoembryonic mango trees, which originated in regions having alternatively wet and dry seasons, have much higher rates of immature fruit drop than polyembryonic mango cultivars (Sturrock, 1961), often are alternate bearing, and generally are more susceptible to anthracnose during flowering and early fruit set (Singh, 1960). The tropical rain forest environment has probably resulted in selection for fruit set under adverse conditions, perhaps by blocking ethylene production during flowering and early fruit set. Ethylene is known to inhibit somatic embryogenesis (Tisserat and Murashige, 1977b), and is also released if S-adenosyl-L-methionine decarboxylase (SAMDC) activity is blocked, resulting in the accumulation of putrescine. Indeed, high concentrations of putrescine also have been observed to accumulate in nucellar callus of monoembryonic mango cultivars in comparison with polyembryonic mangos and in non-embryogenic callus in comparison with embryogenic callus (Litz and Schaffer, 1986). If SAMDC activity in monoembryonic cultivars is more likely to be blocked than in polyembryonic cultivars, it is possible that ethylene production may play the decisive role in regulating adventitious embryogenesis *in ovulo* and somatic embryogenesis *in vitro*.

In contrast with woody tropical fruit trees, herbaceous fruit species have been recovered from various regeneration pathways, including shoot tips and axillary buds and by somatic embryogenesis and organogenesis from callus derived from different tissues (Table 3) (Litz *et al.*, 1985). These procedures are currently being exploited commercially for propagation of banana (Hwang *et al.*, 1984) and in the breeding of papayas (Litz, 1984).

TABLE 3. Regeneration of herbaceous tropical fruit species

Crop	Shoot tip culture	Organogenesis	Somatic embryogenesis	Embryo rescue	Anther culture
Babaco	Cohen & Cooper (1982)	-	-	-	-
Banana	Ma & Shii (1972)	-	-	Cox *et al*. (1960)	-
Papaya	Litz & Conover (1978a)	Litz *et al*. (1974)	DeBruijne *et. al*. (1980)	Khuspe *et al*. (1980)	Litz & Conover (1978b)
Passion fruit	Robles (1978)	Scorza & Janick (1976)	-	-	-
Pineapple	Pannetier & Lanaud (1976)	Mapes (1973)	-	-	-
Plantain	Damasco & Barba (1984)	-	Cronauer & Krikorian (1983)	-	-

A major factor that effectively limits the use of cell and tissue culture for improving important monocotyledonous tropical fruit species, such as banana (Musa gp. AAA), plantain (Musa gp. AAB, ABB) and pineapple (Ananas comosus) is the inconclusive evidence in support of differentiation of plants from callus. Although somatic embryogenesis has reportedly been obtained for Musa (gp. ABB) (Cronauer and Krikorian, 1983), this has been disputed by other studies including those of Srinivasa Rao et al. (1982) and Bakry and Rossignol (1985). Similarly, the report of organogenesis from pineapple tissue cultures (Mapes, 1973) has not been confirmed. On the other hand, Stover and Buddenhagen (1986) have indicated that as many as 20% of banana plants regenerated from shoot tips are deviant to some degree; Wakasa (1979) has also reported very high rates of variability in pineapple plant populations regenerated from shoot tips. Passage of shoot tips through an in vitro cycle may therefore be sufficient to unmask or induce useful variation in this important group of plants.

ACKNOWLEDGEMENTS

The author is grateful for the assistance provided by Rose C. Hendrix, Sara Walker and Callie Sullivan. Florida Agricultural Experiment Stations Journal Series No. 7457.

REFERENCES

Bakry F, Rossignol L (1985). Analyse des capacites de callogenese et d'organogenese obtenues a partir de differents tissus de bananiers (Musa sp., Musacees). Fruits 40:697-708.

Bonga JM (1982). Vegetative propagation in relation to juvenility, maturity, and rejuvenation. In Bonga JM, Durzan DJ (eds): "Tissue Culture in Forestry." The Hague:Martinus Nijhoff/Dr. W. Junk, pp 387-412.

Bradley PM, El-Fiki D, Giles KL (1984). Polyamines and arginine affect somatic embryogenesis of Daucus carota. Plant Sci Lett 34:398-401.

Brown CL, Sommer HE (1982). Vegetative propagation of dicotyledonous trees. In Bonga JM, Durzan DJ (eds): "Tissue Culture in Forestry," The Hague: Martinus Nijhoff/Dr. W. Junk, pp 109-149.

Carron MP, Enjalric F (1985). Somatic embryogenesis from inner integument of the seed of Hevea brasiliensis (Kunth., Mull. Arg.). Comp. Rend. Ser. III 300:653-658.

Cohen D, Cooper PA (1982). Micropropagation of babaco - a Carica hybrid from Ecuador. In Fujiwara A (ed): "Plant Tissue Culture 1982," Tokyo: Japan Assoc Plant Tissue Culture, pp 743-744.

Cox EA, Stotzky G, Goos RD (1960). In vitro culture of Musa balbisiana Colla embryos. Nature 185:403-404.

Cronauer SS, Krikorian AD (1983). Somatic embryos from cultured tissues of triploid plantains (Musa 'ABB'). Plant Cell Rpt 2:289-291.

Damasco OP, Barba RC (1984). In vitro culture of Saba banana [Musa sp. cv. Saba (BBB)]. Phil Agr 67:351-358.

DeBruijne, E, DeLanghe E, Van Rijck R (1974). Action of hormones and embryoid formation in callus cultures of Carica papaya. Intern Symp Fytofarm Fytiat 26:637-645.

Eichholtz D, Robitaille, HA, Hasegawa PM (1979). Adventive embryony in apple. HortScience 14:699-700.

Esan EB (1973). A detailed study of adventive embryogenesis in Rutaceae. Ph.D. Dissertation, Riverside: Univ of California.

Esan EB, Murashige T (1972). Regulation of adventive embryogenesis in the Rutaceae. HortScience 7:208.

Evans DA, Sharp WR, Flick CE (1981). Growth and behavior of cell cultures: embryogenesis and organogenesis. In Thorpe TA (ed): "Plant Tissue Culture Methods and Applications in Agriculture," New York: Academic Press, pp 45-113.

Feirer RP, Mignon G, Litvay JD (1984). Arginine dicarboxylase and polyamines required for embryogenesis in the wild carrot. Science 223:1433-1434.

Frost HB, Soost RK (1968). Seed reproduction: development of gametes and embryos. In Reuther W, Batchelor LD, Webber HJ (eds): "The Citrus Industry, Vol. II," Berkeley: Univ of California Press, pp 290-324.

Halperin W (1967). Population density effects on embryogenesis in carrot cell cultures. Exp Cell Res 48:170-173.

Hwang, SC, Chen CL, Lin JC, Lin AL (1984). Cultivation of banana using plantlets from meristem culture. HortScience 19:231-233.

Khuspe SS, Hendre RR, Mascarenhas AF, Jagannathan V (1980). Utilization of tissue culture to isolate interspecific hybrids in Carica L. In Rao PS, Heble MR, Chadha MS (eds): "Plant Tissue Culture, Genetic Manipulation, and Somatic Hybridization of Plant Cells," Bombay: BARC, pp 198-205.

Krikorian AD, Cronauer SS (1984). In Sharp WR, Evans DA, Ammirato PV, Yamada Y (eds.): "Handbook of Plant Cell Culture, Vol. II. Crop Species," New York: Macmillan, pp 327-348.

Litz RE (1984a). In vitro responses of adventitious embryos of two polyembryonic _Eugenia_ species. HortScience 19:720-722.

Litz RE (1984b). _In vitro_ somatic embryogenesis from nucellar callus of monoembryonic _Mangifera indica_ L. HortScience 19:715-717.

Litz RE (1984c). _In vitro_ somatic embryogenesis from callus of jaboticaba, _Myrciaria cauliflora_. HortScience 19:62-64.

Litz RE (1984d). Papaya. In Sharp WR, Evans DA, Ammirato PV, Yamada Y (eds): "Handbook of Plant Cell Culture, Vol. II. Crop Species," New York: Macmillan, pp 349-368.

Litz RE (1985). Somatic embryogenesis in tropical fruit trees. In Henke RR, Hughes KW, Constantin MP, Hollaender A (eds): "Tissue Culture in Forestry and Agriculture," New York: Plenum Publishing Corp., pp 179-193.

Litz RE (1986). Tissue culture studies with _Theobroma cacao_. In Dimick, PS (ed): "Cacao Biotechnology," University Park: Penn State Univ, pp 111-120.

Litz, RE, Conover RA (1978a). _In vitro_ propagation of papaya. HortScience 13:241-242.

Litz, RE, Conover, RA (1978b). Recent advances in papaya tissue culture. Proc Fla State Hort Soc 91:181-184.

Litz, RE, Knight RJ, Gazit S (1982). Somatic embryos from cultured ovules of polyembryonic _Mangifera indica_ L. Plant Cell Rpt 1:264-266.

Litz, RE, Knight RJ Jr, Gazit S (1984). In vitro somatic embryogenesis from _Mangifera indica_ L. callus. Scientia Hort 22:233-240.

Litz, RE, Moore GA, Srinivasan C (1985). _In vitro_ systems for propagation and improvement of tropical fruits and palms. Hort Rev 7:157-200.

Litz RE, O'Hair SK, Conover, RA (1983). In vitro growth of _Carica papaya_ L. cotyledons. Scientia Hort 19:287-293.

Litz RE, Schaffer B (1986). Polyamines in adventitious and somatic embryogenesis in mango (_Mangifera indica_ L.). (in press)

Ma S, Shu C (1972). _In vitro_ formation of adventitious buds in banana shoot apex following decapitation (in Chinese). J Hort Sci (China) 18:135-142.

Mapes MO (1974). Tissue culture of bromeliads. Intern Plant Prop Comb Soc 23:47-55.

Mhatre M, Bapat VA, Rao PS (1985). Regeneration of plants from the culture of leaves and axillary buds in mulberry (_Morus indica_ L.). Plant Cell Rpt. 4:78-80.

Montague MJ, Armstrong TA, Jaworski, EG (1979). Polyamine metabolism in embryonic cells of _Daucus carota_ II. Changes in arginine decarboxylase activity. Plant Physiol 63:341-345.

Montague MJ, Koppenbrink JW, Jaworski EG (1978). Polyamine

metabolism in embryogenic cells of *Daucus carota*. I. Changes in intracellular content and rates of synthesis. Plant Physiol 62:430-433.

Mullins MG, Srinivasan C (1976). Somatic embryos and plantlets from an ancient clone of the grapevine (cv. Cabernet-Sauvignon) by apomixis in vitro. J Exp Bot 27:1022-1030.

Pijl L van (1934). Uber die Polyembryonie bei *Eugenia*. Rec Trav bot Neerl 31:113-187.

Rangan TS, Murashige T, Bitters WP (1968). *In vitro* initiation of nucellar embryos in monoembryonic *Citrus*. HortScience 3:226-227.

Rangaswamy NS (1982). Nucellus as an experimental system in basic and applied tissue culture research. In Rao AN (ed): "Tissue Culture of Economically Important Plants," Singapore:COSTED, pp 269-286.

Robles MJM (1978). Multiplication vegetative, *in vitro* des bourgeons axillaries de *Passiflora edulis* var. *flavicarpa* Degener et de *P. mollissima* Bailey. Fruits d'Outre Mer 33:693-699.

Scorza R, Janick J (1976). Tissue culture in *Passiflora*. Proc Trop Reg, Amer Soc Hort Sci 20:179-183.

Singh LB (1960). "The Mango," London: Leonard Hill.

Sondahl, MR, Sharp WR (1977). High frequency induction of somatic embryos in cultured leaf explants of *Coffea arabica* L. Z Pflanzenphysiol 81:395-408.

Srinivasa Rao NK, Chacko EK, Doreswamy R, Narayanaswamy N (1982). Induction of growth in explanted infloresence axis of banana. Curr Sci 51:666-667.

Stevenson FF (1956). The behavior of citrus tissues and embryos in vitro. PhD dissertation, Ann Arbor, Univ of Michigan.

Stover RH, Buddenhagen IW (1986) Improvement of the banana tree:polyploid, resistance to diseases and productivity. Fruits 40:175-191.

Sturrock TT (1961). A study of the effect of growth substances on fruit setting of the mango. PhD Dissertation, Gainesville: Univ of Florida.

Tisserat B, Murashige T (1977a). Probable identity of substances in *Citrus* that suppress asexual embryogenesis. In Vitro 13:785-789.

Tisserat B, Murashige T (1977b). Effects of ethephon, ethylene, and 2,4-dichlorophenoxyacetic acid on asexual embryogenesis *in vitro*. Plant Physiol 60:437-439.

Zatyko JM, Simon I, Szabo CS (1975). Induction of polyembryony in cultivated ovules of red currant. Plant Sci Lett 4:822-825.

Plant Tissue and Cell Culture, pages 419–432

APPLICATION OF TISSUE CULTURE TO VEGETABLE CROP IMPROVEMENT

Claire Doré

I.N.R.A., Station de Génétique et d'Amélioration des Plantes, Route de Saint-Cyr, 78000 Versailles, France

INTRODUCTION

Plant improvement consists in gathering or creating many gene combinations, then in selecting and fixing the most agronomically desirable ones and finally in releasing then through commercial varieties after seed multiplication.

This corresponds to first increasing genetic variability and then stabilizing it. To succeed, the plant breeder will require at various stages of the breeding programme the use of many "tools" involving among others, controlled hybridization, selection tests, experimental trials and statistics, quantitative genetics, simulation etc...

More recently, the progress of *in vitro* culture make it reasonable to consider tissue culture as a valuable supplementary tool. It is already available at the various levels of the breeding programme of many crop species such as certain vegetables.

Vegetable is a generic term describing edible plants grown in kitchengardens by amateurs as well as in large fields by industrial farmers. Various parts of the plant are harvested and devoted to human consumption. It probably includes more than one hundred different species of highly various economical importance, belonging to many botanical families and without any biological unity. Potato in this survey will be excluded as it is generally considered as an industrial crop and not a vegetable.

I will try in this paper to establish how far tissue culture is now integrated in vegetable improvement and evaluate the status of practical applications with a few typical examples. (An exhaustive description of the status would be too long and tedious). Its means that theoritical works on vegetables will not be mentioned and also that I will only deal with examples where tissue culture techniques can lead to regeneration of entire plants.

I. INCREASING VARIABILITY

A. Interspecific hybridization

Introduction of valuable new characters is a real necessity for many cultivated vegetables. Conventional methods have provided many desirable interspecific hybrids when crossability has no deep barriers. Embryo culture and protoplast fusion can be very helpfull to overcome such barriers in many cases.

Embryo culture requires more or less sophisticated media depending on the stage where the hybrid embryo aborts. It is often necessary to use this technique to obtain F_1 and BC_1 plants.

Protoplast culture requires conditioning techniques related to microbiology and the control of morphogenesis *in vitro*. Therefore it still remains limited to a few species.

Among many traits, increased disease and pest resistances are probably among the most important breeding objectives for vegetables where quality of the harvested product is so essential. This goal has been achieved for several vegetable species. Interspecific hybrids between Common Bean and *P. acutifolius* led to several Great Northern type Bean cultivars tolerant to common blight (Honma, 1956). Likewise, two alleles Tm_2 and Tm_{22} controlling resistance to tobacco mosaic virus (TMV) were introduced from *Lycopersicon peruvianum* into Tomato using embryo culture techniques and Tm_{22} is currently present in all Tomato cultivars grown in plastic tunnels around the world (Laterrot, 1973). More recently Blackgram (*Vigna mungo*) and Greengram (*Vigna radiata*) were hybridized and embryos cultivated *in vitro*. F_1 and segregating F_2 showed interesting morphological variations some of which are associated with Bean Yellow Mosaïc Virus resistance (Gosal and Bajaj, 1983).

Another interesting possibility concerns the cumulation of several disease resistances. Transfer of cucumber mosaic virus (CMV), watermelon mosaic virus (WMV) and zucchini yellow mosaic virus (ZYMV) resistances present in _Cucurbita ecuadorensis_ has been successfully achieved in Zucchini (_C. pepo_) by embryo culture (Dumas de Vaulx and Pitrat, 1980). This very desirable trait can then be easily introduced in other cultivated cucurbits.

Embryo culture is also at the origin of a new heading type leafy vegetable : Hakuran (_Brassica napus_) was synthetized from Chinese Cabbage (_B. campestris_ var _pekinensis_) with Cabbage (_B. oleracea_) and developped in western Japan since 1966 (Nishi, 1980).

On the other hand the Cabbage nucleus could be introduced into the cytoplasm of male sterile Ogura Radish by repeated backcrosses using embryo culture (Bannerot et al., 1974). This cytoplasmic male sterility could be very useful in _Brassica_ crop improvement for commercial production of F_1 hybrid seeds provided the chlorophyll deficiency associated with this cytoplasm could be corrected. In fusing Rapeseed (_B. napus_) protoplasts, Pelletier et al. (1983) successfully obtained cybrids with a new male sterile cytoplasm relieved of the undesirable trait. This kind of process is now being developed in several countries for Cabbage improvement.

Likewise creation of novel organelle-nuclear combinations by somatic hybridization between Tomato and _Lycopersicon pennellii_, as established by O'Connell and Hanson (1985), represents a very interesting possible towards the research of new cytoplasmic male sterilities for such an economically important crop.

Protoplast fusion may also be an interesting alternative when sexual crosses attempts remain unsuccessful. Such is the case of crosses between Eggplant and the wild species _Solanum sisymbrifolium_. The successful production of somatic hybrids between these two species, recently achieved by Gleddie et al. 1986, is the first step of a long term programme for transferring nematode and mite resistances into eggplant.

Finally I would mention that reproducible totipotency of protoplasts has been demonstrated for several vegetables such as asparagus, Broccoli, Cabbage, Capsicum, Carrot, Hakuran. This result represents the key for future integration of this technique in breeding scheme.

B. Somaclonal variation and in vitro selection

The callus stage is a period of indifferentiated growth during which extensive genetic variability may occur. The degree of variability is generally related to the duration of the callus stage. If the regeneration process is possible to achieve, it is interesting to try to take advantage of this phenomenon, otherwise so undesirable for obtaining cloning.

Somaclonal variation has been proposed by Novak (1984) as a new breeding system for Garlic. Garlic is an obligate apomictic crop without any possibility of genetic improvement through sexual recombination. Polyploids and aneuploids were very frequent in longterm callus cultures and considerable phenotypic variations could be observed in 40 % of the regenerants. Vegetative propagation offers an interesting possibility for stabilizing the worthwhile variants.

Somaclonal variation has also been studied in Tomato by Evans and Sharp (1984) to establish its genetic basis and for evaluation for breeding purposes. They obtained among regenerants many chromosome number variations, several single gene mutations with a very high frequency, and also chloroplast DNA mutations. Several somaclones have been identified which present a few agriculturally important traits such as higher pigment.

Somaclonal variation and in vitro selection are associated in several studies dealing with development of tolerance to mineral stress or herbicide and disease resistance.

The fact that Nyman et al. (1983) succesfully achieved the production of plantlets of Taro (Colocasia esculenta) from salt-tolerant callus is a very encouraging step towards the extension culture of this very important crop in brakish soils.

Celery is the species where genetic instability is being studied by several groups (Orton 1985, Williams and Collin, 1976) and Merrick and Collin (1981) were able to select an Asulam resistant mutant.

Several attempts concerning the use of a pathogenic toxin as a selection system at cell level are being made but so far very few results dealing with vegetables have been published. Indeed many problems arise such as obtention of consistent resistances, persistence of resistance selected in vitro at cell level, in regenerated plants faced to pathogens (Scala et al. 1984, Jullien 1985, Alichio et al. 1984).

C. Haploidization

Haploidy, a special reproduction system which leads to a plantlet from a gametophyte, can bring new genetic variability immediatly fixed at homozygous level after chromosomic stock doubling. Haploid plants can be obtained in particular from in vitro anther culture. These plants are of interest for various reasons, as illustrated by a few examples concerning vegetables.

Haploidization is a very useful method for rapid fixation of characters and creation of new varieties :

- The first application was in Versailles, in 1973, where the use of Asparagus anther culture allowed us to consider it as a source of homozygous genitors of F_1 hybrids. In this dioecious and perennial species haploidy represents the only possible way to obtain homozygous material in a reasonnable time.

In Asparagus the male plants are heterogametic and through anther culture we could obtain supermale plants i.e. male plants homozygous for all characters. This kind of plant used as male plant give an all male F_1 progeny instead of a mixed one. All male varieties have been sought for their yield and earliness.

We have developed this breeding model and many F_1 male hybrids are in experimental trials for evaluation. In our fields some of them appear to be significantly superior in terms of productivity and earliness but cumulation of these desirable traits with a large diameter of spear is sometimes a difficult goal to achieve (Thévenin and Doré, 1976 ; Doré, 1979). This technique has been since applied by several scientists in particular by Falavigna in Italy. His last results confirm the interest of anther culture in Asparagus breeding (Falavigna, 1985).

- With Eggplant the use of high temperature has made it possible to improve the yield of haploids (Dumas de Vaulx and Chambonnet, 1982). The haplomethod is now integrated in a recurrent selection scheme at INRA in France : in each cycle a relatively small number of randomly selected DH lines are tested for rapid evaluation at the homozygous level. When a suitable quality level is reached, a large number of DH lines can be obtained for the creation of new varieties. These varieties are pure lines or F_1 hybrids selected after a reciprocal recurrent scheme (Dumas de Vaulx and Doré, 1984).

In Brassica, Keller has developed a high temperature procedure to significantly enhance embryogenesis from pollen, and pollen plants could be obtained for instance with high efficiency with B. oleracea sp italica, Broccoli cv. Green Mountain (Keller and Armstrong 1983). Several other groups have also described the obtention of large quantities of pollen plants from different subspecies of B. oleracea such as Brussel sprouts, Cabbage, Broccoli (Ockendon, 1984 and 1985, Lelu and Bollon, 1985, Orton and Browers, 1985, Chiang et al. 1985, Doré, 1986) or chinese Cabbage (B. campestris var. pekinensis) (Deng et al., 1982) and proposed to use them in breeding programmes. Very few results concerning their evaluation are yet published.

Nevertheless large differences between genotypes in their responsiveness to anther culture might be a limitating factor to its easy incorporation in breeding schemes for routine production of inbred lines. But it looks very promising for special objectives such as rapid achievement of fixation of resistance to clubroot in Cabbage (Chiang et al. 1985).

When responsiveness to anther culture is relatively high with any genotype such as for Capsicum (Dumas de Vaulx et al. 1981, Vagera and Havranek, 1985), DH lines can be very helpful for analysis of complex inherited disease resistances as there are no allelic interactions. Furthermore, it becomes possible to submit each genotype to a large range of pathogens successively. This kind of approach is now being used in Montfavet (France) (Dumas de Vaulx and Pochard, 1986).

II. STABILIZING VARIABILITY

A. Vegetative multiplication

The interest of *in vitro* micropropagation has already been demonstrated in many circumstances. The control of proliferation and the obtention of true to type plants are essential prerequisites when plant breeders need to establish true clones. This has been achieved with numerous vegetables.

The very first application was the use of meristem culture for virus elimination (Morel and Martin, 1952). The possibility to cure virus-diseased plants and then to multiply the virus-free shoots *in vitro* away from risks of virus recontamination is of considerable interest for many vegetable crops, and vegetatively propagated ones in particular.

Mother stock plants are cured of virus by meristem culture and then either multiplied *in vitro* at industrial scale or by horticultural methods. This has been done for clonal varieties of Strawberry (Boxus, 1976), Garlic (Quiot et al., 1972, Havranek, 1973, Bhojwani et al., 1982), Shallot, Yam (Mantell and Hague, 1979), Sweet Potato (Frison and N_g, 1981), Artichoke (Pécaut and Dumas de Vaulx, 1983), Chinese Artichoke (Boccon-Gibod, 1980) grown in several countries. A sanitary certification system has been established for virus-free plant sale for Strawberry, Garlic and Shallot in a few EEC countries. The quality and the agronomical performance of virus-free plants represent an important economical increase in value.

The micropropagation techniques which are now available are employed by plant breeders in many occasions :

- as a useful safeguard against plant loss ;

- for rapid introduction of new interesting material in recurrent selection (Brussel Chicory, Doré 1984) or elite genitors such as genic male-sterile plants of Lettuce (Doré, 1984) ;

- as an efficient and reliable system of selection of characters whose expression is affected by the environnement. Such is the case of curd quality in Cauliflower as demonstrated by Crisp and Gray, 1979.

- as an economic and easy way of maintaining of genotypes after virus elimination if necessary (Cauliflower, Grout and Crisp, 1977).

Finally it represents an interesting possibility for seed production by cloning the parents of the hybrids. This aspect has been developed for hybrid seed production of Broccoli (Anderson and Carstens, 1977), of Cabbage (Leike et al.) and of Leek (Doré and Schweisguth, 1980). For Asparagus *in vitro* vegetative multiplication of two heterozygous parents made possible the creation a new type of variety : United States with UC 157 and Jersey Giant (Benson and Takatori, 1980 ; Ellison, 1984), France with Aneto, Desto, Cito and Stéline (Doré, 1977) and Italy with Diego (Falavigna et al. 1982) have thus been able to release clonal hybrid varieties.

Finally I would like to mention somatic embryogenesis which is considered by many scientists to be the micropropagation method of the future. Feasibity of the technique, conformity and agronomical behaviour of plants obtained are now being investigated with several vegetables and it is really too early to assess the practical interest of this very powerful system of embryo production.

B. *In vitro storage*

Slow growth process at low temperature (generally a few degrees above 0°C) represents an easy way to maintain *in vitro* plantlets for a few months without any special care.

Many vegetable breeder do use this method to store collection of genotypes, elite genitors within selection process or parents of hybrid varieties. It prevents them to encumber glasshouses at winter time or to risk pest and disease damages.

Long term storage requires special procedures such as cryopreservation but standard procedures should give some success with all species as Withers (1984) asserts it and many vegetables are concerned.

Finally, I should mention in vitro gene banks,(there is one for instance in CYP for Sweet Potato), which are very used to distribute germplasm. They correspond to one aspect of plant resources management where more and more in vitro technology could be used as proposed IBPGR (1984).

CONCLUSION

This rapid survey though obviously incomplete, establishes that many culture techniques are now available and are being used for vegetable breeding purposes and the feasibility of many of them could already be demonstrated.

The number of species mentionned here shows that vegetable studies are very dynamic and that one should probably consider vegetables as pionneer plants. Indeed they are not only pionneers for the application of tissue culture techniques described here, but also for other very recent ones which will probably also have an impact in plant improvement :

- The very first results in cell suspension cryopreservation were achieved with Sweet Potato (Latta, 1971).

- Carrot appears to be an excellent model plant for investigating the development of synthetic seeds (Kitto and Janick, 1985).

- Asparagus is the very first Monocot where genetic transformation of cell by *Agrobacterium tumefaciens* could be obtained (Hernalsteens et al., 1984).

ACKNOWLEDGEMENTS

I thank J.C. Lescure and F. Charlot for excellent technical assistance particularly appreciated during the redaction of this text. I thank also C. Mandon for bibliographical investigation, Dr. Marc Tepfer for the correction of my English and J. Wencek for typing the manuscript.

REFERENCES

Alichio R, Antonioli C, Palenzona D (1984). Karyotypic variability in plants of *Solanum melongena* regenerated from callus grown in presence of culture filtrate of *Verticillium dahliae*, T.A.G., 67: 267-271.

Anderson W, Carstens J (1977). Tissue culture propagation of Broccoli for use of in F_1 hybrid seed production. J. Amer. Soc. Hort. Sci. 102, 1: 69-73.

Bannerot H, Boulidard L, Cauderon Y, Tempé J (1974). Transfer of cytoplasmic male sterility from *Raphanus sativus* to *Brassica oleracea*. Proc. Euc. Meet. Cruciferae, 52-54.

Benson B, Takatori F (1980). Foundation Asparagus Seed program and the release of UC 157 parental clones to the seed industry. Asparagus Research, 1979-1980. Univ. of Calif., Davis.

Bhojwani S, Cohen D, Fry P (1982). Production of virus-free Garlic and field performance of micropropagated plants. Scientia Hort. 18: 39-43.

Boccon-Gibod J (1980). Regeneration du Crosne du Japon par culture de méristèmes. C.R. Réunion Eucarpia Versailles: Applic. de la cult. *in vitro* à l'amélioration des plantes potagères : 31-37.

Boxus P (1976). Rapid production of virus free Strawberry by *in vitro* culture. Acta Hort. 66, 35-38.

Chiang S, Frechette S, Kuo C, Chong C, Delafield S (1985). Embryogenesis and haploid plant production from anther culture of Cabbage. Can. J. Pl. Sci. 65: 1033-1037.

Crisp P, Gray A (1979). Successful selection for curd quality in Cauliflower, using tissue culture. Hort. Res. 19 : 49-53.

Deng L, Zeng Y, Guo Y, Teng Y, Wang W (1982). Induction of pollen plants in Chinese Cabbage. Acta Hortic. Sin. 9,2,: 37-42.

Doré C, 1977. *In vitro* techniques as an efficient tool in Asparagus breeding. Acta Hortic. 78: 89-93.

Doré C (1979). Anther culture as a way towards F_1 hybrids. Proc. 5th Int. Asparagus Symp. Geisenheim, 87-90.

Doré C, Schweisguth B (1980). Multiplication végétative du Poireau pour la production de semences. C.R. Réunion Eucarpia Versailles "Application de la culture *in vitro* à l'amélioration des plantes potagères, 54-57.

Doré C (1984). Multiplication végétative *in vitro* de la Chicorée (*Cichorium intybus*) au service de la sélection. C.R. Réunion Eucarpia "Légumes à feuilles", Versailles, 123-125.

Doré C, Michel H (1984). Multiplication végétative *in vitro* de la Laitue à partir de plantes en floraison. C.R. Réunion "Légumes à feuilles" Versailles, 10-15.

Doré C (1986). Evaluation du niveau de ploïdie des plantes d'une population de Chou de Bruxelles d'origine pollinique. Agronomie 6, n° 9, (under press).

Dumas de Vaulx R, Pitrat M (1980). Application de la culture d'embryons immatures à la réalisation de l'hybridation interspécifique entre *Cucurbita pepo* et *C. ecuadorensis* F_1 et BC_1. C.R. Réunion Eucarpia Versailles : Applications de la culture *in vitro* à l'amélioration des plantes potagères, 126-131.

Dumas de Vaulx R, Chambonnet D, Pochard E (1981). Culture *in vitro* d'anthères de Piment (*Capsicum annuum*). Agronomie 1, 10, 859-864.

Dumas de Vaulx R, Chambonnet D (1982). Culture *in vitro* d'anthères d'Aubergine (*Solanum melongena*). Agronomie 2, 10, 983-988.

Dumas de Vaulx R, Doré C (1985). Applications of *in vitro* techniques to the breeding of some vegetable species. Pro CEC Seminar Braunschweig. In : *In vitro* techniques : Propagation and long term storage, Schäfer-Menuhr (ed), 15-22.

Dumas de Vaulx R, Pochard E (1986). Parthogenèse et androgenèse chez le Piment : rôle actuel dans les programmes de sélection. Le Sélectionneur Français, n° 36 (under press).

Ellison H (1984). Release of Jersey Giant and Greenwich. Asparagus Research Newsletter 2, 1, p. 23.

Evans A (1984). Genetic basis of somaclonal variation in Tomato. Proc. Int. Symp. Pl. Tiss. and Cell Cult. Application to Crop Improvement, Olomouc, 259-265.

Falavigna A, Casali P, Soresi G (1982). Diego : first Italian hybrid of Asparagus from heterozygous clones. Genetica Agraria, 162-163.

Falavigna A, Tacconi M, Casali P (1985). Asparagus breeding through anther culture : eight years of activity. Proc. 6th Int. Asparagus Symp. Guelph, 31-40.

Frison E, Ng S (1981). Elimination of Sweet Potato Virus disease Agents by meristem tip culture. Trop. Pest Management, 27, 4, 452-454.

Gleddie S, Keller W, Scherfield G (1986). Production and characterization of somatic hybrids between Solanum melongena and S. sisymbrifolium T.A.G. 71, 613-621.

Gosal S, Bajaj Y (1983). Interspecific hybridization between Vigna mungo and Vigna radiata through embryo culture. Euphytica, 32, 129-137.

Grout, Crisp P (1977). Practical aspects of the propagation of Cauliflower by meristem culture. Acta Hort. 78, 289-296.

Havranek P (1973). Occurence of viruses in the genus Allium and virus free clones of common Garlic. Proc 7th Conf. Szech. Pl. Virol, 133-138.

Hernalsteens J.P, Thia-Toong L, Schell L, Van Montagu M (1984). An Agrobacterium transformed cell culture from the Monocot Asparagus officinalis. Embo Journ. 3, 13, 3039-3041.

Honma S (1956). A bean interspecific hybrid. J. Hered., 47, 217-220.

IBPGR (1984). The potential for using in vitro techniques for germplasm collection. IBPGR Publication AGP IBPGR/83/108. Int. Board for Pl. Gen. Resources, Rome.

Jullien (1985). Toxicity of fusaric acid on mesophyll cell cultures of Asparagus officinalis : implications for selection to resistance. Proc. 6th Asparagus Symp. Guelph, 238-245.

Keller W, Armstrong K (1983). Production of haploids via anther culture in Brassica oleracea var. italica. Euphytica 32, 151-159.

Kitto S, Janick J (1985). Production of synthetic seeds by encapsulating asexual embryos of Carrot. J. Amer. Soc. Hort. Sci, 110, 277-282.

Laterrot H (1973). Résistance de la Tomate au virus de la mosaïque du Tabac. Ann. Amel. Pl. 23, 4, 287-313.

Latta R (1971). Preservation of suspension cultures of plant cells by freezing. Can. J. Bot. 49, 1253-1254.

Leike H, Jahr W, Köhler H (1984). Micropropagation of hybrid lines in vegetable breeding. Proc. Int. Symp. Pl. Tiss. and Cell Cult. Application to Crop improvement. Olomouc, 432-438.

Lelu M-A, Bollon H, (1985). Obtention d'haploïdes par culture d'anthères de Brassica oleracea var capitata et var gemmifera. C.R. Acad. Sci. 300, 2, 71-76.

Mantell S, Hague S (1979). Disease-free yams : their production, maintenance and performance. Yam virus Bull. CARDI, 2, 22p.

Merrick M, Collin H (1981). Selection for asulam resistance in tissue cultures of Celery. Pl. Sci. Lett. 20, 291-296.

Morel G, Martin C (1952). Guérison de Dahlia atteints d'une maladie à virus. C.R. Acad. Sci. 235, 1324-1325.

Nishi S (1980). Differentiation of Brassica Crops in Asia and the breeding of Hakuran, a newly synthesized leafy vegetable . In : Brassica Crops and wild allies. Tsunoda (ed) 132-150.

Novak F (1984). Somaclonal variation in Garlic Tissue Culture as a new breeding system. 3rd Allium Symp. Eucarpia Wageningen, 39-43.

Nyman L, Gonzales C, Arditti J (1983). In vitro selection for salt tolerance of Taro (Colocasia esculenta). Ann. Bot. 51, 229-236.

Ockendon D (1984). Anther culture of Brussel Sprouts. Ann. Appl. Biol. 105, 285-291.

Ockendon D (1985). Utilization of anther culture in breeding Brussel Sprouts. Int. Symp. Eucarpia Genetic manipulation in plant breeding- Horn (ed) -on printing-

O'Connell M, Hanson M, (1985). Somatic hybridization between Lycopersicon esculentum and L. pennellii. T.A.G. 70, 1-12.

Orton T (1985). Genetic instability during embryogenic cloning of Celery. Pl. Cell Tissue Organ. Cult. 4, 159-169.

Orton T., Browers M (1985). Segregation of genetic markers among plants regenerated from cultured anthers of Broccoli. T.A.G., 69, 637-643.

Pécaut P., Dumas de Vaulx R, Lot H (1983). Virus free clones of globe Artichoke obtained after in vitro propagation. Acta Horticulturae, 131, 303-309.

Pelletier G, Primard C, Vedel F, Chetrit P, Remy R, Rousselle P, Renard M (1983). Intergeneric cytoplasmic hybridization in Cruciferae by protoplast fusion. M.G.G. 191, 244-250.

Quiot J-B, Messiaen C-M, Marrou J, Leroux J.P (1972). Régénération par culture de méristèmes de clones d'Aïl infectés de façon chronique par le virus de la Mosaïque de l'Aïl. Act. 3è Cong. U.P.M, Oeiras (Portugal), 429-433.

Scala A, Bettini P, Binatti M, Bogani P, Pellegrini G, Tognoni F (1984) In vitro analysis of the tomato Fusarium oxysporum system and selection experiments. Int. Symp. Pl. Tissue and Cell Cult. Application to Crop Improvement, Olomouc, 361-362.

Thevenin L, Doré C (1976). L'amélioration de l'Asperge et son atout majeur, la culture in vitro. Ann. Amél. Pl., 26, 4, 655-674.

Vagera J, Havranek P (1985). In vitro production of androgenesis in Capsicum annuum and its genetic aspects. Biol. Plant 27, 1, 10-21.

Williams L, Collin H (1976). Embryogenesis and plantlet formation in tissue cultures of Celery. Ann. Bot. 40, 325-332.

Withers L (1984). Long term storage of in vitro cultures Proc. C.E.C. Seminar Braunschweig. In : In vitro techniques : Propagation and long term storage, Schäfer-Menuhr (ed), 137-148.

Section VII.

Technology Development

Plant Tissue and Cell Culture, pages 435–446

TRANSFORMATION OF HIGHER PLANT CELLS BY INTRANUCLEAR MICROINJECTION

Brian L. Miki[1], V.N. Iyer[2] and Terry J. Reich[2]

[1]Plant Research Centre, Agriculture Canada, Ottawa, Ontario, Canada, K1A 0C6
[2]Biology Department, Carleton University, Ottawa, Ontario, Canada, K1S 5B6

INTRODUCTION

Several techniques have been developed for introducing cloned genes into plant cells. Artificial delivery systems such as direct DNA uptake (Krens et al. 1982; Paszkowski et al. 1984), electroporation (Fromm et al. 1986) and fusion with liposomes (Deshayes et al. 1985) have been substantiated with protoplasts only. For cells with intact cell walls (Pollock et al. 1985) and tissue explants (Horsch et al. 1985) the biological delivery system of virulent _Agrobacterium_ species has been essential; however, the dependence on effective but non-destructive plant-microbe interactions has also restricted the use of this approach from many plant species, most notably the cereals. Although artificial delivery systems are being developed for intact cells, their effectiveness has yet to be demonstrated. All of these methods depend on selection systems for the recovery of transformants. Several selectable marker genes have been constructed for this purpose (Fraley et al. 1983; Herrera-Estrella et al. 1983; DeBlock et al. 1984; Waldron et al. 1985). Their effectiveness varies with plant species, tissue source and culture conditions. The adaptation of these technologies for the transformation of plant species that have not been well studied _in vitro_, such as crop species, may be very difficult.

Intranuclear microinjection is a direct physical approach that overcomes many of these biological obstacles

(Reich et al. 1986c); however, this technology is just emerging in the plant sciences. Whereas the genetic transformation of animals (Brinster et al. 1981; Hammer et al. 1985), frogs (Rusconi and Schaffner, 1981) and insects (Rubin and Spradling, 1982) by the microinjection of embryos has been well established, only recently has the transformation of plant protoplasts by microinjection been verified (Crossway et al. 1986; Reich et al. 1986c). In this report we will describe aspects of this technology and speculate on its relevance to the general areas of plant cell and molecular biology.

BACKGROUND

Microinjection is the most precise way for depositing substances into specific cells, yet early attempts to employ the technique were seldom sustained. The microinjection of fluorescent dyes into differentiated cells contributed to the understanding of intercellular transport in plants (Goodwin and Erwee, 1985). The transfer of microorganisms into plasmolyzed cells was demonstrated (Bradley, 1979). Techniques for the microinjection of virus particles into single tobacco cells from suspension culture were developed and used to study virus infection (Nims et al. 1967a,b; Halliwell and Gazaway, 1975). From these studies, it was clear that special methods were needed for holding cells and penetrating their cell walls without inflicting irreversible damage. The early solutions to these problems supported the microinjection of only small numbers of cells (Nims et al. 1967a,b). For genetic transformation it was assumed that the microinjection of large numbers of proliferating cells was a minimum requirement.

Protoplasts were an obvious source of dividing single cells without cell walls. A variety of methods for holding, injecting and culture of protoplasts were developed (Griesbach, 1983; Steinbiss and Stabel, 1983; Lawrence and Davies, 1985; Morikawa and Yamada, 1985; Crossway et al. 1986; Reich et al. 1986a,b,c). Only those procedures that ensured delivery of DNA into the nucleus yielded efficient transformation (Crossway et al. 1986; Reich et al. 1986a,b,c).

MICROINJECTION TECHNIQUES

For transformation, microinjection systems are designed for speed and operational simplicity because these factors govern the total number of cells that can be manipulated. In each experiment it is now possible to microinject 150-300 protoplasts intranuclearly (Reich et al. 1986a). Since transformation frequencies of 15-66% have been demonstrated (Table 1), it would be reasonable to expect that 22-180 would express the microinjected gene; however, a plating efficiency of at least 5% would be necessary to recover one colony. Microinjection systems that are less efficient would be restricted in application and could be used with only well developed protoplast systems such as tobacco (*Nicotiana tabacum*) mesophyll protoplasts.

To facilitate rapid injection rates, protoplasts with partially reformed cell walls can be attached to glass coverslips with poly-L-lysine without damaging the cells (Steinbiss and Stabel, 1983; Reich et al. 1986a). After one day in culture the cell walls of protoplasts are firm enough to provide support and protection yet they are easily penetrated with micropipettes. At the same time DNA synthesis has reinitiated in alfalfa (*Medicago sativa*) protoplasts (Holbrook et al. 1985) but it is not known if this influenced DNA integration. By means of reference markings on the coverslip, such as microscopic coded grids, the location of each microinjected protoplast can be recorded with a video camera (Reich et al. 1986a). In this way the operator is free to move from one microinjected cell to the next one of choice without interuption. At the end of an experiment or the next day, the microinjected cells can be segregated and transferred to optimal conditions for low density cell culture. For alfalfa protoplasts, the double filter paper technique (Horsch and Jones, 1980) has been an extremely efficient method for rescuing calli from individual protoplasts (Reich et al. 1986c). This approach has yielded large numbers of transformed lines of alfalfa and oilseed rape (*Brassica napus*) (Table 1 and unpublished data).

An alternative method employed a blunt holding pipette for holding each protoplast and segregating it directly into a microdrop culture after each microinjection

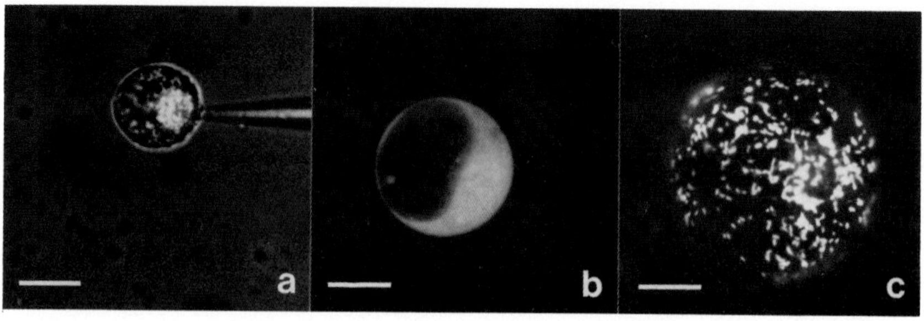

Figure 1. Micrographs of alfalfa protoplasts showing selective staining of the nuclear (a), cytoplasm (b) and mitochondria (c). Protoplasts nuclei were partially stained with Hoechst 33258 and examined by fluorescence microscopy with Leitz filter block A combined with a low level of bright-field illumination to reveal unstained portions of the protoplast and the position of the micropipette (a). Reproduced from Reich et al (1986c). The cytoplasm was selectively stained with fluorescein diacetate and examined by fluorescence microscopy using Leitz filter block I2 (b). Reproduced from Reich et al. (1986b). The mitochondria were selectively stained with $DiOC_2$ (3,3'-diethyloxadicarbocyanine iodide) and examined with Leitz filter block N_2. Reproduced from Reich et al. (1986b). Bar=25μm.

(Crossway et al. 1986). This yielded smaller numbers of tobacco transformants (Table 1). Holding pipettes or embedding in gels (Lawrence and Davies, 1985) may become essential for cell types that do not readily attach to coverslips with agents that bind to their surfaces.

INTRANUCLEAR MICROINJECTION

Studies with somatic animal cells (Capecchi, 1980) and embryos (Brinster et al. 1985) revealed the need for intranuclear delivery of DNA for transformation by microinjection. This was also confirmed with tobacco mesophyll protoplasts (Crossway et al. 1986). Generally, the nucleus in plant cells has been extremely difficult to resolve with conventional light microscope optics (Reich et al. 1986b). With protoplasts from several species this can be improved considerably by differential interference contrast microscopy (Gordon-Kamm and Bushnell, 1985). Alternatively, partial staining with fluorescent dyes and partial fluorescence microscopy have been very effective (Fig. 1a, Reich et al. 1986b). DNA-specific dyes such as Hoechst 33258 and DAPI (4',6-diamidino-2-phenylindole) specifically and transiently stained nuclei in permealized plant cells (Meadows and Potrykus, 1981; Reich et al. 1986b). Intranuclear microinjection can be verified by monitoring the disturbance in the fluorescence as the solution is ejected from the pipette tip. The fluorescence disappears within a day and no effect on DNA synthesis, mitosis or sustained cell divisions has been observed (Reich et al. 1986b).

Other non-toxic fluorescent dyes stain the cytoplasm and mitochondria selectively (Figs. 1b and 1c; Reich et al. 1986b). These may be useful for monitoring cytoplasmic microinjections and organelle transplantations.

GENETIC TRANSFORMATION

Transformation by intranuclear microinjection of protoplast-derived cells is a recent accomplishment. In all cases, it was attained without selection and with extremely high transformation frequencies, ranging from 14% to 66% (Table 1). Since large numbers of transformants have been recovered with various plasmid types and different plant species it is fair to conclude that this approach is very efficient. The methodology ensures the delivery of 10^1-10^3 copies of plasmid DNA to the nucleus of specific cells. The high transformation frequencies that follow support the hypothesis that delivery of DNA to the nucleus is the key factor in all of the transformation strategies for plants. Other factors such as the state of differentiation or the metabolic conditions of the recipient cells may also be important but they have not yet been examined. Due to the unique experimental capabilities offered by microinjection, it may become useful in the study of detailed events associated with fundamental genetic processes such as DNA replication, integration, recombination or transposition.

The mechanism by which microinjected plasmid DNA is integrated into the plant chromosome is poorly understood.

TABLE 1. Transformation of Protoplasts by Intranuclear Microinjection of Plasmid DNA

Species	Plasmids injected	No. of experiments	No. of protoplasts injected	No. of calli recovered	No. of calli transformed	Transformation frequency	References
Nicotiana tabacum	pCGN561[1]	5	113	56	8	14%	Crossway et al. (1986)
Medicago sativa	pTiC58[2]	3	523	65	17	26%	Reich et al. (1986c)
	pMON120[3]	3	813	116	17	15%	Reich et al. (1986c)
	pMON8015[4]	3	492	90	18	20%	Reich et al. (1986c)
Brassica napus		1	42	6	4	66%	
	pMON38[5]	1	107	21	12	57%	Reich et al. (unpublished)

[1] pCGN561 is a 29 kb plasmid that does not carry genes for expression in plants but possess T-DNA borders (Crossway et al. (1986)).

[2] pTiC58 is a wild-type nopaline plasmid.

[3] pMON120 is an 8 kb intermediate vector carrying a functional T-DNA right border and *nos* gene. (Fraley et al. (1983)).

[4] pMON8015 is a 9 kb vector carrying a *nos* gene and chimeric *nptII* gene but lacks functional T-DNA borders (Fraley et al. (1985)).

[5] pMON38 is a 6.7 kb plasmid carrying a *nos* gene and T-DNA right border. This plasmid was linearized with a restriction endonuclease prior to microinjection.

Integration appears to occur before cell division since subcloning of transformed lines through protoplasts did not uncover untransformed cells (Reich et al., unpublished data). The presence of T-DNA borders on the plasmids and the coinjection of plasmids carrying the vir genes did not enhance the transformation frequencies (Table 1). Among the transformants examined utilization of the T-DNA borders could not be detected following microinjection of Ti plasmids (Reich et al. 1986c) or smaller vectors carrying T-DNA borders (Crossway et al. 1986). These data and similar observations on the integration of plasmids delivered by direct DNA uptake (Krens et al. 1985) clearly showed that the mechanisms governing transformation by A. tumefaciens do not operate with artificial methods of transformation.

The structure of the integrated DNA is very dependent on the method of delivery. On direct DNA uptake, plasmid DNA is highly rearranged and the integration pattern is very complex (Krens et al. 1985; Czernilofsky et al. 1986). Microinjected plasmid DNA does not seem to be rearranged; however, limited, nonrandom changes in specific Ti plasmid sequences have been found (Reich et al. 1986c). At this time transformation by A. tumefaciens seems to yield the simplest and most predictable pattern of integration.

APPLICATION

The immediate contributions that can be made to the field of plant genetic manipulation by microinjection technology are partly illustrated by the experiments shown in Table 1. Alfalfa is difficult to infect with A. tumefaciens (Mariotti et al. 1984) and oilseed rape exhibits such a strong hypersensitive response to the bacteria that transformants are difficult to recover. Although conditions have been developed to overcome these problems (Deak et al. 1986; Holbrook et al. 1985b), transformation of these same species by microinjection was quickly achieved in only a few experiments because it bypassed the dependence on effective plant-microbe interactions. Furthermore, the delivery of DNA directly to the nucleus eliminated the need to determine the optimal electrochemical conditions for achieving DNA uptake. The obvious extension of this approach is the transformation of the cereals. To provide a major advantage over other methods for cereal protoplast transformation (Lorz et al. 1985; Fromm et al. 1986), it is important to demonstrate the microinjection of regenerable cells with intact cell walls. This would provide transformation technology for cereal species which cannot yet be regenerated from protoplasts. Preliminary experiments indicate that the microinjection of embryogenic microspores of oilseed rape may provide a model for such manipulations (Pechan and Keller, unpublished results). The major restriction on the application of microinjection for transformation remains the limited number of cell culture systems that support efficient recovery of plants from single cells or possibly small cell aggregates.

The recovery of large numbers of transformants clonally and without selection by microinjection is very significant for many plant species. Alfalfa calli from suspension culture-derived protoplasts were inherently tolerant to kanamycin, geneticin, hygromycin, chloramphenicol and methotrexate. The selectable marker genes commonly used for plants (Fraley et al. 1983; Herrera-Estrella et al. 1983; Waldron et al. 1985) were therefore inapplicable. Without good selection systems direct DNA uptake and A. tumefaciens - mediated transformtion can be difficult to adapt in practice. Furthermore the expression of the gene being studied by transformation may impart a growth disadvantage to the cell. This can be avoided following microinjection, by maintaining each manipulated cell as an independent clone from the outset.

Microinjection technology for plant cells has advanced to a level of sophistication that should allow its application in diverse areas of cell biology and genetics. Proteins or their antibodies can be introduced into specific cell compartments to study post-synthetic processing, intracellular transport and influence on cell architecture or function. In the past it was necessary to use frog ocytes for comparable studies (Matthews et al. 1981). In many cases it may be desirable to examine the influence of a gene product in the cell by microinjection before transformation with the gene. It seems reasonable to consider the transplantation of organelles such as mitochondria, in order to study cytoplasmic genetics. Clearly, the potential of microinjection as an experimental research tool is just beginning to be explored in the plant sciences.

ACKNOWLEDGEMENTS

The authors wish to thank M. Haffner, L. Holbrook, D. Brown, D. Simmonds, W. Keller and B. Scobie who contributed both guidance and research support for the development of these technologies. T. Reich was on educational leave from Agriculture Canada and supported in part by grants to V.N. Iyer.

REFERENCES

Bradley PM (1979). Micromanipulation of cyanelles and a cyanobacterium into higher plant cells. Physiol Plant 47: 293-298.

Brinster RL, Chen HY, Trumbauer M, Senear AW, Warren R, Palmiter RD (1981). Somatic expression of herpes thymidine kinase in mice following injection of a fusion gene into eggs. Cell 27: 223-231.

Brinster RL, Chen HY, Trumbauer ME, Yagle MK, Palmiter RD (1985). Factors affecting the efficiency of introducing foreign DNA into mice by microinjecting eggs. Proc Natl Acad Sci USA 82: 4438-4442.

Capecchi MR (1980). High efficiency transformation by direct microinjection of DNA into cultured mammalian cells. Cell 22: 479-488.

Crossway A, Oakes JV, Irvine JM, Ward B, Knauf VC, Shewmaker CK (1986). Integration of foreign DNA following microinjection of tobacco mesophyll protoplasts. Mol Gen Genet 202: 179-185.

Czernilofsky AP, Hain R, Herrera-Estrella L, Lorz H, Goyraerts E, Baker B, Schell J (1986). Fate of selectable marker DNA integrated into the genome of Nicotiana tabacum. DNA 5: 101-113.

Deak M, Kiss GB, Koncz C, Dudits D (1986). Transformation of Medicago by Agrobacterium mediated gene transfer. Plant Cell Rep 5: 97-100.

DeBlock M, Herrera-Estrella L, VanMontagu M, Schell J, Zambryski P (1984). Expression of foreign genes in regenerated plants and in their progeny. EMBO J 3: 1681-1689.

Deshayes A, Herrera-Estrella L, Caboche M (1985). Liposome-mediated transformation of tobacco mesophyll protoplasts by an Escherichia coli plasmid. EMBO J 4: 2731-2737.

Fraley RT, Rogers SG, Horsch RB, Sanders PR, Flick JS, Adams SP, Bittner ML, Brand LA, Fink CL, Fry JS, Gallupi GR, Goldberg SB, Hoffman NL, Woo SC (1983). Expression of bacterial genes in plant cells. Proc Natl Acad Sci USA 80: 4803-4807.

Fraley RT, Rogers SG, Horsch RB, Eichholtz DA, Flick JS, Fink CL, Hoffman NL, Sanders PR (1985). The sev system: A new disarmed Ti plasmid vector system for plant transformation. Bio/Technology 3: 629-635.

Fromm ME, Taylor LP, Walbot V (1986). Stable transformation of maize after gene transfer by electroporation. Nature 319: 791-793.

Goodwin PB, Erwee MG (1985). Intercellular transport studied by micro-injection methods. In Robards (ed): "Botanical Microscopy 1985", Oxford New York Tokyo: Oxford University Press, pp. 335-358.

Gordon-Kamm WJ, Bushnell WR (1985). Intranuclear microinjection of isolated protoplasts using differential interference contrast microscopy. In "First International Congress of Plant Molecular Biology, Savannah, 1985", Athens: The University of Gerogia Center for Continuing Education, p. 109.

Griesbach RJ (1983). Protoplast microinjection. Plant Mol Biol Reporter 1: 32-37.

Halliwell RS, Gazaway WS (1975). Quantity of microinjected tobacco mosaic virus required for infection of single cultured tobacco cells. Virology 65: 583-587.

Hammer RE, Pursel VG, Caird ER, Wall RJ, Bolt DJ, Ebert KM, Palmiter RD, Brinster RL (1985). Production of transgenic rabbits, sheep and pigs by microinjection. Nature 315: 680-683.

Herrera-Estrella L, DeBlock M, Messens E, Hernalsteens J-P, VanMontagu M, Schell J (1983). Chimeric genes as dominant selectable markers in plant cells. EMBO J 2: 987-995.

Holbrook LA, Reich TJ, Iyer VN, Haffner M, Miki BL (1985). Induction of efficient cell division in alfalfa protoplasts. Plant Cell Rep 4: 229-232.

Holbrook L, Charest P, Grainger J, Miki B (1985). Transformation of Brassica protoplasts and thin cell layers by Agrobacterium tumefaciens. In "First

International Congress of Plant Molecular Biology, Savannah, 1985", Athens: The University of Georgia Center for Continuing Education. p. 28.

Horsch RB, Fry JE, Hoffman NL, Eichholtz D, Rogers SG, Fraley RT (1985). A simple and general method for transferring genes into plants. Science 227: 1229-1231.

Horsch RB, Jones GE (1980). A double filter paper technique for plating cultured plant cells. In Vitro 16: 103-108.

Krens FA, Mans RMW, VanSlogteren TMS, Hoge JHC, Wullems GJ, Schilperoort RA (1985). Structure and expression of DNA transferred to tobacco via transformation of protoplasts with Ti-plasmid DNA: co-transfer of T-DNA and non T-DNA sequences. Plant Mol Biol 5: 223-234.

Krens FA, Molendijk L, Wullems GJ, Schilperoort RA (1982). In vitro transformation of plant protoplasts with Ti-plasmid DNA. Nature 296: 72-74.

Lawrence WA, Davies DR (1985). A method for the microinjection and culture of protoplasts at very low densities. Plant Cell Rep 4: 33-35.

Lorz H, Baker B, Schell J (1985). Gene transfer to cereals cells mediated by protoplast transformation. Mol Gen Genet 119: 178-182.

Mariotti D, Davey MR, Draper J, Freeman JP, Cocking EC (1984). Crown gall Tumorigenesis in the forage legume Medicago sativa L. Plant Cell Physiol 25: 473-482.

Matthews JA, Brown JWS, Hall TC (1981). Phaseolin mRNA is translated to yield glycosylated polypeptides in Xenopus oocytes. Nature 294: 175-176.

Meadows MG, Potrykus I (1981). Hoechst 33258 as a vital stain for plant cell protoplasts. Plant Cell Rep. 1: 77-79.

Morikawa H, Yamada Y (1985). Capillary microinjection into protoplasts and intranuclear localization of injected materials. Plant Cell Physiol 26: 229-236.

Nims RC, Halliwell RS, Rosberg DW (1967a). Disease development in cultured cells of Nicotiana tabacam L var Samsun NN injected with tobacco mosaic virus. Cytologia 32: 224-235.

Nims RC, Halliwell RS, Rosberg DW (1967b). Wound healing in cultured tobacco cells following microinjection. Protoplasma 64: 305-314.

Paszkowski J, Shillito RD, Saul M, Mandák V, Hohn T, Hohn B, Potrykus I (1984). Direct gene transfer to plants. EMBO J 3: 2717-2722.

Pollock K, Barfield DG, Robinson SJ, Shields R (1985). Transformation of protoplast-derived cell colonies and suspension cultures by Agrobacterium tumefaciens. Plant Cell Rep 4: 202-205.

Reich TJ, Iyer VN, Scobie B, Miki BL (1986a). A detailed procedure for the intranuclear microinjection of plant protoplasts. Can J Bot In press.

Reich TJ, Iyer VN, Haffner M, Holbrook LA, Miki BL (1986b). The use of fluorescent dyes in the microinjection of alfalfa protoplasts. Can J Bot In press.

Reich TJ, Iyer VN, Miki BL (1986c). Efficient transformation of alfalfa protoplasts by the intranuclear microinjection of Ti plasmids Bio/Technology In press.

Rubin GM, Spradling AC (1982). Genetic transformation of Drosophila with transposable element vectors. Science 218: 348-353.

Rusconi A, Schaffner W (1981). Transformation of frog embryos with a rabbit β-globin gene. Proc Natl Acad Sci USA 78: 5051-5055.

Steinbiss H-H, Stabel P (1983). Protoplast derived tobacco cells can survive capillary microinjection of the fluorescent dye lucifer yellow. Protoplasma 116: 223-227.

Waldron C, Murphy EB, Roberts JL, Gustafson GD, Armour SL, Malcolm SK (1985). Resistance to hygromycin B. A new marker for plant transformation studies. Plant Mol Biol 5: 103-108.

Plant Tissue and Cell Culture, pages 447–458

ADVANCES IN THE CRYOPRESERVATION TECHNOLOGY OF PLANT CELLS AND ORGANS

K.K. Kartha

Plant Biotechnology Institute, National Research Council, Saskatoon, Saskatchewan, S7N 0W9, Canada

NRCC No. 26013

INTRODUCTION

Cryopreservation (preservation in the frozen state) is based on the reduction and subsequent arrest of metabolic functions of biological material by imposition of ultra-low temperature. At the temperature of liquid nitrogen (-196°C) almost all the metabolic activities of cells are at a standstill and the specimens can then be preserved in such a state for extended periods. However, only few biological material, in their natural state, can be frozen to sub-freezing temperatures without adversely affecting the cell viability. The discovery of chemicals with cryoprotective properties paved the way for the development of effective cryopreservation technique; the first chemical to be discovered in this category being glycerol (Polge et al., 1949) followed by dimethyl sulfoxide (DMSO or Me_2SO) (Lovelock and Bishop, 1959).

One of the earliest reports on cryopreservation attempts with plant material was made by Sun (1958) who achieved partial success with desicated seedlings of *Pisum sativum*. This indicated the possibility that plant materials could be subjected to ultra-low temperature exposure. However, cryopreservation of plant cells and organs received widespread acceptance mainly due to the emphasis placed, in recent years, on plant biotechnology wherein *in vitro* manipulations with cultured cells, tissues, and organs became the *modi operandi*. This article is not intended to cover the subject matter in any great detail, but rather to point out some of the basic essentials involved in cryopreservation technology. For a detailed treatment of the subject matter, the readers are referred to a recently published book "Cryopreservation of Plant Cells and Organs" (Kartha, 1985a) and other review articles (Kartha, 1985b; Withers, 1985a).

NEED FOR CRYOPRESERVATION

The first and foremost application of cryopreservation is in the area of preservation of valuable genetic resources, especially of vegetatively propagated material and also of those species with short-lived seeds. Before cryogenic storage strategies could be envisaged, it becomes imperative to develop appropriate tissue culture techniques for reliable and faithful regeneration of plants. In this context, use of meristems is highly recommended because the technique of plant regeneration from meristems is simple and straightforward and the regenerants, in most cases, are genetically identical to the parents. The latter criterion is essential for the purpose of germplasm preservation. However, the possibility of genetic aberration induced through adventitious bud proliferation, although of negligible magnitude, still exists in the propagules.

With the emphasis placed on biotechnology, callus and cell cultures are being widely used, in recent years, to create additional genetic variability for various biotechnological applications related to agriculture and industry. When occasion warrants, such induced variability would be preserved infinitely through cryogenic methods. It is a well recognized phenomenon that cell or callus cultures are prone to inadvertent genetic changes resulting from the selection pressure imposed on the system through various culture maintenance procedures including routine subculture. In certain cell lines which possess specific biosynthetic capabilities such as production of value-added chemicals, secondary metabolites etc., such changes could seriously result in alteration or inhibition of the product formation. In such specific instances, cryopreservation could play a significant role in preserving the original reference inoculum.

APPROACH TO CRYOPRESERVATION

1) Pretreatments and Cryoprotection

Some form of cryoprotection is necessary for cryopreservation of plant material unless they are naturally dehydrated, as in the case of winter vegetative buds, or articially cold acclimated. A number of compounds such as DMSO, glycerol, various sugars and sugar alcohols protect living cells against damage during freezing and thawing. Such compounds can lower the temperature at which freezing first occurs and can alter the crystal habit of ice when it separates. The colligative properties of the cryoprotectants minimize the deleterious action of electrolyte concentration resulting from conversion of water to ice (Nash, 1966). Higher solubility and low toxicity are the two essential characteristics these compounds should possess. The cryoprotectants generally used for freezing biological material fall into two cateogies, permeating and non-permeating. The former includes DMSO, methanol, glycerol, while the latter, various sugars and sugar alcohols and high molecular weight additives such as polyvinylpyrolidone, dextran, hydroxyethyl starch, etc. Among the permeating category of compounds, DMSO enters the cells more rapidly than glycerol, and therefore, requires shorter treatment duration. Most of the cryoprotectants exhibit varying degrees of cytotoxicity at higher concentrations. Generally a concentration of 5 to 10% for DMSO and 10 to 20% for glycerol is adequate for most material. In instances where application of a single cryoprotectant does not result in high survival, a mixture of cryoprotectants has been beneficial (Finkle and Ulrich, 1979; Hauptman and Widholm, 1982; Chen et al., 1984a, b; Finkle et al., 1985; Withers, 1985b).

Where cryoprotectants at non-toxic levels do not afford sufficient cryoprotection, certain physiological modifiations are to be made to the experimental material in order to increase their post-freezing survival. This kind of treatment now has become a routine protocol, especially for meristems and cell cultures. Preconditioning of cells prior to cryopreservation has been beneficial in increasing the freezing resistance of cells of certain species. Supplementation of culture medium with such osmotically active compounds as mannitol, sorbitol, sucrose and proline, has resulted in increasing the freezing resistance. These compounds mainly act by their dehydration effect. Mannitol has been found to be beneficial in reducing the mean cell volume of cells of *Acer pseudoplatanus* and *Capsicum annuum* and increase their post-freezing survival (Withers and Street, 1977). However, preculturing in mannitol-supplemented medium did not enhance the survival of cryopreserved cells of *Nicotiana sylvestris* (Maddox et al., 1983), and alkaloid producing cell cultures of *Catharanthus roseus* (Chen et al., 1984a). All these osmotically active compounds also possess cryoprotective properties. Although the exact mechanism of action of these compounds are unclear, in a recent study, Pritchard et al. (1986) observed that sycomore cells, grown to exponential phase in either mannitol or sorbitol supplemted media exhibited increased levels of total and soluble protein and respiratory activity, but decreased amounts of free proline. Similarly, preculturing of cells or meristems in DMSO-supplemented medium for periods of 24 to 48 hours prior to freezing has resulted in increased survival (Kartha et al., 1979, 1980, 1982; Manzhulin et al., 1983).

Sorbitol has been used as an osmotic agent in the preculture medium as well as the sole cryoprotectant during the freezing process. Weber et al. (1983) successfully cryopreserved cell cultures of *Glycine max*, *Datura innoxia*, *Brassica napus* and *Daucus carota* and cell culture derived protoplasts of *G. max* using this procedure. While cell cultures of alkaloid producing *Catharanthus roseus* could not be cryopreserved using sorbitol as the sole cryoprotectant, they could be successfully cryopreserved by preculturing the cells in nutrient medium containing 1M sorbitol for 6-20 hours followed by freezing using 1M sorbitol + 5% DMSO (Chen et al., 1984a). Equimolar concentrations of $CaCl_2$ and KCl, in addition to being toxic to cells, were most ineffective in their cryoprotective efficacy. Although considerable survival was noticed subsequent to preculturing the cells in media supplemented with 1M concentrations of glucose, trehalose, and sucrose, the overall survival and recovery was much superior when sorbitol was used as the preculturing agent (Chen et al., 1984a). In a detailed study using nuclear magnetic resonance (NMR) and differential thermal analysis (DTA), it was found that incorporation of DMSO or sorbitol into the liquid medium had a significant effect in the temperature range for initiation to completion of ice crystallization and the amount of liquid water present in the cell at various subzero temperatures. There was a close association between the percent unfrozen water at -40°C and the percent cell survival after freezing in liquid nitrogen. These studies revealed that in the alkaloid producing *C. roseus* cells, the amount of liquid water at -40°C is critical for successful cryopreservation and the combination of DMSO and sorbitol was the most effective in preventing water from freezing (Chen et al., 1984b). These results may explain the cryoprotective properties of DMSO and sorbitol and why DMSO and sorbitol in combination are most effective than when used alone.

The benefical effect of using a mixture of cryoprotectants involving polyethylene glycol (PEG), glucose, and DMSO (10-8-10%) was demonstrated with sugarcane callus by Ulrich et al. (1979). In subsequent studies several other cell or callus lines have been cryopreserved using the same mixture (Finkle et al., 1985). Prior to the discovery of sorbitol-DMSO system in the author's laboratory (Chen et al., 1984a), the most effective cryoprotectant mixture applicable to a variety of cell cultures was either 0.5M DMSO + 0.5M glycerol + 1M sucrose, or 0.5M DMSO + 0.5M glycerol + 1M proline (Withers, 1985b).

Cold hardening or ABA treatment of wheat suspension culture increased the freezing resistance and survival following cryopreservation (Chen et al., 1985). Similarly cold hardening of *Panax* suspension cultures at 2°C in a medium containing 3 to 25% sucrose also increased the survival of cells following storage in liquid nitrogen (Butenko et al., 1984). Incorporation of amino acids such as asparagine, alanine, proline and serine in the preculture medium also enhanced the survival rate of *Diascoria deltoidea* cells subsequent to cryostorage (Butenko et al., 1984).

From the foregoing examples, it becomes evident that various preculturing strategies prior to freezing enhance the post-freezing survival.

2) Freezing Methods

Freezing methods employed in cryopreservation attempts with various types of plant materials generally fall into the following categories: a) slow freezing, b) rapid freezing, and c) droplet-freezing.

a) Slow freezing. Methods of freezing biological specimens slowly are based on the physicochemical events occuring during freezing. When plant cells are exposed to progressive temperature reduction, ice formation is usually initiated extracellularly. It is generally believed that plasma membrane acts as a barrier that prevents the ice from seeding the cytoplasm. In absence of nucleators, the cytoplasm remains supercooled. Following further reduction in temperature, the concentration of extracellular liquid increases as more water is converted to ice. Since the vapor pressure of the frozen solution is lower than the same concentration of the supercooled liquid, the vapor pressure of slowly cooled cells reach equilibrium with external ice by efflux of water. Slow freezing is thought to prevent the intracellular ice formation, thus preventing freezing injury. It is generally agreed that upon extracellular freezing the cytoplasm will be effectively concentrated and plant cells survive better when adequately dehydrated. There are different ways of achieving protective dehydration, including slow cooling at constant or varying cooling rates, or holding the samples at one or more intermediate subzero temperatures.

The success of slow freezing depends upon a number of factors including cooling rates, pretreatments and cryoprotection, type and physiological states of the material, and the temperature prior to immersion in liquid nitrogen. The most commonly used methods for the cryopreservation of plant cells generally involve regulated slow cooling at a constant rate of 0.5 to 2°C/min to terminal temperature between -30°C to -40°C followed by storage in liquid nitrogen. Each specimen is characterized by its unique narrow range of optimal cooling rates. Most reports on cryopreservation of cell cultures involve slow freezing (for a review see Withers, 1985b). Successful examples of cryopreservation of meristems include such diverse species as pea, strawberry, carnation, potato, cassava, etc. (for a review see Kartha, 1985).

b) Rapid freezing. While rapid freezing has been unsuitable for the cryopreservation of cell cultures, it has been employed to cryopreserve shoot-tips of a few species such as Carnation (Siebert, 1976), potato (Grout and Henshaw, 1978; Bajaj, 1977), *Arachis hypogaea*, *Cicer arietinum* (Bajaj, 1979), strawberry (Sakai et al., 1978) and *Brassica napus* (Withers, 1982) and somatic embryos (e.g. oil palm, Engelmann et al., 1985).

Rapid freezing is accomplished by direct immersion of the cryoprotectant-treated specimens in liquid nitrogen. The cooling rate is in the order of several hundred degrees per minute. At such rapid cooling rates, the intracellular fluids do not have sufficient time to equilibrate with the external ice with the possibility of intracellular ice formation which is considered to be lethal. At the same time, it is also possible that ultra-rapid cooling may prevent the growth of intracellular ice crystals by rapidly passing the cells through the temperature zone in which lethal ice crystal growth occurs.

c) Droplet-freezing. The droplet freezing was originally developed by Kartha et al. (1982b) for the preservation of cassava meristems. In this method, the cryoprotectant-treated meristems are dispensed in droplets of 2-3 μl on an aluminum foil contained in a petri dish. Each specimen is then transferred to a droplet and frozen by slow cooling (0.5°C/min) to a sub-zero temperature between -20°C, and -40°C prior to immersion in liquid nitrogen. Probably, freezing plant specimens in droplets on aluminum foil, which has an efficient thermal conductivity, results in more homogenous cooling of the sample. Also, since a very small amount of cryoprotectant solution is used for each meristem, a uniform freezing of the cell occurs and lesser a amount of ice forms, thus avoiding the excess mechanical pressure exerted on the cell wall as the ice propagates. The droplet freezing technique has been applied to the cryopreservation of cassava meristems and cultured cells (Chen and Kartha, unpublished results) with better results than using conventional slow freezing in glass or plastic cryogenic ampoules.

3) Storage, Thawing and Regrowth

Storage of material is best accomplished in liquid nitrogen immersion phase (-196°C) on in its vapor phase (ca -150°C). Rapid thawing is recommended for most cryopreservation protocols. The basis of applying rapid thawing is to avoid the damaging ice recrystallization which may occur during slow warming. In general, thawing is carried out by removing the sample from liquid nitrogen storage and immersing in a 34-40°C water bath for about 1-2 minutes or until the phase change occurs and ice is transformed into water.

Regrowth of cryopreserved specimens is the most convincing and accurate estimate of viability, while other techniques such as the use of vital or fluorescent stains, triphenyltetrazolium chloride assay etc. might provide immediate preliminary assays of viability. However regrowth of cryopreserved materials would depend upon the manner the specimens are handled such as plating the cells without washing away the cryoprotectants, incorporation of special additives in regrowth medium and the physical environmental conditions the cells are exposed to during the early stages of regrowth. The reader is referred to Withers (1985b) for a detailed account of these variables.

TABLE 1. List of Cryopreserved Cell, Callus, Protoplasts and Meristems/Shoot-tip Cultures.*

A. CELL CULTURES

Acer pseudoplatanus	_Malus domestica_
Atropa belladona	_Medicago sativa_
Brassica napus	_Nicotiana plumbaginifolia_
Capsicum annum	_N. sylvestris_
Catharanthus roseus	_N. tabacum_
Datura innoxia	_Oryza sativa_
D. stramonium	_Panax ginseng_
Diascorea deltoidea	_Populus_ sp.
Digitalis lanata	_Pseudotsuga menziesii_
Daucus carota	Rosa Paul's Scarlet
Glycine max	_Saccharum_ sp.
Hordeum vulgare	_Sorghum bicolor_
Hyoscyamus muticus	_Zea mays_

Table 1 (continued)

B. CALLUS CULTURES

Lavandula vera
Medicago sativa
Phoenix dactylifera
Populus sp.
Saccharum sp.
Triticum aestivum
Ulmus americana

C. PROTOPLASTS

Bromus inermis
Datura innoxia
Daucus carota
Glycine max
Hordeum vulgare
Merchantia polymorpha
Triticum aestivum
Zea mays

D. MERISTEMS/SHOOT-TIPS

Arachis hypogea
Brassica napus
Cicer arietinum
Dianthus caryophyllus
Fragaria x *ananassa*
Lycopersicon esculentum
Malus domestica
Manihot esculenta
Pisum sativum
Solanum etuberosum
Solarum goniocalyx
Solarum tuberosum

E. OTHERS

Somatic and zygotic embryos
Anthers/pollen
Whole seeds

---

* For references see Kartha (1985b), Withers (1985b).

EXAMPLES OF SUCCESSFUL CRYOPRESERVATION

Various types of plant cells and organs have been cryopreserved to date (Table 1); these include meristems or shoot-tips, zygotic and somatic embryos, cultured cells, callus, protoplasts, pollen/anthers and whole seeds. For a detailed account of the techniques involved in the cryopreservation of these diverse type of plant materials, see Kartha (1985a).

SUMMARY

With the recent development in preculture techniques using various additives such as sorbitol, mannitol, sucrose and amino acids coupled with the judicious use of cryoprotectants, either alone or in combination and regrowth techniques, many types of plant material have been successfully cryopreserved. Cryopreservation techniques are very valuable for the long-term preservation of cell, callus and protoplast cultures, meristems/shoot-tips, zygotic or somatic embryos and whole seeds. These materials serve definite experimental purposes aimed at crop improvement through various biotechnological approaches. Cryopreservation could be successfully employed both for the preservation of genetic stability and induced variability. Cell cultures are more amenable to cryopreservation than meristems or shoot-tips. Due to space limitation, only some of the parameters that affect the viability of plant material could be discussed in this article.

REFERENCES

Bajaj YPS (1977). Initiation of shoots and callus from potato-tuber sprouts and axillary buds frozen at -196°C. Crop Improv 4: 48-53

Bajaj YPS (1979). Freeze-preservation of _Arachis hypogaea_ and _Ciar arietinum_. Indian J Exp Biol 17: 1405-1407

Butenko RG, Popov AS, Volkova LA, Chernyak ND, Nosov AM (1984). Recovery of cell cultures and their biosynthetic capacity after storage of _Dioscorea deltoidea_ and _Panax ginseng_ cells in liquid nitrogen. Plant Sci Lett 33: 285-292

Chen THH, Kartha KK, Leung NL, Kurz GWG, Chatson KB, Constabel F (1984a). Cryopreservation of alkaloid-producing cell cultures of periwinkle (_Catharanthus roseus_). Plant Physiol 75: 726-731

Chen THH, Kartha KK, Constabel F, Gusta LV (1984b). Freezing characteristics of cultured _Catharanthus roseus_ (L.) G. Don cells treated with dimethylsulfoxide and sorbitol in relation to cryopreservation. Plant Physiol 75: 720-725

Chen THH, Kartha KK, Gusta LV (1985). Cryopreservation of wheat suspension culture and regenerable callus. Plant Cell Tissue Organ Cult 4: 101-109

Engelmann F, Duval Y, Dereuddre J (1985). Survie et prolifération d'embryons somatiques de Palmier à huile (Elaeis guineensis Jacq) après congélation dans l'azote liquide. C R Acad Sc Paris 301: 111-116.

Finkle BJ, Ulrich J (1979). Effects of cryoprotectants in combination on the survival of frozen sugarcane cells. Plant Physiol 63: 598-604.

Finkle BJ, Zavala ME, Ulrich JM (1985). Cryoprotective compounds in the viable freezing of plant tissues. In Kartha KK (eds.) "Cryopreservation of Plant Cells and Organs", Boca Raton: CRC Press Inc., pp 75-113.

Grout BWW, Henshaw GG (1978). Freeze-preservation of potato shoot-tip cultures. Ann Bot 42: 1227-1229.

Hauptman RH, Widholm JM (1982). Cryostorage of cloned aminoacid analog-resistant carrot and tobacco suspension cultures. Plant Physiol 70: 30-34.

Kartha KK (1985a). "Cryopreservation of Plant Cells and Organs." Boca Raton: CRC Press Inc., pp 276.

Kartha KK (1985b). Cryopreservation of plant cells and organs. Newsletter IAPTC #45 (March issue) 2-15.

Kartha KK, Leung NL, Gamborg OL (1979). Freeze-preservation of pea meristems in liquid nitrogen and subsequent plant regeneration. Plant Sci Lett 15: 7-15.

Kartha KK, Leung NL, Pahl K (1980. Cryopreservation of strawberry meristems and mass propagation of plants. J Amer Soc Hort Sci 105: 481-484.

Kartha KK, Leung NL, Gaudet-LaPrairie P, Constabel F (1982a). Cryopreservation of periwinkle Catharanthus roseus cells cultured in vitro. Plant Cell Repts 1: 135-138.

Kartha KK, Leung NL, Mroginski LA (1982b). In vitro growth responses and plant regeneration from cryopreserved meristems of cassava (Manihot esculenta Crantz). Z Pflanzenphysiol 107: 133-140.

Lovelock JE, Bishop MWH (1959). Prevention of freezing damage to living cells by dimethylsulfoxide. Nature 183: 1394-1395.

Maddox AD, Gonsalves FG, Shields R (1983). Successful cryopreservation of suspension cultures of three Nicotiana species at the temperature of liquid nitrogen. Plant Sci Lett 28: 157-162

Manzhulin AV, Butenko RG, Popov AS (1983). Effect of pretreatments of potato apices surviving after deep freezing. Soviet Plant Physiol 30: 1188-1193 (English summary).

Nash T (1966). Chemical constitution and physical properties of compounds able to protect living cells against damage due to freezing and thawing. In Meryman (ed.) "Cryobiology," New York: Academic Press, pp 197-211.
Polge C, Smith AU, Parkes AS (1949). Revival of spermatozoa after vitrification and dehydration at low temperature. Nature 164: 666.
Pritchard HW, Grout BWW, Short KC (1986). Osmotic stress as a pregrowth procedure for cryopreservation 3. Cryobiology of sycomore and soybean cell suspensions. Ann Bot 57: 379-387.
Sakai A, Yamakawa M, Sakata D, Harada T, Yakuwa T (1978). Development of a whole plant from an excised strawberry runner apex frozen to -196°C. Low Temp Sci Ser Bull 36: 31-38.
Seibert M (1976). Shoot initiation from carnation shoot apices frozen to -196°C. Science 191: 1178-1179.
Sun CN (1958). The survival of excised pea seedlings after drying and freezing in liquid nitrogen. Bot Gaz 19: 234-236.
Ulrich JM, Finkle B, Moore PH, Ginoza H (1979). Effect of a mixture of cryoprotectatns in attaining liquid nitrogen survival of callus cultures of a tropical plant. Cryobiology 16: 550-556.
Weber G, Roth EJ, Schweiger HG (1983). Storage of cell suspensions and protoplasts of _Glycine max_ (L.) Merr, _Brassica napus_ (L.), _Datura innoxia_ (Mill.) and _Daucus carota_ (L.) by freezing. Z Pflanzenphysiol 109: 29-39.
Withers LA (1982). The development of cryopreservation techniques for plant cell tissue and organ cultures. In Fujiwara (ed.) "Plant Tissue Culture," Tokyo: Proc 5th Inter Cong Plant Tissue Cell Cult, pp 793-794.
Withers LA (1985a). Cryopreservation of cultured cells and meristems. In Vasil IK (ed.). "Cell Culture and Somatic Cell Genetics of Plants", vol. 2, New York: Academic Press, pp 253-316.
Withers LA (1985b). Cryopreservation of cultured plant cells and protoplasts. In Kartha KK (ed.) "Cryopreservation of Plant Cells and Organs," Boca Raton: CRC Press Inc., pp 243-267.
Withers LA, Street HE (1977). Freeze-preservation of plant cell cultures. In Barz W, Reinhard E, Zenk MH (eds.) "Tissue Culture and its Bio-technological Application," Berlin: Springer pp 226-244.

Plant Tissue and Cell Culture, pages 459–471

PROCESS SYSTEMS AND APPROACHES FOR LARGE SCALE PLANT CELL CULTURE

Michael W. Fowler

The Wolfson Institute of Biotechnology
University of Sheffield, Sheffield. S10 2TN U.K.

INTRODUCTION

One of the more intriguing features of plant cell culture is that although there are comparatively few processes operational at either an industrial scale or Pilot Plant (up to ca. 1000 litres) level, the range of process options and bioreactor systems which are available to those concerned with the development of plant cell culture processes is quite wide (Fowler, 1985). It is the purpose of this short review to examine some of these options and to assess their potential for use in natural product synthesis from plant cell cultures on an industrial scale.

LARGE SCALE CELL CULTURE

At the heart of any process system based upon the use of plant cells or their components is the need at an early stage to produce varying quantities of biomass. This may of itself comprise the process system, or it may be a prelude to a further production stage based, for instance, on immobilised cell or enzyme technology.

The growth of plant cells on a large scale has received a great deal of attention in recent years (see for example reviews by Martin, 1980 and Scragg and Fowler, 1985). Although quite a number of the early large scale experiments were carried out with classical stirred tank reactor systems used for microbial fermentations (see for example Routier and Nickell, 1956 and Martin, 1980), it was generally thought that such reactor conformations were not suitable for plant cells, because of the rather different physiological and structural properties encountered as compared with microbial systems. Specifically plant cells are much larger than microbial cells, have much lower basal metabolic rates, and are surrounded by a cellulose based cell wall. This latter is generally considered to have a high tensile strength but low shear resistance. In consequence the high shear conditions developed in microbial tank reactors were seen as a major restriction in the use of such bioreactor systems for plant cell growth. A number of groups presented data confirming the problem (Mandels, 1972, Wagner and Vogelman 1974 and Fowler 1982). In attempts to overcome this difficulty a range of different bioreactor systems have been investigated, of which perhaps the best known are based upon air-driven reactors in three major configurations, bubble columns, draught tubes and air-lift loops (Smart and Fowler, 1984a,b). These air driven systems are particularly characterised by low shear levels, and a range of plant cell types have now been grown successfully in them (Scragg and Fowler, 1985). However, while good growth and productivity appears achievable up to biomass levels of about 30g dry weight per litre, data presented by Tanaka (1981) indicates that at biomass levels much above this, mixing and aeration of the culture becomes incomplete and culture performance falls off. Unfortunately Goldstein and colleagues (1980), in a detailed process cost analysis, indicated that bioreactor systems which could cope with biomass loadings of around 100g per litre dry weight may well be required before plant cell systems achieve commercial viability. On present evidence this would appear not to be feasible with conventional air-driven designs and may well require novel reactor systems. A solution however might lie in the nature of the cells themselves.

In recent years we have carried out an extensive programme into the effects of shear on the viability of plant cell cultures. This has involved exposing plant cell cultures to differing levels of shear for varying periods of time in conventional flat blade impellar stirred tank reactors. Viability has then been assessed by transferring the sheared cells to new medium in a low shear culture regime and assessing growth performance under standard conditions. Given the previously held view that plant cells are highly shear sensitive we were somewhat surprised to find that the test cells did in fact remain viable under very high shear loadings. We have now observed this situation with quite a wide range of cell cultures. Table 1 (Scragg et al., 1986) is illustrative of our general observations, in this case with cells of *Catharanthus roseus* as the test system. It is difficult

Table 1

The effect of shear on the viability of C.roseus suspension cultures

Time of exposure to shear (hr)	Dry and fresh weights of cultures after 9 days incubation	
	Dry wt.mg.ml^{-1}	Fresh wt.mg.ml^{-1}
Expt. 1.		
0	7.7 ± 0.2	171.9 ± 10
1	8.5 ± 2.8	164.7 ± 56
2	8.2 ± 1.9	230.7 ± 81
3	7.7 ± 1.2	198.4 ± 35
4	9.8 ± 1.5	206.1 ± 33
5	10.3 ± 1.0	233.0 ± 35
Expt. 2.	Dry and fresh weights after 14 days incubation	
0	7.2 ± 0.2	195.5 ± 6.9
2.5	7.9 ± 0.5	222.3 ± 2.1
5	7.7 ± 0.1	214.4 ± 6.1

All cultures were subjected to shear at 166 sec^{-1} at a stirring rate of 1000 rpm, with a flat blade impellar.

to provide an adequate explanation of these results; one possibility is that over time our own cell cultures have been subjected to a moderately high shear regime in the conditions under which we normally maintain them, and that in consequence we have unwittingly selected high shear resistant cells by the very nature of our culture regime and practice. Against this is the observation that found certain newly initiated cultures also exhibit high levels of shear tolerance.

As a final general comment to this section, it is perhaps instructive to note that the Mitsui Petrochemical process for shikonin production from *Lithosperum* cell cultures (Curtin, 1983) uses conventional stirred tank reactors. In addition Noguchi and his colleagues (Naguchi et al, 1977) working with tobacco, Alfermann et al., 1983, working with *Digitalis lanata* cells and Berlin (1986) working with *Catharanthus* cells, have all reported good cell growth in conventional stirred tank reactors.

Such observations are of course important against the general background of concern over the economics of cell culture. If it proves possible to use even slightly modified stirred tank reactors this should leave the way open to take advantage of already available reactor capacity, thereby cutting investment costs in new plant and equipment. More recently Ulbrich and co-workers have reported on the use of a type of Archimedian screw reactor the *Coleus* cells synthesising rosmarinic acid. This reactor system pumps rather than stirs the cells and can operate at high biomass levels, towards those suggested by Goldstein et al., (1980). It may therefore provide yet a further generally applicable bioreactor configuration.

Another key factor in the large scale growth of plant cells is the nature and level of the gassing regime applied to the culture. There are now a number of reports indicating that a carefully balanced gas regime is vital to optimal culture performance; while low venting rates may limit growth, high rates may also be deleterious to the culture (Smart and Fowler, 1981, 1984; Tanaka 1982; Scragg and Fowler, 1984; Breuling et al., 1985; Scragg et al, 1986). Current data suggest that at high ventilation rates carbon dioxide may be 'stripped off' from the system which in turn may lead to carbon limitation of anaplerotic reactions supporting biosynthesis. To a degree this may be overcome by supplementing the carbon dioxide supply to the system, while maintaining the same gas flow rate. An effect of high oxygen levels on culture performance has been more difficult to establish. There are indications that increasing the dissolved oxygen level may increase product yield (Breuling et al., 1985). Equally there are also concerns that at the higher gassing rates required in air driven reactors, high dissolved oxygen levels may be deleterious (Tanaka, 1982; Scragg and Fowler, 1985).

Such observations on the impact of gassing regimes are important, not only because of their effects on cell physiology, but also because they have implications for the type of bioreactor configuration which may be used in the large scale growth of a particular cell line. Comment was earlier passed about the problems of maintaining adequate mixing in a plant cell culture at high biomass levels in air driven reactors. If high venting rates are used in an attempt to improve mixing characteristics then it is possible that this could lead to difficulties with overgassing. Quite obviously a delicate balance has to be struck between achieving adequate mixing of the culture without engendering deleterious effects by upsetting the gas balance and dissolved gas levels.

PROCESS SYSTEMS

The ability of plant cells to divide, grow and synthesise products under a variety of different culture regimes has been amply demonstrated in recent years by a number of research groups (for review see Fowler, 1985). Plant cells have been grown in batch, semi-continuous/fed batch and continuous culture and have been maintained in a viable state for extended periods of time in a variety of immobilised reactor systems. Out of all this has come a variety of process options, involving both single and two stage formats (Table 2).

There is however, one key facet of secondary metabolite synthesis by plant cell cultures which very much limits the options available. With very few exceptions secondary metabolite synthesis is dissociated from growth and cell division, a phenomenon well established for microbial cells and earlier termed trophophase-idiophase development by Bu'Lock (1975). Such a situation severely restricts the use of batch or continuous culture in a single step operation for product synthesis. With batch culture the result is usually a compromise between optimising growth to achieve short run times and getting the maximum product yield possible to balance this. With continuous culture and cells in a continuous state of growth and division, secondary metabolite yield is typically very low, a situation exemplified by the work of Tal Goldberg (1982) who compared batch and continuous culture for the synthesis of diosgenin by *Dioscorea* cell cultures. Because of this it is generally considered that for many systems the

Table 2

Process format options for plant cell culture systems[a]

Single-stage operations	Two-stage operations	
	First Stage (biomass production)	Second Stage (product synthesis)
(1) Batch culture	(1) Batch culture	Batch culture
(2) Fed batch semi-continuous culture	(2) Continuous culture	Batch culture
(3) Continuous culture culture	(3) Continuous/batch immobilization	
	(4) Fed batch culture	Fed batch culture

[a] These options do not take into account the need to grow up inocula into a first stage operation.

development of a two-stage process option may be the most fruitful approach (Alfermann and Reinhard, 1978; Noguchi et al., 1977; Fowler, 1983), where the first stage is directed towards rapid growth and biomass accumulation and the second stage is focussed upon product synthesis with little if any cell division or growth. In the case of an immobilised cell second stage the aim is of course for zero growth and cell division.

Where a two stage option is selected the nature of the second stage to a large degree depends upon the fate of the secondary metabolite on synthesis, and whether it is released from the cells or not. If it is released from the cells, or a mechanism can be found which achieves this without reducing the viability of the immobilised culture (Brodelius, 1985; Rosevere and Lambe, 1985), then an immobilised cell second stage is probably the most effective way of product synthesis, maximising as it does the use of cell biomass and achieving a high utilisation rate of bioreactor capacity. Perhaps the most effective process format in economic terms would be a continuous culture first stage followed by an immobilised cell second stage. However, such a format also rests on the proviso of technical feasibility.

At this point it is perhaps useful to review the process systems which have so far been either brought on stream at an industrial scale, or are at a Pilot Plant scale (Table 3). Apart from those concerned with biomass

Table 3

Process formats for plant cell culture systems which are either operational or at pilot plant stage

Product	Process format	Single or first stage	Second stage	Origin
Shikonin	Two-stage	Batch culture Air-lift reactor	Batch culture Air-lift reactor	Mitsui Petrochemical Industries: Curtin, 1983
Rosmarinic	Two-stage	Fed-batch culture Spiral reactor	Batch culture Spiral reactor	Ulbrich et al., 1985
Digoxin	Two-stage	(1) Continuous culture Air-lift reactor (2) Batch culture Air-lift reactor	Batch culture Air-lift reactor Immobilized cell culture	Reinhard & Alfermann, 1980 Alfermann et al., 1983
Tobacco	Two-stage	Batch culture Stirred tank reactor	Batch culture Stirred tank reactor	Noguchi et al., 1977
Geranoil	Two-stage	Batch culture Stirred tank reactor	Batch culture Stirred tank reactor	Kanebo (Japan)
Ginseng	Single-stage	Batch culture Stirred tank reactor		Furuya, 1982
Cinnamoyl putrescines	Single-stage	Fed batch culture Stirred tank reactor		Schiel et al., 1984

production all the systems so far developed are two stage in nature, and reply principally on batch culture. The reason that immobilised cell systems do not feature as the second stage is presumably because of lack of product release from the cells or a mechanism for achieving this without losing cell viability. It is also interesting to note that almost all the systems are based on conventional stirred tank reactors, again adding weight to earlier comments that it may well be possible to think of using established fermentation equipment for plant cell processes.

SCALE UP

At some point in the consideration of any process system thought has to be given to the manner in which a process is developed from bench to Pilot Plant to industrial scale. Although a limited number of processes are operational at full or Pilot Plant scale little information has been published on scale-up parameters. However, in our own experience, and reflecting the situation with microbial systems, a linear scale-up profile is rarely achieved. Figure 1 illustrates the situation with a Catharanthus roseus culture where we have scaled up from a 250 ml shake flask to an 80 litre air-lift draught tube bioreactor. As can be seen from the

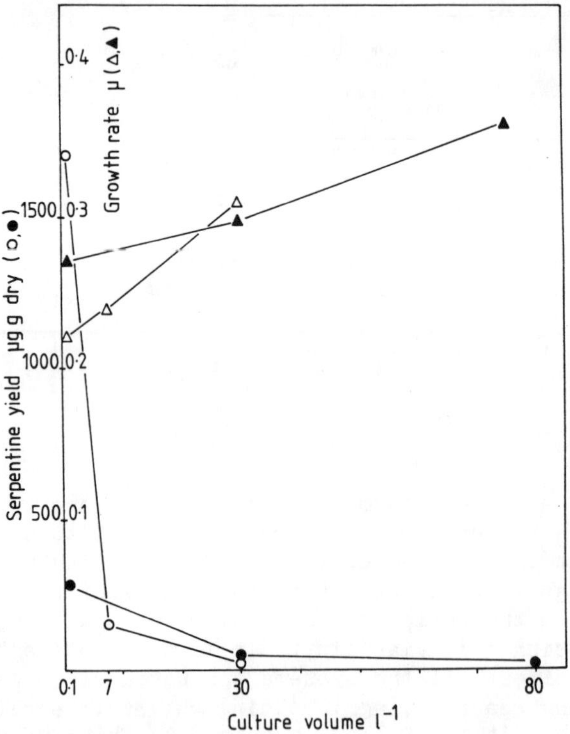

Figure 1. The growth rate and serpentine yields of C.roseus cell lines C87, and C87N grown in vessels of various sizes. Growth rate u (day^{-1}), of C.roseus C87 (Δ-Δ); C.roseus C87N (▲-▲); serpentine yield ug.gram^{-1} dry weight C.roseus C87 (O-O); C.roseus C87N (●-●).

figure, increasing the volume of culture and nature of the culture vessel has resulted in an increased biomass yield but with a much reduced product yield. Loss of product on scale-up is a fairly common occurrence. However, that this can be overcome is well exemplified by both the process for shikonin production (Curtin, 1985) and rosmarinic acid synthesis (Ulbrich et al., 1985). Possible causes of the problem lie in altered and perhaps inadequate mixing of the nutrient and cells at high reactor volumes, and lowered dissolved oxygen levels (Breuling et al., 1985). Data on scale-up of plant cell cultures is very much in need of urgent development.

Another aspect of scale-up which is important in the totality of the process is the time taken to grow up the culture from a 'seed inoculum' to full scale; what might be referred to as 'the pre-production phase'. While this is obviously a necessary part of the whole, it should nonetheless occupy as little of the whole as possible, given that it is non-productive in terms of secondary metabolite synthesis. Typically most research groups work on an inoculum ratio of about 1:10, old culture to new medium. Starting at a 1 litre level and working up to say 1000 litre, and a culture cycle time of about 10 days a minimum period of something like 30 to 40 days is involved in this pre-production phase. Such a length of time reinforces the need for consideration to be given not only to developing cultures with a high product yield but also which have a high growth rate in the pre-production period.

DOWN STREAM RECOVERY

The area of down stream recovery is one which has so far received comparatively little attention in plant cell culture circles, the general assumption having been that traditional techniques of organic extraction or steam distillation would be appropriate. There are however new approaches being developed for whole plant systems which are applicable to cell cultures. In particular the use of enzyme cocktails, containing a range of hydrolase activities, has been found in a number of instances to be more effective at releasing phytochemicals from a range of tissues than traditional techniques. Table 4 is illustrative of typical results obtained in our own

Table 4

Effects of Enzyme Cocktails on release of secondary products from plant tissues and cells,

A. Yield of Vanillin precursors from micropropagated seedlings of Vanillin planifolia : effect of predigest with Depol NCC

Treatment	Tissue	Yield phenolic µg.g^{-1} fresh weight ($\bar{x}$) Coumaric acid	Ferulic acid
Buffer	Stem	2.69 (± 1.59)	5.89 (± 0.93)
	Leaf	1.37 (± 0.92)	3.05 (± 0.58)
Buffer	Stem	38.48 (± 10.74)	1.35 (± 0.88)
Depol NCC	Leaf	101.66 (± 16.41)	18.74 (± 1.54)
Acid (HCl)	Stem	31.39 (± 11.40)	5.25 (± 6.45)
Hydrolysis	Leaf	20.67 (± 0.40)	1.26 (± 0.84)

B. Yield of unknown phenolic from tissues of hop as affected by a predigest with Depol NCC

Tissue	µg.g.$^{-1}$ dry weight Compound X* Ether extraction	pre-digest with Depol NCC
Stem	9	76
Leaf	81	96
Suspension culture (a)	1	188
(b)	5	218

*Compound X elutes on HPLC with α-acids but has a different UV spectrum. Comparative quantitation given here has been made using the α-acid response factor.

laboratory where we have compared enzyme cocktails with traditional techniques for the extraction of materials from both whole plant and cell culture material. Where we

are working at marginal product yields and economics extraction of the maximum amount of product from the tissue is of course vital. Such enzyme extraction techniques may therefore have much to offer plant cell technology.

FINAL COMMENTS

To do justice to such a large area as process development in such a short space is of course extremely difficult. That this is so is however indicative of the progress being made in this area of plant cell biotechnology and hopefully presages a period of strong development in the future.

ACKNOWLEDGEMENTS

I would like to express my thanks to Dr. Alan Scragg, Dr. Angela Stafford and Miss Pamela Bond for providing data relating to our own work in the Wolfson Institute. We would wish to express our gratitude to the U.K. Science and Engineering Research Council for supporting our work on shear, gassing regimes and scale-up, and to Biocatalysts Ltd. for providing test samples of enzyme cocktails for product extraction.

REFERENCES

Alfermann, A.W. and Reinhard, E. (1978) Possibilities and problems in production of natural compounds by cell cultures methods. In Production of Natural Compounds by Cell Culture Methods. (Alfermann, A.W. and Reinhard, E. eds). pp.3-15. Gesellschaft fur Strahlen und Umweltforschung, Munich.

Alfermann, A.W., Bergmann, W., Figur, C., Helmbold, U., Schwantag, D., Schulter, I. and Reinhard, E. (1983) Biotransformation of -methyldigitoxin to -methyldigoxin by cell cultures of *Digitalis lanata*. In: Plant Biotechnology (Eds. Mantell S.H. and Smith, H.), pp.67-74. Cambridge University Press, Cambridge.

Berlin, J. (1985) Immobilised plant cells. IAPTC News letter 46, 8-14.

Breuling, M., Alfermann, A.W. and Reinhard, E. (1985) Cultivation of cell cultures of *Berberis wilsonae* in 20-l airlift bioreactors. Plant Cell Reports 4, 220-223.

Brodelius, P. (1985) The potential role of immobilisation in plant cell biotechnology. Trends Biotechnol. 3, 280-285.

Curtin, M.E. (1983) Harvesting profitable products from plant tissue culture. Biotechnology 1, 649-57.

Fowler, M.W. (1982) Substrate utilization by plant cell cultures. J. chem. Tech. Biotech. 32, 338-46.

Fowler, M.W. (1983) Commercial applications and economic aspects of mass plant cell culture. In Plant Biotechnology (Eds. Mantell S.H. and Smith H.), pp.3-38. Cambridge University Press, Cambridge.

Fowler, M.W. (1986) Industrial Applications of Plant Cell Culture. In Plant Cell Culture Technology (Yeoman, M.M. ed.) Blackwell Scientific Publications, Oxford. pp. 202-227.

Goldstein, W.E., Ingle, M.B. and Lasure, L. (1980) Product cost analysis. In: Plant Tissue Culture as a Source of Biochemicals (Ed. Staba E.J.), pp.191-134. CRC Press, Boca Raton, Florida.

Mandels, M. (1972) The Culture of Plant Cells. Advances in Biochemical Engineering, 2, 201-215.

Martin, S.M. (1980) Mass culture systems for plant cell suspensions. In: Plant Tissue Cultures as a Source of Biochemicals (Ed. Staba E.J.), pp.146-66. CRC Press, Boca Raton, Florida.

Noguchi, M., Matsumoto, Hirata, Y., Yansamoto, K., Katsuyama, A., Kato, A., Azechi, A. and Kato, K. (1977) Improvement of growth rates of plant cell cultures. In: Plant Tissue Culture and its Biotechnological Application (Eds. Barz W., Reinhard E. and Zenk M.H.), pp.85-94. Springer-Verlag, Berlin.

Rosevere, A. and Lambe, C.A. (1985) Immobilised Plant Cells. Advances in Biochemical Engineering 31, 37-58.

Routier, J.B. and Nickell, L.G. (1956) Cultivation of plant tissue. US patent 2, 747, 334.

Scragg, A.H. and Fowler, M.W. (1984) The mass culture of plant cells. In Cell Culture and Somatic Cell Genetics of Plants (Ed. Vasil I.K.). Academic Press, New York.

Scragg, A.H., Allan, E.J., Bond, P.A. and Smart, N.J. (1986a) The rheology of Catharanthus roseus cell suspensions. In: Secondary Metabolism in Plant Cell Cultures (Eds. P. Morris, A.H. Scragg, A. Stafford, M.W. Fowler) pp.178-194. Cambridge University Press.

Scragg, A.H., Morris, P., Allan, E.J., Bond, P., Hegarty, P., Smart, N.J. and Fowler, M.W. (1986b) The effect of scale up on plant cell culture performance. In: Process Possibilities for Animal and Plant Cell Culture (Eds. C. Webb and F. Mavituna). In press.

Smart, N.J. and Fowler, M.W. (1981) Effect of aeration on large-scale cultures of plant cells. Biotech. Lett. 2, 171-6.

Smart, N.J. and Fowler, M.W. (1984a) An airlift column bioreactor suitable for large scale cultivation of plant cell suspensions. J. exp. Bot. 35, 521-7.

Smart, N.J. and Fowler, M.W. (1984b) Mass cultivation of *Catharanthus roseus* cells using a non-mechanically agitate bioreactor. Appl. Biochem. Biotech. 9, 209-16.

Tal, B. and Goldberg, I. (1982) Growth and diosgenin production by *Dioscorea deltoidea* cells in batch and continuous culture. Planta Med. 44, 107-10.

Tanaka, H. (1981) Technological problems in cultivation of plant cells at high density. Biotech. Bioeng. 23, 1203-18.

Tanaka, H. (1982) Oxygen Transfer in Broths of Plant Cells at High Density. Biotechnology and Bioengineering. Vol. XX1V, 425-42.

Ulbrich, B., Weisner, W. and Arens, H. (1985) Large-scale production of rosmarinic acid from plant cell cultures of *Coleus blumei* Benth. In primary and secondary metabolism of plant cell cultures (Eds. Neumann, K.-H., Barz, W. and Reinhard, E.). pp.293-303. Springer-Verlag.

Wagner, F. and Vogelmann, H. (1977) Cultivation of plant tissue cultures in bioreactors and formation of secondary metabolites. In Plant Tissue and Its Biotechnological Application (Eds. Barz, W., Reinhard, E. and Zenk, M.H.), pp.245-52. Springer-Verlag, Berlin.

Plant Tissue and Cell Culture, pages 473–493

SCALE-UP: ARTIFICIAL SEEDS

Keith Redenbaugh, Peter Viss, David Slade, and Jo Ann Fujii

Plant Genetics, Inc., 1930 Fifth Street, Davis, CA 95616 USA

CONCEPT OF ARTIFICIAL SEEDS

The origin of the idea of an artificial seed is difficult to determine. Certainly, those who first produced somatic embryos may have considered such an application (Steward, *et al.*, 1958 and Reinert, 1958). However, it was not until the early 1970's that the concept of using somatic embryos began to be presented as a potential propagation system for seed-sown crops. Toshio Murashige gave a number of seminars on tissue culture propagation where he concluded with this concept. For a period of time he conducted research in his laboratory that was focused on the developmental physiology of somatic embryos which he felt to be the limiting factor for large-scale propagation (Murashige, personal communication). He formally presented his ideas on artificial seeds at the *Symposium on Tissue Culture for Horticultural Purposes* in Ghent, Belgium, September 6-9, 1977. His terse comments in the proceedings, however, were "But to be applicable, the cloning method must be extremely rapid, capable of generating several million plants daily, and competitive economically with the seed method" (Murashige, 1977).

During the mid-1970's, two separate research groups began work on somatic embryogenesis for crop propagation. Keith Walker, then at Monsanto

Company, directed a group of scientists that identified basic concepts of delivery of cloned, agricultural crops. Since the focus was to develop thrifty somatic embryo systems that would recapitulate zygotic embryogenesis, their choice was the advanced system developed for *Medicago sativa* L. (alfalfa) using a line (Regen S) identified by Bingham, *et al.* (1975). Soybean and vegetable crops were also of interest to them (Walker, personal communication). Walker cited two reports that had a strong impact on their thinking about the use of somatic embryos for crop propagation. Street (1977) discussed the problem of reliability in embryogenesis. He wrote "morphogenic competence is determined from the time of culture initiation" such that there is "the need to have an initiation medium that will ensure that the competent cells are involved in callus formation." The second report (Sunderland, 1977) demonstrated that the production of hundreds of morphologically uniform embryos from *Datura* and *Nicotiana* pollen was possible.

At the same time a second group at Union Carbide led by Robert Lawrence began to consider various methods for cloning forest trees and difficult-to-produce hybrids for crops such as celery and lettuce. This group focused on delivery of somatic embryos using fluid drilling technology (Lawrence, 1981) and using polyoxyethylene to form seed tapes or sheets (Lawrence, personal communication). Lawrence and Walker's groups came together at a symposium workshop, *Advances in Methods of In Vitro Cloning for Large Scale Propagation of Plants*, held September 21-22, 1981 at the W. Alton Jones Cell Science Center in Lake Placid, New York. Discussions at that time centered on the concept of a low-cost, high-volume propagation system for vegetable and agronomic crops using somatic embryos delivered by fluid drilling, in a seed tape, or as an artificial seed.

During the late 1970's and early 1980's, other groups were active in developing methods to commercially propagate crops using somatic embryos. Drew (1979) suggested delivering carrot somatic embryos in a fluid drilling system, but was able to produce only three plants from carrot embryos on a carbohydrate-free medium. With his system he was not able to produce very many plants and found that the crucial problem was "the very slow rate of development of plantlets derived from culture." Kitto and Janick (1982) coated clumps of carrot embryos, roots, and callus with polyoxyethylene. Some embryos survived the coating process as well as a desiccation step (Kitto and Janick, 1985a and 1985b).

Early in 1980, Walker moved to Plant Genetics, Inc. where Redenbaugh, *et al.* (1984 and 1986) discovered that hydrogels such as sodium alginate could be used to produce single-embryo artificial seeds. In a few experiments, the artificial seeds were planted in the greenhouse with plant production (7% for alfalfa and 10% for celery). The Lawrence group (then at Agrigenetics) subsequently began to use alginate for encapsulating carrot and celery somatic embryos (Lutz, *et al.* 1985). Because their encapsulation results with alginate were similar to those at Plant Genetics, they focused on the problems of somatic embryo physiology. In this they were successful in obtaining germination of carrot somatic embryos in vermiculite in a growth chamber (James Wong, personal communication).

The early assessments of Murashige (personal communication) and Street (1977) on the difficulty of somatic embryogeny are still valid today. The quality and fidelity of somatic embryos are the limiting factors for development and scale-up of artificial seeds. In this report, results on encapsulation, scale-up research, and improving the maturation of alfalfa somatic embryos will be presented. Projected costs and potential crop targets also will be discussed.

HYDROGEL ENCAPSULATION OF ARTIFICIAL SEEDS

The potential uses of artificial seeds are twofold: as a delivery system and as an analytical tool (Table 1). Artificial seeds offer the possibility of a low-cost, high-volume propagation system that will compete with true seeds and transplants. Most likely, the technology will first be used with hybrid vegetable crops such as celery or with high-value flower and ornamental species. Because of the relative ease of producing large numbers of somatic embryos, artificial seeds will be applicable for monoculture as well as mixed genotype methods of agriculture. The artificial seed coating also has the potential to hold and deliver beneficial adjuvants such as growth-promoting rhizobacteria, plant nutrients and growth control agents, and pesticides for

Table 1. Potential uses of artificial seeds

Delivery Systems
- Reduced costs of transplants
- Direct greenhouse and field delivery of:
  - elite, select genotypes
  - hand-pollinated hybrids
  - genetically engineered plants
  - sterile and unstable genotypes
- Large-scale monocultures
- Mixed-genotype plantations
- Carrier for adjuvants such as microorganisms, plant growth regulators, and pesticides
- Protection of meiotically-unstable, elite genotypes

Analytical Tools
- Comparative aid for zygotic embryogeny
- Production of large numbers of identical embryos
- Determine role of endosperm in embryo development and germination
- Study of seed coat formation
- Study of somaclonal variation

precise placement. Although researchers currently turn to zygotic embryos for improving the initiation and maturation of somatic embryos, in the future somatic embryos will aid the study of embryogeny because of their ready availability in large numbers. Other possible uses as analytical tools are listed in Table 1.

Water soluble hydrogels have been found suitable for making artificial seeds (Table 2). Two methods have been used to coat somatic embryos: gel complexation via a dropping procedure and molding. Redenbaugh *et al.* (1986) mixed alfalfa somatic embryos with sodium alginate (2% w/v) and dropped them into a calcium nitrate solution (100mM). Surface complexation began immediately and the drops were gelled completely in 30 minutes. Alternatively, the embryos could be mixed in a temperature-dependent gel such as Gel-rite™, placed in the well of a microtiter plate, and gelled as the temperature was lowered. Somatic embryos from several crops have been encapsulated in alginate with plants recovered *in vitro* (Table 3).

Table 2. Useful gels for encapsulation of somatic embryos

Gel	Conc. % w/v	Complexing Agent	Conc. mM
Sodium Alginate	0.5-5.0	Calcium Salts	30-100
Sodium Alginate with Gelatin	2.0 5.0	Calcium Chloride	30-100
Carrageenan with Locust Bean Gum	0.2-0.8 0.4-1.0	Potassium or Ammonium Chloride	500
Gelrite	0.25	Temperature lowered	

Table 3. Crops encapsulated in calcium alginate beads

Species	Common Name
Apium graveolens L.	Celery
Brassica species	Rapid-cycling *Brassica*
Daucus carota L.	Carrot
Gossypium hirsutum L.	Cotton
Lactuca sativa L.	Lettuce
Medicago sativa L.	Alfalfa
Zea mays L.	Corn

Initially, the effect of encapsulation was difficult to assess because of the overall poor quality of the somatic embryos. Although visually normal embryos were produced (i.e. bipolar, root and shoot axis, cotyledons), the germination and continued development of the embryos was very inconsistent. In fact, the use of germination (root elongation and emergence) as an efficacy assay was found to be unsuitable when shoot production and further growth was not observed concomitantly. Consequently, the approach for developing artificial seeds was shifted away from one focused on somatic embryogenesis (initiation of embryo formation) to one of somatic embryogeny (initiation, development, and maturation of embryos). The concept of somatic embryogeny and the production of high quality embryos is not one that is widely followed by many researchers who have either focused on the production of a somatic embryogenesis system that results in some plant recovery or who have interest only in the study of the early stages of embryogenesis. This focus separated from an emphasis on producing mature, true-to-type, high quality embryos can possibly lead to conclusions based on abnormal somatic embryogenesis. To overcome this problem and to achieve high quality embryo production for scale-up of artificial seeds, we measured embryo response in terms of embryo-to-plant development or "conversion" (Table 4). Essentially, embryo conversion frequency is the percent of the somatic embryos that produce green plants having a normal phenotype. This assay has been critical for developing conditions and media that select for uniform plant production. Over the past 4 1/2 years, the conversion frequency for alfalfa somatic embryos was increased from less than 0.5% to 50-70% using a random population of embryos 1 mm in diameter and greater (Table 5).

Table 4. Embryo-to-plant conversion includes the following events:

1. Germination (radicle elongation)
2. Development of a vigorous root system
3. Growth and development of the shoot meristem
4. Production of true leaves
5. A direct shoot-to-root connection
6. Absence of hypertrophy of the hypocotyl
7. Minimization of callus growth in the hypocotyl
8. A green plant with a normal phenotype

Table 5. Progress in alfalfa somatic embryo conversion

Time	% *in vitro* conversion of random SE
1982	<0.5%
1983	8.0
1984	30.0
1985	40-50
1986	50-70

SCALE-UP: ALFALFA SOMATIC EMBRYOGENY IN LIQUID CULTURE

Scale-up research is required in several areas (Figure 1). Somatic embryos need to be produced on a large scale, matured to a stage where germination will be at a high rate and frequency, and encapsulated in a suitable material. The encapsulated embryos will probably need to be coated to prevent capsule desiccation and allow for singulation during planting. Then the artificial seeds will require a shelf life specific for the planting practices of the crop. The final assessment will be the harvest and yield.

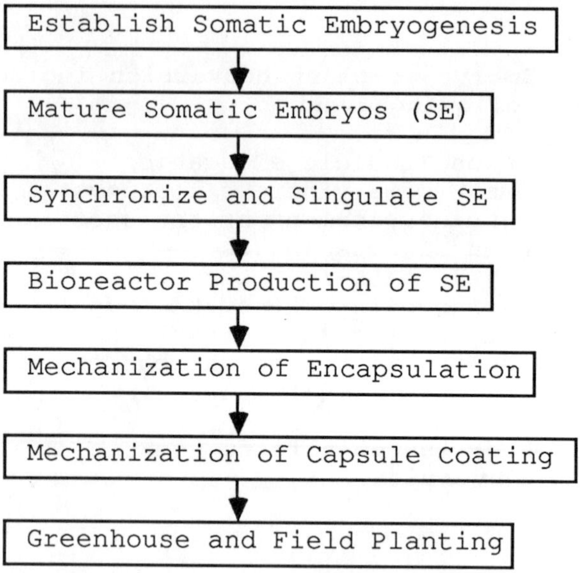

Figure 1. Production of artificial seeds

For medium-scale purposes, a liquid flask system (Figure 2) was developed for alfalfa based on the agar methods previously reported (Walker and Sato, 1981 and Stuart and Strickland, 1984a). The agar method (Redenbaugh, *et al.* 1986) was modified to include a liquid induction stage of three days. The induced cells were collected on a 60 mesh stainless steel sieve and placed in liquid regeneration medium consisting of Schenk and Hildebrandt (1972) salts (SH), 100 mM proline and 25 mM ammonium. The agar system had a regeneration phase of 3 weeks followed by embryo harvest. However, in liquid, the embryos began to degrade in quality at 3 weeks. Consequently, it was necessary to collect the embryos on a 40 mesh sieve after 2 weeks and resuspend them in fresh medium for a second 2 week period. An additional improvement in embryo quality was to wash the embryos with sterile water and store them at 4C for a maturation period. A final modification was based on the discovery that maltose was effective for agar-based regeneration (Strickland *et al.*, 1986). In the conversion medium (half-strength SH medium), 1.5% maltose was substituted for sucrose.

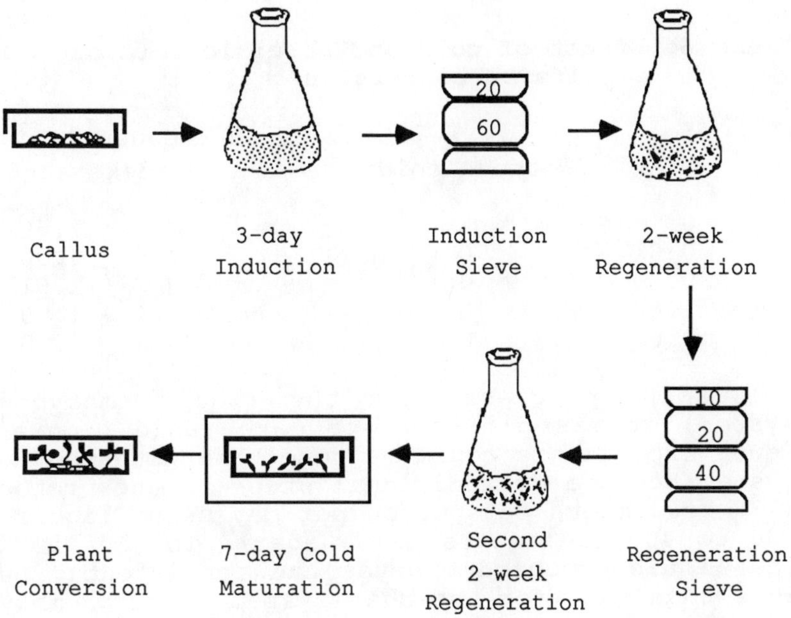

Figure 2. Liquid production of alfalfa somatic embryos

Using these conditions, 10 to 20 somatic embryos per mg induced callus tissue were produced. These embryos had a conversion frequency of 20-30% (using a random population greater than 1 mm length). The cold maturation treatment was critical for obtaining these conversion frequencies (Table 6). Interestingly, the use of abscisic acid during the second regeneration stage had no consistent effect.

Table 6. Effect of cold and abscisic acid on alfalfa SE conversion

ABA Conc. µM	% Conversion Frequency	
	Without cold	3 weeks at 4°C
0	1	17 ± 2.60
0.01	0	16 ± 4.00
0.10	0	7 ± 2.61
1.0	0	24 ± 4.99
10.0	0	13 ± 3.60

Despite progress with the liquid embryogeny system, it was difficult to make rapid progress. Fewer experiments could be done than with the agar system because of additional steps in the protocol (125 treatments per person per day using liquid vs 200 to 400 treatments using agar). The quality of the liquid-produced embryos had consistently been less than that for the agar system. Liquid embryos normally had a somewhat swollen appearance while agar embryos had a waxy tone and a conversion frequency usually 10 percentage points higher. Consequently, to develop artificial seeds more rapidly, concerted efforts were made with the agar system.

MATURATION: IMPROVING EMBRYOGENY OF ALFALFA EMBRYOS ON AN AGAR BASED CULTURE

The uses of a cold maturation treatment and maltose in the conversion medium were tested with the agar protocol (Table 7). The cold treatment was further examined to identify optimum storage conditions and duration (Figure 3). Embryos stored on the original regeneration plates did not show a benefit from cold storage, while embryos transferred to agar water or liquid SH medium (embryos were rinsed in the medium and stored under high humidity) resulted in an increase in conversion. However, this increase declined at 14 days in cold storage. As a consequence, the protocol was modified to mature the embryos for 7-10 days at 4C.

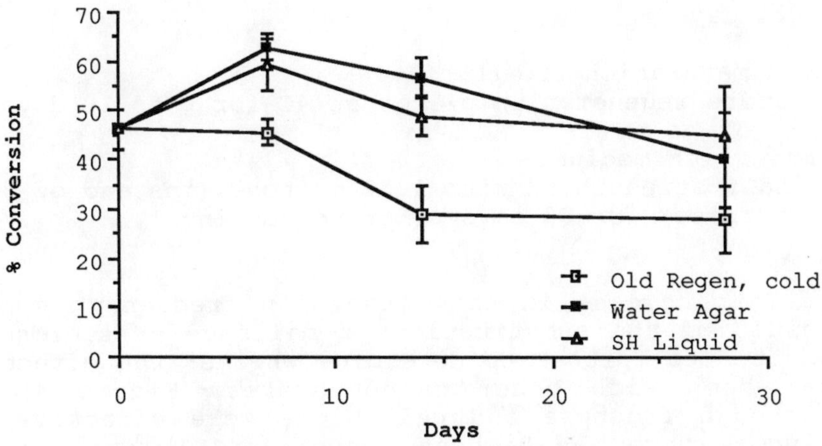

Figure 3. Effect of cold storage on embryo conversion

Table 7. Alfalfa regeneration protocol on agar

Initiation and maintenance medium:
Schenk and Hildebrandt (SH) medium with
25 μM naphthaleneacetic acid and 10 μM kinetin
subcultured every 21 days

Induction medium:
SH with 50 μM 2,4-dichlorophenoxyacetic acid and
5 μM kinetin for 3 days

Regeneration medium:
SH with 10 mM ammonium and 30 mM proline for
21 days

Cold maturation treatment:
store regeneration plates at 4C for 7 to 10 days

Conversion medium:
half strength SH with 1.5% maltose (instead of
sucrose) for 28 days prior to scoring

The increase in the conversion frequency as a result of the substitution of maltose for sucrose was tested further to determine whether the effect was seen with other carbohydrates (Figure 4). Although fructose and cellobiose were effective, glucose gave the highest conversion. Because the salts and sugars in the medium were autoclaved, filter-sterilized maltose was used to determine if the conversion effect was due to a breakdown of maltose to glucose. As seen in Figure 5, the maltose effect does appear to be due to the glucose moiety. In general, when conversion was done on a sucrose medium, the conversion frequency was the same whether the plants were scored after 4 or 6 weeks. With maltose, however, there was an increase in the conversion frequency if the plants were scored after 6 weeks. This indicates that maltose serves as a maturation factor for inducing further development of those embryos that lag behind more mature embryos and that would not convert on a sucrose medium.

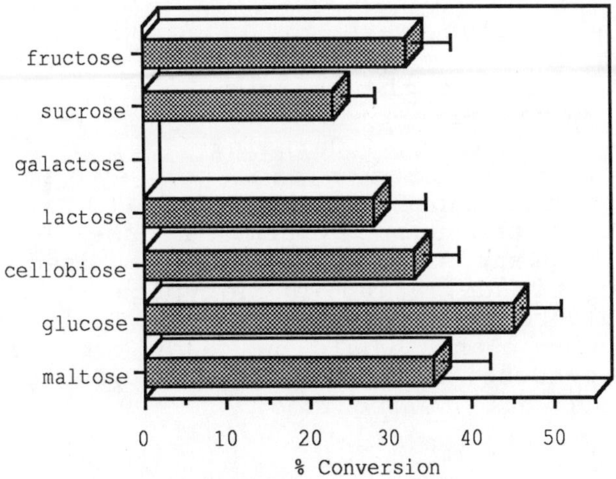

Figure 4. Carbohydrate effect on embryo conversion

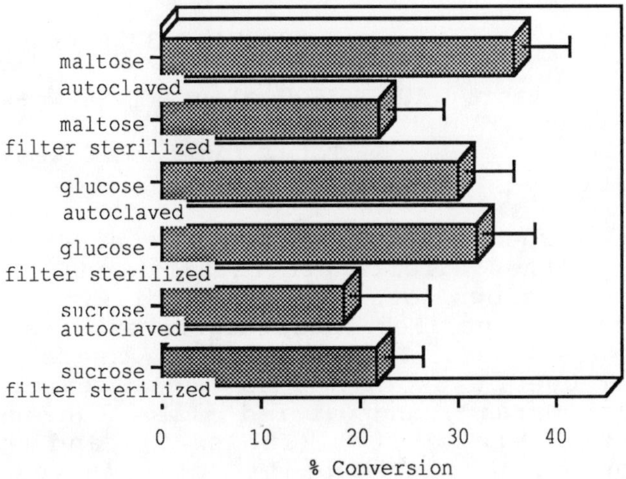

Figure 5. Effect of sterilization treatment on embryo conversion

Using the protocol in Table 7, 2-6 embryos per mg induced callus were produced which had a conversion frequency of 40-50%. For experimental use, hand selecting the embryos from agar plates was not more time-consuming than selecting liquid-produced embryos. Certainly, for production purposes a mechanized bioreactor system with minimum operator input will be required. Using the numbers from the agar system, a baseline mass balance for plant production was determined. From 1g callus tissue, 3400 phenotypically normal plants could be produced in 10 weeks (Table 8). At this time the plants would be ready for transfer to the greenhouse. For producing 1,000,000 plants, 300 g callus would be required.

Table 8. Mass balance for alfalfa SE production

Start	1 g fwt callus
Maintenance	4.0 g tissue in 3 weeks
Regeneration	8500 SE in 3 weeks
Conversion *in vitro* (40%)	3400 plants in 4 weeks
	Total time is 10 weeks

Initial experiments have been conducted to determine whether alfalfa artificial seeds can be planted and grown outside a sterile, tissue culture environment. For these experiments, select embryos were used. The artificial seeds were planted in one of two soilless media, partially covered with Saran™, and watered daily. Conversion frequencies were 2-10% (Table 9) and gave indication that the quality of embryos is sufficient to achieve greenhouse conversion even though the *in vitro* conversion was not consistent.

Table 9. In vitro and greenhouse conversion of encapsulated somatic embryos

% Conversion *In Vitro*		% Conversion in Greenhouse	
Ave.	Std. Error	Ave.	Std. Error
24.0	2.5	2.0[a]	2.0
18.0	3.7	8.0[a]	2.0
81.5	4.4	10.0[a]	4.5
86.0	6.4	4.0[b]	4.0
38.7	10.1	6.0[b]	4.0

[a]artificial seeds planted in peat in aluminum trays
[b]artificial seeds planted in vermiculite in aluminum trays under mist

ARTIFICIAL SEEDS POTENTIAL

Somatic embryogenesis has been observed in a great many species to date, which indicates that it may be possible to produce embryos in almost any desired crop. As more experience is gained with embryogenesis, common features begin to emerge. Examples are the discovery that somatic embryos could be produced for the first time in cereals using immature zygotic embryos (Lu and Vasil, 1981). The same explant yielded the first embryos in conifers (Hakman, *et al.* 1985). The use of 2,4-dichlorophenoxyacetic acid for inducing cells to develop into somatic embryos is a widespread phenomenon for most species. Proline and other reduced nitrogen sources have been found useful for some species such as corn (Armstrong and Green, 1985) and alfalfa (Stuart and Strickland, 1984b). It is anticipated that greater universality will be found in somatic embryogenesis.

More attention must be given to somatic embryogeny so that fully mature embryos are produced. Already, abscisic acid has been found to be critical for embryo maturation in the Umbelliferae (Ammirato, 1985). The use of maltose for increasing the conversion of alfalfa embryos is another example of embryo maturation. One limitation that must be overcome for developing artificial seeds is storage time. Currently, the conversion frequency of encapsulated alfalfa somatic embryos declines sharply when the artificial seeds are stored at 4C. Naked embryos show a characteristic slight drop in conversion that then remains stable over a 1-2 month period. Based on experience with true seed encapsulation in alginate capsules, the limited storage time of artificial seeds is probably due to an anaerobic environment in the capsules. This is a problem for somatic embryos because they are not developmentally arrested and continue very active respiration. Two possible solutions are to have a smaller ratio of capsule volume to embryo volume so that gas diffusion can readily occur or to induce an arrested state in the embryos.

The costs of artificial seeds (based on the alfalfa system) was calculated to be $261.36 per 1,000,000 units or 0.026¢ each. This cost includes labor and material for the tissue culture processes, encapsulation (machinery and labor), and overhead (Table 10). A cost of 0.026¢ compares favorably with such crops as hybrid vegetables and flower crops (Table 11). The costs for celery and lettuce are based not on hybrid seed (which is not available commercially), but on self-pollinated seed. Although the cost for alfalfa seed is considerably less than that for artificial seeds, there may be advantages for use of artificial seeds in inhouse breeding programs.

Table 10. Calculated costs for producing 1,000,000 artificial seeds based on alfalfa model system[1]

Production stage	Labor cost	Materials cost
Callus growth (21 days)	$4.00	$1.36
Induction (3 days)	4.00	2.00
Embryo regeneration (28 days)	4.00	2.00
Maturation treatment (21 days)	4.00	2.00
Encapsulation (1 day)	112.00	28.00
Subtotal	128.00	35.36
Combined totals	$163.36	
Overhead (60%)	98.00	
Total cost per 1,000,000 artificial seeds	$261.36	
Cost per unit	0.026¢	

[1]Calculations, in part, provided by C. Sluis

Ten stages of commercialization of artificial seeds can be identified. The first is to select candidate crops based on both technological and commercial potential. A crop that only fits one potential should probably be reconsidered. Then a somatic embryogeny system must be established or adapted, followed by stage 3 which is to optimize the system, including embryo maturation. These three stages need to be developed for each crop and constitute the bulk of time required for developing artificial seeds. The fourth stage, embryo automation production, may require the development of unique production systems. Not enough research has been done in this area. From our experience with alfalfa, a liquid production method produces embryos of lower quality than those produced on agar. The final embryogeny stage is to produce mature, synchronized embryos that have been developmentally arrested. This may involve a desiccation stage to produce somatic embryos that are very similar to zygotic embryos.

Table 11. Comparison of costs for artificial seeds and true seeds[1]

Crop	Cost per unit
Artificial seeds	
alfalfa	0.026¢
True seeds	
alfalfa	0.00066
broccoli, hybrid	0.085
cauliflower, open pollinated	0.04
cauliflower, hybrid	0.14
celery	0.0052
lettuce	0.012
begonia, annual	0.3
geranium, hybrid	6.0
gerbera daisy	1.4
impatiens	0.13
petunia	0.35

[1]Calculations, in part, provided by C. Sluis

The next two stages involve encapsulation. Once the encapsulation and coating systems are optimized and automated, they should be sufficient for all crops. Greenhouse and field conversion will also need to be optimized for each species. Watering, fertilizer, transplanting, and other requirements will need to be developed. The ninth stage involves identification and control of any pest and disease problems that may be unique to artificial seeds. The final stage is perhaps the most important: determining whether the artificial seed delivery system works in terms of increased productivity for a specific crop.

ACKNOWLEDGMENTS

We wish to thank Ian Sussex for stressing the concept of somatic embryogeny and Carolyn Sluis for developing the cost analysis for artificial seeds. We also thank Toshio Murashige, Keith Walker, Robert Lawrence, and James Wong for their perspectives on the conceptional development and early work on artificial seeds.

REFERENCES

Ammirato, P. (1985). Patterns of development in culture. In Henke, R., K. Hughes, M. Constantin, and A. Hollaender (eds.): *Tissue Culture in Forestry and Agriculture,* Plenum Press, NY. p. 9-62.

Armstrong, C. and C. Green (1985). Establishment and maintenance of friable embryogenic maize callus and the involvement of L-proline. Planta 164:207-214.

Bingham, E.T., L.V. Hurley, D.M. Koutz, and J.W. Saunders (1975). Breeding alfalfa which regenerates from callus tissue in culture. Crop Sci. 15:719-721.

Drew, R.L.K. (1979). The development of carrot (*Daucus carota* L.) embryoids (derived from cell suspension culture) into plantlets on a sugar-free basal medium. Hort. Res. 19:79-84.

Hakman, I., L. Fowke, S. von Arnold, and T. Eriksson (1985). The development of somatic embryos in tissue culture initiated from immature embryos of *Picea abies* (Norway spruce). Plant Sci. Lett. 38:53-63.

Kitto, S. and J. Janick (1981). Polyox as an artificial seed coat for asexual embryos. Hort. Sci. 17:448 (abst.).

Kitto, S. and J. Janick (1985). Production of synthetic seeds by encapsulating asexual embryos of carrot. J. Amer. Soc. Hort. Sci. 110:277-282.

Kitto, S. and J. Janick (1985). Hardening treatments increase survival of synthetically-coated asexual embryos of carrot. J. Amer. Soc. Hort. Sci. 110:283-286.

Lawrence, R.H., Jr. (1981). *In vitro* cloning systems. Environ. and Exp. Bot. 21:289-300.

Lu, C. and I. Vasil (1981). Somatic embryogenesis and plant regeneration from leaf tissues of *Panicum maximum* Jacq. Theor. Appl. Genet. 59:275-280.

Lutz, J., J. Wong, J. Rowe, D. Tricoli, and R. Lawrence, Jr. (1985). Somatic embryogenesis for mass cloning of crop plants. In Henke, R., K. Hughes, M. Constantin, and A. Hollaender (eds.): *Tissue Culture in Forestry and Agriculture,* Plenum Press, NY. p. 105-116.

Murashige, T. (1977). Plant cell and organ cultures as horticultural practices. Acta Hort. 78:17-30.

Redenbaugh, K., J. Nichol, M. Kossler, and B. Paasch (1984). Encapsulation of somatic embryos for artificial seed production. In Vitro 20:256-257 (abst.).

Redenbaugh, K., B. Paasch, J. Nichol, M. Kossler, P. Viss, and K. Walker (1986). Somatic seeds: encapsulation of asexual plant embryos. Biotech. (in press).

Reinert, J. (1958). Morphogenese und ihre Kontrolle an Gewebekulturen aus Carotten. Naturwissenschaft 45:344-345.

Schenk, R. and A. Hildebrandt (1972). Medium and techniques for induction and growth of monocotyledons and dicotyledons cell cultures. Can J. Bot. 50:199-204.

Steward, F., M. Mapes, and K. Mears (1958). Growth and organized development of cultured cells II. Organization in cultures grown from freely suspended cells. Am. J. Bot. 45:705-708.

Street, H.E. (1979). Embryogenesis and chemically induced organogenesis. In Sharp, W., P. Larsen, E. Paddock, and V. Raghavan (eds.), *Plant Cell and Tissue Culture Principles and Applications*, Ohio State University Press, Columbus, Ohio. p. 123-153.

Strickland, S., C. McCall, J. Nichol, and D. Stuart (1986). Enhanced somatic embryogenesis in *Medicago sativa* by addition of maltose and its higher homologs to the culture medium. In Somers, D.A., *et al.* (eds.), *VI International Congress of Plant Tissue and Cell Culture Abstracts*, University of Minnesota, Minneapolis, MN. p.188.

Stuart, D. and S. Strickland (1984a). Somatic embryogenesis from cell cultures of *Medicago sativa* L. I. The role of amino acid additions to the regeneration medium. Plant Sci. Lett. 34:165-174.

Stuart, D. and S. Strickland (1984b). Somatic embryogenesis from cell cultures of *Medicago sativa* L. II. The interaction of amino acids with ammonium. Plant Sci. Lett. 34:175-181.

Sunderland, N. (1977). Comparative studies of anther and pollen culture. In Sharp, W., P. Larsen, E. Paddock, and V. Raghavan (eds.), *Plant Cell and Tissue Culture Principles and Applications*, Ohio State University Press, Columbus, Ohio. p. 203-219.

Walker, K. and S. Sato (1981). Morphogenesis in callus tissue of *Medicago sativa* : the role of ammonium ion in somatic embryogenesis. Plant Cell Tiss. Org. Cult. 1:109-121.

Index